Electronic Media

(formerly *Telecommunications*)

Tenth Edition

Electronic Media

An Introduction

Lynne Schafer Gross

California State University, Fullerton

 **Higher Education**

Boston   Burr Ridge, IL   Dubuque, IA   New York   San Francisco   St. Louis
Bangkok   Bogotá   Caracas   Kuala Lumpur   Lisbon   London   Madrid   Mexico City
Milan   Montreal   New Delhi   Santiago   Seoul   Singapore   Sydney   Taipei   Toronto

 **Higher Education**

Published by McGraw-Hill, an imprint of The McGraw-Hill Companies, Inc., 1221 Avenue of the Americas, New York, NY 10020. Copyright © 2010, 2006, 2003, 2000, 1997, 1995, 1992, 1988, 1986, 1983. All rights reserved. No part of this publication may be reproduced or distributed in any form or by any means, or stored in a database or retrieval system, without the prior written consent of The McGraw-Hill Companies, Inc., including, but not limited to, in any network or other electronic storage or transmission, or broadcast for distance learning.

This book is printed on acid-free paper.

3 4 5 6 7 8 9 0 DOC/DOC 0

ISBN: 978-0-07-337886-2
MHID: 0-07-337886-0

Vice President, Editor in Chief: *Michael Ryan*
Publisher: *Frank Mortimer*
Executive Editor: *Katie Stevens*
Managing Editor: *Meghan Campbell*
Executive Marketing Manager: *Pamela Cooper*
Developmental Editor: *Phillip Butcher*
Editorial Coordinator: *Erika Lake*
Senior Production Editor: *Karol Jurado*
Production Service: *Anne Draus, Scratchgravel Publishing Services*
Manuscript Editor: *Margaret Tropp*
Cover Designer: *Laurie Entringer*
Cover Images: *© The McGraw-Hill Companies, Lars Niki, photographer (top left); © Digital Vision Ltd./SuperStock (top right); © The McGraw-Hill Companies, David Planchet, photographer (bottom)*
Production Supervisor: *Louis Swaim*
Media Project Manager: *Tom Brierly*
Composition: *10/12 Times Roman by Macmillan Publishing Solutions*
Printing: *50# New Era Matte Plus, RR Donnelley & Sons*

Library of Congress Cataloging-in-Publication Data

Gross, Lynne S.
    Electronic media : an introduction / Lynne Schafer Gross. — 10th ed.
        p. cm.
Rev. ed. of: Telecommunications. 9th ed. c2006.
    Includes bibliographical references and index.
    ISBN-13: 978-0-07-337886-2 (acid-free paper)
    ISBN-10: 0-07-337886-0 (acid-free paper)
1. Telecommunication. 2. Broadcasting. I. Gross, Lynne S. Telecommunications. II. Title.
    TK5105.G759 2010
    384—dc22   2009001127

www.mhhe.com

ABOUT THE AUTHOR

Lynne S. Gross has taught radio-television-film production and theory courses at California State University, Fullerton (where she was Vice Chair of the Communications Department), Pepperdine University, Loyola Marymount University, UCLA, USC, and Long Beach City College. She is the author of 11 other books and numerous journal and magazine articles.

She is currently Associate Producer for the instructional video series *Journeys Below the Line* and in the past was Program Director for Valley Cable TV. She has served as Producer for several hundred television programs, including the series *From Chant to Chance* for public television, *Effective Living* for KABC, and *Surveying the Universe* for KHJ-TV.

Her consulting work includes projects for Children's Broadcasting Corporation, RKO, KCET, CBS, the Olympics, Visa, and the Iowa State Board of Regents. It has also taken her to Malaysia, Swaziland, Estonia, Russia, Australia, New Zealand, and Guyana, where she has taught radio and television production and consulted on film planning and postproduction equipment.

She is active in many professional organizations, serving as a Governor of the Academy of Television Arts and Sciences and President of the Broadcast Education Association. Awards she has received include the Rosebud Award for Outstanding Media Arts Professor in the California State University System, the Frank Stanton Fellow for Distinguished Contribution to Electronic Media Education from the International Radio and Television Society, and the Distinguished Education Service Award from the BEA.

PREFACE

Purpose

The electronic media constitute one of the most potent forces in the world today. They influence society as a whole, and they influence every one of us as an individual. As each year passes, the media grow in scope. The early pioneers of radio would never recognize today's vast array of electronic media—broadcast television, cable TV, satellite radio and television, the internet, DVDs, DVRs, video games, cell phones—just to name a few. Neither would they recognize the structure that has evolved in such areas as regulation, advertising, and audience measurement. They would marvel that their early concepts of equipment have led to such developments as audio recorders, cameras, video recorders, digital effects generators, nonlinear editors, fiber optics, and satellites. If they could see the quantity and variety of programming available today, they might not recognize that it all began with amateurs listening for radio signals on their primitive crystal sets.

All indications are that telecommunications will continue to change at a rapid pace. As it does, it will further affect society. All people, whether individuals working in the communications field or individual members of society, have a right to become involved with media and have an obligation to understand why people need to interact with the media. Some knowledge of the background and structure of the industry is an essential basis for this understanding. A major goal of this book is to provide just that kind of knowledge so that intelligent decisions about the role of media can be made both by practitioners in the field and by members of the general society.

Organization of the Book

This is the tenth edition of a book that first appeared in 1983 with the title of *Telecommunications*. The title has been changed for this edition to more accurately reflect the scope of the contents. The book has also been reorganized and updated to include the many changes that have occurred in the media field in the last few years. The current edition takes into account the growing importance of the internet, the advent of portable devices, the conversion to digital TV, the effects of DVRs, the expansion of the telephone business, the development of satellite radio, the march toward digital cinema, ever-evolving programming strategies, alterations in advertising and ratings as the audience further fractures, new laws and ethical challenges posed by a changing media world, rapid technological changes, and the implications of digital technologies for worldwide communication. All of these areas and more have been updated in the book.

Electronic Media Forms

The book is divided into two parts: Electronic Media Forms and Electronic Media Functions. The opening chapter is an expansion, suggested by the book's reviewers, of material on the Significance of Electronic Media that was previously in a Prologue.

Chapter 2 covers The Internet, Portable Devices, and Video Games. These topics were moved forward in the book, again at the suggestion of reviewers, to introduce electronic media by engaging students in the media they use extensively.

Television is now viewed as a single entity without regard for whether the source is broadcast, public, cable, satellite, computer, or some other form. For that reason, television has been broken into two chapters, Chapter 3: Early Television and Chapter 4: Modern Television, with the various distribution forms discussed in parallel. Chapter 5 deals with Radio and Chapter 6 with Movies, completing the survey of the electronic media forms.

Electronic Media Functions

The first chapter in the Functions section is another new chapter—Chapter 7: Careers in Electronic Media. Although the previous edition provided some material about careers in an Epilogue, reviewers felt that this subject is so important to students that it deserves an entire chapter.

Chapters 8 (Programming), 9 (Sales and Advertising), and 10 (Promotion and Audience Feedback) have been significantly restructured to incorporate the newer media, to acknowledge the role that the public plays in creating material for the media, to discuss the rethinking the advertising world is going through to counter people's ability to ignore ads, and to cover the new methods of measurement needed to augment what used to be a fairly simple method of determining ratings.

In Chapter 11 on Laws and Regulations, some of the older regulations that no longer apply have been weeded out and the discussion of current laws has been updated and expanded. Ethics and Effects (Chapter 12) are with us just as they always have been, but many of the examples in the chapter have been freshened. Chapter 13 covers the Technical Underpinnings that are important to understand as the electronic media evolve, and the final chapter deals with the all-encompassing influence of The International Scene.

Features of the Book

Exhibits

The book contains many photos, drawings, and charts to enhance the reader's understanding of concepts and ideas. More than 70 of these visual exhibits are new to this edition.

Issues and the Future Boxes

All the chapters should lead the reader to assess the strengths and weaknesses of the particular subject being discussed. Toward this end, each chapter contains an

"Issues and the Future" box. This feature should help prepare readers for fast-changing events that they will read about in newspapers and magazines.

Zoom In Boxes

Each chapter includes several "Zoom In" boxes. Because of the popularity of this feature, the number of boxes has been increased. Many of these boxes are designed to increase critical thinking. They discuss current and controversial issues, often of an ethical nature, and end with a series of questions designed to stimulate the reader's thinking. Other boxes relate historical information to help readers understand the genesis of today's media events.

Review Guides

Marginal notes appear in each chapter. These notes highlight the main subjects being discussed in the adjacent paragraphs. Taken together, these notes serve as review points for the reader. Throughout the text, important words are boldfaced. These terms are all defined in the glossary and should serve as another aid to learning.

Suggested Websites

Each chapter lists several pertinent websites that readers may wish to visit to supplement their knowledge. Websites are also mentioned in chapter notes and within the text itself.

Further Study

Extensive notes at the end of each chapter provide many sources for further study of particular subjects.

Beginnings and Endings

Each chapter begins with a pertinent quote and a short introduction. At the end of each chapter, a summary outlines major points in a manner slightly different from that given within the chapter. For example, if the chapter is ordered chronologically, the summary may be organized in a topical manner. This should help the reader form a gestalt of the material presented.

Flexible Chapter Sequence

The chapters may be read in any sequence. Some of the terms that are defined early in the book, however, may be unfamiliar to people who read later chapters first. The glossary can help overcome this problem. It includes important technical terms that the reader may want to review from time to time, as well as terms that are not necessary to an understanding of the text but may be of interest to the reader.

Supplementary Materials

This edition of the text, like the former, includes a valuable website at www.mhhe.com/gross10e. Among other items, the site contains useful practice quizzes for the student and an instructor's manual and testbank.

Acknowledgments

This book represents the combined efforts of many people, including the following reviewers who offered excellent suggestions:

Tammy Trujillo, Mt. San Antonio College

Craig Breit, Cerritos College

John MacKerron, Towson University

B. William Silcock, the Walter Cronkite School of Arizona State University

In addition, I would like to thank the book team at McGraw-Hill, especially Phil Butcher for his guidance and suggestions.

LYNNE SCHAFER GROSS

BRIEF CONTENTS

CONTENTS

Electronic Media

ELECTRONIC MEDIA FORMS

A wide variety of electronic media exists for the dissemination of entertainment and information. Within less than the average lifetime, these media have proliferated into more than a dozen forms. The media both complement and compete with one another, experiencing the slings and security of the free enterprise system. Some people wonder how many forms of media the market can bear; others marvel at how many it does bear. As these media develop, change is inevitable, brought about by both external and internal forces. Although all the media forms are relatively young, they are already rich in history and adaptation. The first chapter of this section deals with the significance of the various media; the succeeding chapters handle such media forms as the internet, portable devices, video games, broadcast TV, cable TV, satellite TV, home video, radio, and movies.

Chapter 1

THE SIGNIFICANCE OF ELECTRONIC MEDIA

Y ou would be hard-pressed in modern American society to find someone who does not interact with electronic media on a daily basis—someone who does not surf the internet, watch TV, listen to the radio, download music to an iPod, go to a movie, play a video game, send an email, talk on a cell phone, watch a DVD. In fact, many people, in this age of multitasking, interact with several media at the same time. They watch a movie on a DVD while listening to music on an iPod, playing a video game, checking email, and telling a friend all about it over a cell phone.

In addition, everyone has opinions that relate to media. Our opinions decide whether a movie or a TV series is a hit or a dud. We vote for our favorite *American Idol* singer. We argue about the media's coverage of political candidates. We talk about the tactics used in commercials and the sex and violence seen on TV and heard in radio song lyrics. We worry about the predators on the internet and whether or not our cell phones will cause cancer. Parents and

> Television is less a means of communication [the imparting or interchange of thoughts, opinions, and information by speech, writing, or signs] than it is a form of communion [act of sharing or holding in common; participation, association; fellowship].
>
> **Richard Schickel, *The Urban Review***

children clash over the amount of time the latter spend playing video games, watching TV, or text messaging.

The media are constantly changing. Not very long ago there was no internet, no iPod, no Wii. Within the past few decades, a number of technologies that were highly touted have gone by the wayside—subscription TV, multichannel multipoint distribution service, teletext. The way people used electronic media and how they perceived the media were quite different ten years ago from what they are today.

1.1 A Rationale for Study

Even though everyone has a basic knowledge of the field of electronic media, there are many reasons to study it. Anyone who is aiming toward a career in this area will profit from an intimate knowledge of the history and organization of the industry. Those armed with knowledge have a greater chance for career survival   **career options** than those who are naive about the inner workings and interrelationships of networks, stations, cable TV facilities, the internet, movie studios, advertisers, unions, telephone companies, the government, and a host of other organizations that affect the actions and programming of the industry. As the various forms of communications expand, they create new and exciting jobs (see Exhibit 1.1). Knowledge of the past will help people predict the direction of their future jobs. Knowing about media can also help practitioners set their own personal values so that they can help mold the industry into a form that they feel is effective in a positive way.

Exhibit 1.1
The jobs pictured here were created recently when digital technologies allowed the number of cable TV channels to grow, and this network, TVG, was formed to specialize in horse racing.

(Courtesy of TVG)

At a broader level, individuals owe it to themselves to understand the messages, tools, and communication facilities that belong to our society because they are so crucial in shaping our lives. Rare is the individual who has not been emotionally touched or repulsed by a scene in a movie. Rare, too, is the individual who has never formed, reinforced, or changed an opinion on the basis of information heard on radio or seen on TV or the internet. Knowledge of the communications industry and its related areas can lead to a greater understanding of how these forces can influence and affect both individual lives and the structure of society as a whole. It can also teach each individual the most effective methods for interacting with media and affecting programs and services.

understanding

In addition, the electronic media are fascinating and worthy of study in their own right. Those in the field are associated with glamour and excitement (and power and greed), both on-screen and off. Although the day-to-day workings of the industry can be as mundane as they are in any other field, the fact that it is a popular art that includes the rich and famous makes this industry of special interest. The ramifications of the power that the electronic media exert over society are most deserving of study.

fascination

1.2 A Matter of Terms

Before embarking on a study of electronic media, it is wise to acquaint yourself with some of the terms related to the industry. Through the years the terms have changed in meaning and use, and new technologies bring forth new terms. You may be familiar with many of the terms listed in the following paragraphs; you will learn more about them and the terms you don't know in the remaining chapters of this book.

The widespread study of radio, television, and film at the university level did not begin until the 1960s. At that time, film was considered to be separate from radio and television and often was not in the same department. Movies were a chemical medium, and films were projected only in theaters, not in homes. There were two electronic media—radio and television—and together they were called **broadcasting.** Radio consisted of a fairly large number of local stations with specific formats. Television was dominated by three commercial networks—ABC, CBS, and NBC—and their **affiliated** stations. A few stations were **independent** and did not broadcast material from any of the three big networks, but they were considered second-class.

1960s

By the late 1960s, broadcasting was divided into two categories—**commercial broadcasting** supported by advertising and **public broadcasting** supported primarily by government funds and personal donations. These two coexisted harmoniously because public broadcasting was small and not a threat to its commercial kin. In fact, it often relieved commercial broadcasting of its more onerous public-service requirements because the commercial broadcasters could point out that public broadcasting, with its educationally oriented programming, served that interest.

1970s

In the mid-1970s, a number of other media challenged radio and broadcast TV, creating an alphabet soup that included **CATV** (**community antenna TV,** which became known as **cable TV**), **VCRs** (**videocassette recorders**), **DBS** (**direct broadcast satellite,** later referred to as **satellite TV**), and **LPTV** (**low-power TV**). **Video games** were becoming popular in arcades. Corporations began using television for training, referring to this as **industrial TV**; later the name was changed to **corporate video** or **organizational TV** because industrial sounded too grimy. The word *broadcasting* no longer seemed to apply because that word implied a wide dissemination of information through the airwaves. Many of these other media were sending information through wires, and cable TV was even touting its **narrowcasting** because its programs were intended for specific audience groups.

The whole concept of television as a form of mass communication began to change during the 1980s. Before that, most people in the country watched the same programming at the same relative time. The three networks competed fiercely, but, on the average, each garnered almost 30 percent of the available audience. They were mass communication systems sending out programming to be viewed by generally passive masses. With the introduction of a variety of delivery systems, TV became a more fractionalized medium that appealed to smaller groups of consumers. No longer did three network programming chiefs call the shots on what people would watch and when they would watch it. People could videotape programs off the air to watch whenever they wanted, or they could visit the local **video store** and rent a movie, thus bringing the film industry into the electronic realm and enabling people to watch movies in their homes. Instead of watching NBC, CBS, or ABC, people could watch one of the many cable or satellite TV channels, the newly formed Fox network, public broadcasting, one of the independent TV stations, or one of the other alternative media forms. They could also use their newly invented **remote controls** to switch from one program to another anytime they became slightly dissatisfied, without taking a long walk from the easy chair to the TV set to change the channel. The share of audience that the "big three" networks had attracted plummeted dramatically (see Exhibit 1.2).[1]

1980s

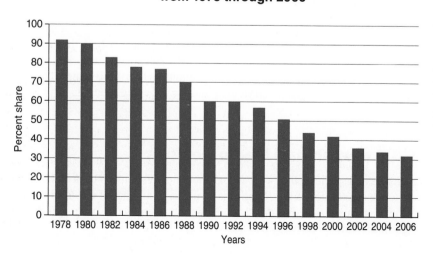

The Prime-Time Share of Audience of ABC, CBS, and NBC from 1978 through 2006

Exhibit 1.2

The share of audience of the original TV networks (ABC, CBS, and NBC) has dipped continuously since the advent of cable TV networks and other new technologies.

In the 1990s, the field continued to broaden. Telephone companies developed the **cell phone** and also entered program distribution areas that had traditionally been reserved for broadcasters and cablecasters. The once lowly phone line allied itself with the computer and the **modem,** spawning a new array of **digital** interactive services, including **electronic mail (email)** and access to information provided on the **internet.** Although at first this information was provided only by text and rudimentary graphics, it did include news, stock market quotes, sports, and other information traditionally provided by radio and TV, as well as newspapers and magazines. By the mid-1990s, the internet was flexing its muscle to become a prime electronic media distribution system, and along with it came many new terms: **webcasting, browsers, ISPs, search engines, blogging, chatrooms.** On the home video front, DVDs began replacing **analog** VCRs.

As digital technology improved in the 21st century, so did the quality and quantity of internet offerings. High-speed (**broadband**) modems, offered by phone companies and cable TV companies, and larger-capacity hard drives accommodated improved graphics, audio, and video. Anyone, not just the powerful networks, could place program material on the internet that could be accessed by anyone else connected to the internet. High-quality cameras became so portable and inexpensive that individuals could use them (and their lower-quality cell phone cameras) to produce material. The passive masses became interactive individuals.

In addition, **digital video recorders (DVRs),** commonly known as TiVos, made it easier for people to record TV programs and watch them when and how they want. **Satellite radio** (originally called **digital audio radio service,** or **DARS**) allowed for radio programming that could cover the entire nation, and terrestrial radio stations started switching from analog to **HD radio,** also known as **in-band on-channel (IBOC).** Television outlets switched from analog to digital and began emphasizing **high-definition TV.** The DVD industry adopted a new high-definition format, and movie theaters started switching from film projection to electronic based **digital cinema.** Apple introduced the iPod, and **podcasting** entered the vernacular. Cell phones, **pods, MP3** players, and **PDAs** became additional avenues for accessing and using the internet and also brought forth concepts such as **text messaging.** In addition, internet sites arose that enabled individuals to engage in **social networking.**

The terms will not stop coming. As technology improves, companies will invent new ways to enhance media. The one element that runs through all the terms listed and all the ones to come is the need for content. Regardless of the delivery system, there must be something worth delivering. There will always be a need for good storytelling and for engaging methods of presenting information.

1.3 Ownership of Media Devices

Statistics and studies tell us a great deal about the relationship of people and society to the electronic media. Americans own a large number of electronic communication devices (see Exhibit 1.3). However, there have been a number of significant shifts in ownership recently.[2]

Percentage of American Households with Various Electronic Media Devices and Services

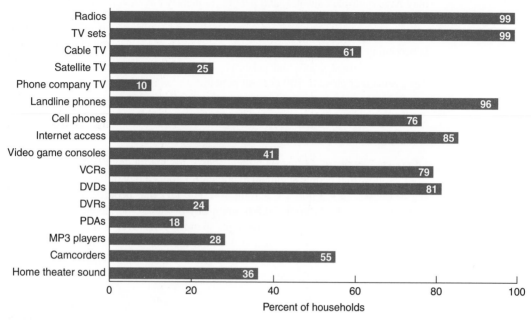

Exhibit 1.3

Americans own many devices that help them keep entertained, informed, and connected. How many of these do you own?

Internet access, which didn't even register on the charts until the 1990s, has grown at a phenomenal rate and should soon qualify as a universal service. Cell phone penetration continues to grow while **landlines,** which were almost 100 percent for many years, are shrinking. The reason is fairly obvious—the recent trend for people to use their cell phones as their only phones, eliminating the need for a phone attached to a home base. The chart shows households with cell phones, but there is a definite trend for multiple people within one household to own phones. In fact, there currently are more cell phones than people in the United States—as well as in many other countries.

internet and phones

Cable TV penetration is down, having at one point reached 70 percent. This is due in part to the rise in satellite TV subscribers, and some households have switched to a new source for their TV signals—FiOs provided by Verizon telephone company. Studies show that 96 percent of people are satisfied with the TV service given by the phone company, 86 percent are satisfied with satellite, while only 71 percent are satisfied with cable. No doubt this discontent is part of the reason for cable TV's decline.

cable, satellite, and FiOs

VCR penetration is down from about 90 percent to 79 percent and many of the 79 percent are probably units people still have but rarely use. Households have switched to DVDs, which have superior quality. The number of households with high-definition DVDs is too small to measure, but that may be something that increases in the future.

VCRs and DVDs

video games and DVRs

The percentage of homes with video game consoles has grown about 8 percent from the beginning of the 21st century to over 40 percent. A single household may have multiple consoles, and people frequently buy new models. The number of digital video recorders is growing at a rapid rate, allowing more people to shift their viewing to when they want it rather than when it is aired—a phenomenon that is becoming known as "me TV."

radio and TV

The two most stable and pervasive media devices are radios and TV sets. They remain at close to 100 percent penetration, where they have been for many years. The average household owns 6 radios and 2.4 TV sets.

1.4 Use of Electronic Media

Of course, people don't just own these electronic media devices; they use them. Studies show that people spend an average of 3 hours a day online, 3 hours listening to radio, and 4.5 hours watching TV. The fact that people multitask means that this does not add up to 10.5 hours a day of electronic media use. The best

hours spent

guesstimate is that people, on average, spend a little over 4 hours a day with radio, TV, and the internet, but that does not include the amount of time people talk on the phone, play video games, or go to the movies. Regardless what the exact number is, people spend a great deal of time with media, making it a very significant part of their lives.

internet and TV

Interestingly, although many people thought the internet would cut down on the amount of time people spent watching TV, this apparently has not been the case. In fact, in one survey, more heavy internet users said their TV viewing had gone up than said it had gone down. The number of hours that the TV set is on in the average household has remained stable (see Exhibit 1.4), although there has been a definite trend away from viewing broadcast television (NBC, CBS, ABC, Fox, etc.) to watching cable TV (MTV, USA, Lifetime, ESPN, etc.). "Watching TV" is becoming a complicated concept because people can watch traditional TV

Exhibit 1.4

The number of hours that the TV set is on continues to grow, but at a slower rate than in the past.

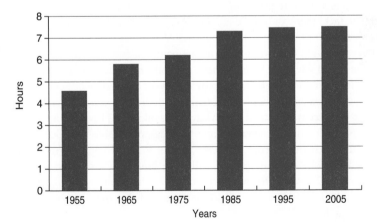

Average Hours a TV Set Is On in Households

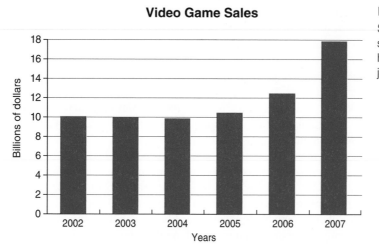

Video Game Sales

Exhibit 1.5
Sales of video game software and hardware took a jump in 2007.

programs on the internet and can also watch programming fare that is being created specifically for the internet.[3]

    Video games are an area of growth, with the amount of money spent on game software and hardware spiking in the mid-2000s (see Exhibit 1.5). *Grand Theft Auto IV* grossed $500 million in its first week of sales in 2008, and the Wii accessories rack up sales and profits. Video games have become a bigger industry than domestic movies.[4] **video games**

    Since 2003, the Online Publishers Association in cooperation with Nielsen/Net Ratings has measured internet use in terms of categories: commerce, communications, community, content, and search (see Exhibit 1.6). "Commerce" includes **internet categories**

Average Time Spent on Online Activities

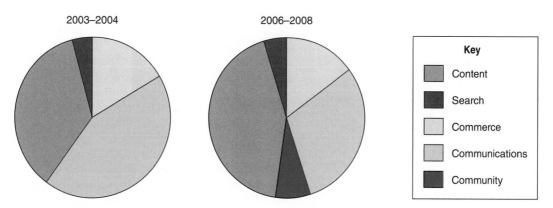

Exhibit 1.6
This chart shows some of the changes that have occurred in online activities between 2003 and 2008.

online shopping at such sites as amazon.com, eBay, and shopping.com. "Communications" originally just meant email but now also includes instant messaging and some types of group communication. "Community" was not added until 2008; it refers to sites such as Facebook and MySpace that combine user-generated content with communications in order to foster relationships. "Content" involves sites that provide news, information, and entertainment, such as MapQuest, ask.com, and CNN.com. "Search" means using the internet to look for material by requesting certain information; examples would be Google and Yahoo's search engine.

The most interesting change between 2003 and 2008 was the exchange of position between content and communications. Whereas communication was first (44 percent) and content second (36 percent) in 2003, by 2008 the numbers had almost reversed, with content at 43 percent and communication at 30 percent. This change may be due, on the one hand, to a dramatic increase in content of a reliable nature on the internet and, on the other, to the movement of some communication to the category of community.[5]

In general, people like the electronic media. Only 6 percent of people would even consider giving up television or the internet. Younger people are more likely to prefer the internet over TV, but many also report watching TV and surfing the internet simultaneously (see "Zoom In" box).

 ZOOM IN: **Extra Credit**

For those of you who like statistics, here are a few other miscellaneous facts about the electronic media.

1. More than 50 percent of teens do not buy any CDs; they download, mostly from iTunes.
2. Within the last 24 hours, about 75 percent of the population will have listened to radio and 90 percent will have watched TV.
3. Between 2006 and 2007, people reduced their time spent with broadcast TV and newspapers by 6.3 percent and increased their time spent with cable TV and video games by 19.8 percent.
4. In 2002 *Monsters, Inc.* became the first movie to sell more DVDs (7 million) than VHS tapes (4 million). DVDs have outsold VHS tapes ever since.

5. Of people who have broadband access to the internet, 30 percent view live streaming video, 20 percent watch saved video files, and 46 percent have listened to online radio.
6. Women are more likely than men (47 percent to 43 percent) to be concerned about the quality of material downloaded to portable devices.
7. Some 89 percent of teens say the internet and portable devices make their life easier; 71 percent of their parents agree.

Can you explain these facts? Also, why do you think the internet apparently has not affected TV viewing? To what do you attribute the fact that content is now the leading use of the internet?

1.5 The Functions of Media

Several decades ago, theoreticians claimed that the main functions of media were to entertain, inform, and persuade. As electronic media expanded and became more important to the social fabric, so did their functions. Today one way to look at the functions is to consider that the electronic media are involved with (1) presenting entertainment, (2) disseminating news and information, (3) aiding commerce, (4) transmitting culture and customs, (5) acting as a watchdog, (6) providing relaxation and companionship, and (7) connecting people to each other. These purposes are not mutually exclusive. It is entirely possible that one television program, for example, could serve all these functions.[6]

list of functions

1.5a Presenting Entertainment

Entertainment occupies the lion's share of time or space for many forms of electronic media. Movies, for example, are almost totally entertainment based, as are video games. Approximately three-fourths of the programming on TV qualifies as entertainment; for radio stations with a music format, the percentage is even higher. The internet started out delivering primarily information, but in recent years its capacity to deliver audio and video has improved, and now many people use the internet for entertainment purposes. Portable devices, such as an iPod filled with music, also provide entertainment.

major function

Electronic media provide so much entertainment because that is what people seem to want. Ratings for dramas are almost always higher than ratings for documentaries. Individuals, however, certainly have the ability to choose not only how much entertainment they want, but also what type of entertainment they desire. Sports fans can always find something on some medium devoted to sports, and even a group as narrow as *Law and Order* aficionados can get their fill by using broadcast TV, cable TV, DVDs, internet downloads, and other media forms.

options

For many years entertainment was a passive activity; people watching TV were often referred to as "couch potatoes" because they just sat on their couches and watched what someone else produced. Today watching passively is still the major mode for entertainment, but there are more opportunities than ever before for active participation in entertainment. Anyone can set up a website to program audio or video material. Sites such as YouTube, MySpace, and Facebook encourage people to program their own creative ideas. Some movies allow viewers to select the ending, and the user has great control over how video games progress.

passivity

The media do not always fulfill their entertainment function to the satisfaction of everyone. Politicians, citizens groups, parents, and others complain about the amount of sex and violence in TV programs, movies, and music lyrics on the radio. Although the electronic media do provide excellent entertainment, some of it is trite or just plain bad. When some idea catches on for one network, others quickly imitate it, and the imitations are usually not as good as the original. Some

criticisms   believe that this is particularly true of the reality genre. Started because it was inexpensive, it has now spawned across many networks. However, entertainment will no doubt continue to be the dominant function of electronic media. People like to be entertained, and the electronic media are well equipped to provide entertainment.[7]

1.5b Disseminating News and Information

importance   Although entertainment may rank as the top function of electronic media in terms of quantity and accessibility, there are many times when the importance of news or information transcends anything of an entertainment nature. When

routine
information   there is a terrorist attack, a fire, a tornado, or some other imminent danger, people turn to the media. This certainly happened on September 11, 2001, when

terrorists crashed airplanes into the World Trade Center (see Exhibit 1.7); internet sites were overloaded, and people were glued to their TV sets. Radio is particularly valuable in emergency circumstances because most people have battery-operated radios that continue to work when other media do not. This happened when Hurricane Katrina hit New Orleans in 2005; other media forms failed, but radio stations stayed on the air. The telephone also becomes important, especially now that emergency agencies can use reverse 911 to call many people to give them information. During the 2007 Virginia Tech shootings, students kept in touch with each other and exchanged information through cell phones and internet blogs.[8]

It does not take an emergency for the information function to be valuable, however. If you want to know whether to carry an umbrella or what route you should take to work, the media can instruct you. They can also provide the latest economic information (unemployment statistics, stock market results), as well as information about the media (what's playing at the local theater, what's on TV tonight). During political campaigns, most people learn about the candidates through the media. Both radio and television are valuable for providing useful general information, but the internet is proving even more valuable because the information you need is always available. You don't have to wait for your radio or TV station to broadcast the weather; you can access your favorite internet weather site and obtain the information instantaneously.

Most people like to stay up-to-date on the news of the day, and the electronic media provide ways of

Exhibit 1.7

The electronic media were instrumental in disseminating the news of the terrorist attacks on the New York World Trade Center on September 11, 2001.

(Amy Sancetta/AP Wide World)

doing that. For many years, the broadcast network evening news was the main source of news for many people. Those newscasts still lead in number of viewers, but the numbers have eroded greatly from the 1980s as audience members have turned to cable TV 24-hour-a-day news services, all-news radio, the internet—and *The Daily Show with Jon Stewart* (see "Zoom In" box).[9]

news

In addition to information that you need, the media also provide information that is interesting but not crucial. Although you can get through life without knowing about the mating patterns of the cockroach, a well-presented chronicle on the subject can convey provocative, and sometimes useful, information. Many TV channels provide informational programming ranging from cooking shows to science experiments to history reenactments, and the amount of information available on the internet is growing exponentially.

noncrucial information

People who use the internet for news can take an active role by personalizing their internet pages with news stories from around the world that contain specific

personalized news

 ZOOM IN: **News and *The Daily Show***

Rumors (and some research results) indicate that college-age people are using *The Daily Show* as one of their main sources of news. To oldsters, it seems preposterous that anyone would use a frothy comedy show that pokes fun at the news as an actual source for learning about the news. Jon Stewart himself admits that he is a not a journalist but rather a comedian.

However, several researchers from Indiana University conducted an academic study comparing coverage related to the 2004 presidential debates and conventions on *The Daily Show* and on the broadcast networks' nightly newscasts. They looked at NBC, CBS, and ABC newscasts and *The Daily Show* on specific days in July, August, and September when this subject was covered on all the programs. They then coded the material in terms of seconds devoted to hype (horse-race comments and hoopla), substance (campaign issues and candidate qualifications), and humor (joking and laughter). What they found was that the network shows had more hype than substance and *The Daily Show* had more humor than substance, but the amount of substance on both was the same. In other words, *The Daily Show* had just as much substance as the network newscasts.

Do you watch *The Daily Show*? If so, do you look at it as a source of news? Would you consider *The Daily Show* to be an example of how program material can be both entertaining and informative?

Jon Stewart of *The Daily Show*.

(Courtesy of the Academy of Television Arts and Sciences)

words or phrases—Egypt, SAT scores, bipolar disorder. Some services automatically email you on a portable device such as a BlackBerry any time there is a new story with the words or phrases you have indicated.

criticisms

There are many complaints about news and information, ranging from "it's boring" to "it's sensationalized." Because news departments of media entities need to make money, they sometimes resort to titillating stories that will attract audience members rather than hard news. Reporters don't always get it right. In the effort to be fast, they sometimes broadcast incorrect information. Often news is capsulated, with very little content for any one story. Most people think the news is biased—against whatever their point of view is. On the other hand, there are so many news venues now that individuals can find some source that agrees with their political leanings and thus can avoid being exposed to the views of others—a danger in a democracy. Documentaries become controversial when they present a point of view while programs that are purely informational, such as nature shows, do not attract a large audience. Although there is a great deal of information on the internet, anyone can place it there, so its accuracy is suspect.[10]

1.5c Aiding Commerce

Like it or not, advertising is a major part of the electronic media business. In the past, advertising was considered part of the persuasion function of the media because commercials were geared primarily to persuade consumers to buy products. Advertisers still hope purchases will be the outcome of most commercials, but commercials have changed, and the media now aid commerce in other ways.

persuasion

humor and effects

Research has shown that humor is effective in helping people remember commercials and the products they advertise. Also visual effects have improved greatly from what they were in the early days of electronic media. As a result, some commercials are not as hard sell as they used to be; they may contain only a short reference to the product while emphasizing humor and effects.

tailoring

Another trend is to match consumers with products that interest them. In some cable TV service areas, not everyone sees the same ads; they are tailored to the demographics of a particular area, and switching equipment at the cable headquarters sends different ads to different homes. Google has made important inroads into matching consumers and advertisers. When you type in the name of a product, the screen displays places where you can buy that product. Companies have paid to have their links included, but generally these links are helpful to the person doing the search, and they are unobtrusive.

shopping

Media, the internet in particular, enable commerce between individuals through sites such as eBay and craigslist that help people sell possessions they no longer want (such as a 1972 encyclopedia) or find people to provide services they desire (such as walking a dog). The internet has also changed shopping patterns. Many people order products through the internet without setting foot in a store; the products are delivered to their homes via one of the truck delivery services. Some products, such as music, are delivered over the internet and do not involve any form of conventional transportation.

In many ways, advertisers need to work harder to gain the attention of poten-   **gaining attention**
tial customers. People with DVRs easily skip through commercials, and the clut-
ter of commercials on radio and TV means that any one commercial can get lost
in the crowd. Advertisers have responded, in part, by placing their products within
TV programs, movies, and video games, a practice known as **product placement.**
They also create entire programs about single products (**infomercials**) or multiple
products (**home shopping** channels).

Hardly anyone will admit to liking commercials—except, perhaps, the ones   **criticisms**
that accompany the Super Bowl. Commercials can be loud, irritating, and deceptive. They are particularly controversial when aimed at young children who are
not sophisticated enough to sort out the hype. They are so numerous that they
really don't aid commerce; viewer fatigue sets in by the time the fourth ad in a
commercial break airs. Some people take offense at the personal products that
certain networks and stations air, such as feminine hygiene and condoms. Companies are constantly rethinking their marketing strategies, hoping that new ideas
will be more acceptable and more productive. Any changes they make are likely
to involve the electronic media in a major way.[11]

1.5d Transmitting Culture and Customs

The electronic media transmit culture and customs without even trying. A TV   **unintentional**
program about lifestyles of the rich and famous may not show the customs most
Americans are used to, but it does show a slice of Americana. American sitcom
mothers are usually portrayed as loving, cheerful, and devoted to their children—
cultural values held high for mothers. A home show such as *Martha Stewart Living* conveys a message about what constitutes the good life. A documentary about
Ethiopia details that country, but if that documentary is made by a French company, it may have a slant that reflects France as much as Ethiopia. A dramatic
movie about ancient China will reveal the history and customs of that time, but
will no doubt also reveal traits of modern times. People talking on cell phones
convey their attitudes and values. The music that pours from an iPod is part of
culture.

Sometimes the transmitting of culture and customs is intentional. Spanish-   **intentional**
language networks, Vietnamese-language stations, and similar entities have as
part of their mission transmitting the values of the culture related to their
languages. Religious programming is intent on disseminating the beliefs of a
particular religion. Public service campaigns often try to enforce or change customs—fasten your seat belt, don't smoke, don't drink and drive. People who
interpret the news, such as conservative radio talk show hosts, liberal TV commentators, or bloggers, are trying to transmit their ideas of what society should
be like.

By transmitting culture and customs, people involved with the media also   **criticisms**
create **stereotypes** and give false impressions. Sitcoms, in order to garner a
laugh, often exaggerate negative traits of a particular group of people. A movie
that casts a particular minority as villains can be accused of creating a false
stereotype. If programming from one culture is shown in another culture, this
can result in culture clash. Governments have tried (largely unsuccessfully) to

prevent programming from reaching their citizens because it shows customs that are not accepted in that country. Leaders in poor countries worry that if their citizens see lifestyles and customs of wealthier countries, they will become discontent and dissatisfied with the leaders.[12]

1.5e Acting as a Watchdog

In the United States, one of the functions of the media is to watch over aspects of society, particularly the government. Media have a responsibility to report defects, not ignore them. Most of this watchdog function falls to journalists.

politicians   They keep a close eye on the actions of politicians and report indiscretions—large or small. Sometimes this reporting creates an adversarial relationship between the government and the press. This is not the case in many other countries; in other societies, the press is a public relations arm of the government and primarily keeps people informed about the good things the government is doing for them.

election campaigns   During election campaigns, the American press reports on the candidates and the election process, informing people so that they can make more intelligent voting decisions (see Exhibit 1.8). As part of their watchdog function, journalists

interpretation   also interpret. For example, after the president gives a speech, they tell us what the most important points were. They try to explain why the price of gasoline is rising or why unemployment is on the increase.

Sometimes various media entities just present material and let members of the public draw their own conclusions. C-SPAN, for example, cablecasts the

Exhibit 1.8

The media covered all aspects of the 2008 election campaign.

proceedings of various government bodies, including the House of Representatives. People can watch, without commentary from journalists, and make their own observations. On the internet are political sites of every flavor that people can examine to form (or reinforce) their points of view.

public conclusions

The government is not the only body that the media watch over. They keep an eye on business, reporting misuse of funds and executive improprieties. The media also report on failures and successes in education, recreation, religion, and other aspects of society.

The people being watched do not look favorably upon the watchdog function of the media. They feel they are being treated unfairly by overzealous journalists looking to blow any minor indiscretion into a major story. Some people who might make good public servants will not enter politics because they do not want to put themselves and their families through the brash scrutiny of the press.[13]

criticisms

1.5f Providing Relaxation and Companionship

Electronic media can fulfill the function of providing relaxation and companionship just by being there. For example, TV can be an excellent companion for older people who live alone. A simple turn of a radio switch can bring you relaxing music. Some people find relaxation in sorting through their email, playing video games, or talking to friends on the phone.

companion

Part of the reason that media can be relaxing is that they provide vicarious experiences. After a hard day at work, many people want to forget their own problems and become wrapped up in the lives of fictional characters whose problems far surpass theirs. Vicarious experiences can also serve as a release from boredom, or they can simply help pass the time. They can serve as an emotional release. Horror movies allow people to scream in a theater, and tear-jerkers allow them to cry without other people looking at them strangely.

vicarious experience

To some, relaxation is a synonym for laziness, and they do not buy into the value of media as an agent of relaxation. And, indeed, some people become so narcotized by media that they spend hours in front of a TV set or computer screen when they would be better off going out and socializing or exercising. Although media can provide relaxation and companionship, they can also waste a lot of time. Television, in particular, is criticized for being a "boob tube" and delivering content that requires no thinking or action on the part of the viewer. It is blamed for obesity because people tend to eat while watching, and they watch passively rather than engaging in any physical activity that might shed a few pounds. Video games receive similar criticism, but at least they involve hand-eye coordination, and some of them require other bodily movement.[14]

criticisms

1.5g Connecting People to Each Other

Electronic media have always provided material that enables people to have something in common. When a movie comes out, people talk about it—and producers hope the word-of-mouth buzz will sell more tickets. In the heyday of radio drama, large numbers of people listened to the same program at the same time and then talked about it with their friends and co-workers the next day. For many years, network TV served the same "watercooler" function. Today viewing and listening are more fragmented, so people are less likely to have conversations in which everyone has seen or heard the same material, but even if a few people haven't seen a particular show, they have probably heard about it—through the media.

conversations

In addition to providing content that connects people who have all experienced it, many media forms provide for direct contact between people. Email over the internet is one example; people who would never have bothered to write a letter to someone will dash off an email or a text message. The current generation of high school and college students is often referred to as the "connected generation," in part because of their high use of cell phones and other communication devices (see Exhibit 1.9). But they aren't the only ones who use media to connect. Their parents find old high school and college classmates through internet sites. Talent agents use the internet to scout for promising actors and musicians. Bloggers connect with people they have never met. People who play video games or have social networking sites on the net make new friends they never would have met otherwise. Internet dating services bring people together, and the most common form of dating is going to the movies. Chat groups act as support for people with various diseases, people who have been victims of abuse, or people who simply share common interests.

direct contact

Exhibit 1.9

Text messaging has become an important form of communication, especially for young people.

Media may help people to connect, but they also help them to disconnect—
the boy who is too immersed in a video game to talk to his parents, the teenager
who drowns out the world with an iPod, the father who won't answer his daughter's question because he is watching a TV sports event. Also, the ease of connecting makes it easier for people to express hate through emails or phone
messages and to prey upon unsuspecting people such as children and the
elderly.[15]

criticisms

1.6 The Democratization of Media

One major change in electronic media in recent times is that people in general
are now much more in control of the media. This is a major **paradigm shift**
from earlier times when **gatekeepers** controlled most the public's use of media.
Gatekeepers included network executives who chose the series to be aired and
decided what time they could be watched; news producers who selected which
stories would make it to the evening newscast; radio music directors who
selected the music their station would play; and movie studio executives who
gave the green light to film projects and then decided when and where they
would open.

gatekeepers

Today those people still exist, but they have much less power than in the
past. People with DVRs record programs and watch them when they want to.
There are so many news outlets that people can choose which they want to frequent. In addition, people with camcorders or cell phone cameras often supply
material for the newscast if they happened to be where major news breaks (see
"Zoom In" box on page 20). Individuals select their own music, filling their iPods
or MP3 players with what they like, not necessarily what a radio station music
director might select. People have many more options for watching movies than
just going to a theater, and they can create their own little movies and exhibit
them on the web.

supplying news

User-generated content has greatly democratized electronic media.
Much of what happens in the world does not get reported or programmed by
the major media companies. That leaves plenty of room for individuals to
report on events of importance to them. Parents can distribute an announcement about the birth of their baby over the internet. People can post facts,
anecdotes, thoughts, photos, videos, and just about anything else on social networking sites. Music lovers can create and operate their own radio stations.
It's also easy for people to burn CDs of music they perform or to make a DVD
of a video project. Any number of people have had success marketing their
materials through the internet or by getting video stores to carry videos they
create.

*user-generated
content*

Cable TV facilities have **public access** channels, and although they are
somewhat "yesterday," they are another outlet through which people can produce
video material and gain an audience. Callers to radio talk shows play a more
important role than they used to, often setting the agenda for the show. All these
examples and many more have changed the nature and basic structure of the
media, making them more accessible and democratic.[16]

other aspects

ZOOM IN: We're All Reporters

A large number of people carry news recording equipment with them all the time. The first time this became obvious was in July 2005 when bombs ripped through London's subway system, and victims trapped below ground used the cameras built into their cell phones to record the events and email the images to British TV networks. Because of the dangerous conditions, members of the press were not allowed near the places where the explosions had occurred, so the original reports were based on these citizens' footage.

The London incident was far from the first time that ordinary people had contributed to the news. Much of what was seen of the devastation of the December 2004 tsunami in South Asia was taken by tourists with camcorders who were vacationing in Thailand. As far back as 1991, George Holliday used his camcorder to record the arrest and beating of Rodney King by the Los Angeles police—a tape that was played repeatedly by news organizations and used in the trials of the police officers involved. Earthquakes, airplane near misses, and many other events have been reported first by people who were in the vicinity with a recording device.

At another level, there are nonjournalists who purposely gather news and then try to sell the footage to news organizations. Others put together news stories for their blogs or websites. Some

local stations and networks encourage amateurs to submit footage. Local TV stations, especially, do not have staff to cover all that happens in the community and welcome outside contributions.

This "citizen journalism" has its problems. One is determining the authenticity of the footage. Unscrupulous people seeking to dupe a news organization may send anonymous footage that is fictional. Another problem is that overzealous amateur reporters who have not been trained in journalistic procedures may get in the way of fire or rescue crews or put themselves in harm's way. The quality of the footage is often poor—shaky, with too many zooms.

To cope with these problems, some media outlets do not air footage sent anonymously and also make contributors sign documents attesting to the authenticity of the material. Some stations offer courses to train people who wish to contribute news footage. This is win-win because the independent "reporters" make contact with the stations and can then sell their wares more easily, and the stations receive better footage from people who they know have had some training.

If you were a news director of a TV station, under what circumstances would you air amateur footage? Can you think of a time when you might have been able to capture footage that could be used by a news organization?

A scene from the 1991 tape a citizen made of the Los Angeles police arresting Rodney King.

(AP/Wide World Photos)

1.7 Convergence, Proliferation, and Resilience

Convergence, proliferation, and resilience are in some ways contradictory and in other ways complementary. They all affect the degree of significance that media have in our lives. **Convergence** refers to bringing together various media forms and facilities. **Proliferation** refers to distributing media content to a wide variety of venues through an ever-increasing number of distribution platforms. **Resilience** refers to the ability of various forms of electronic media to adapt and continue to exist. Interestingly, although different forms are brought together, they are also more widespread, and they seem to be able to withstand a changing landscape.

One aspect of convergence involves the blending of separate media to create new types of information and entertainment suppliers. Radio stations stream programming over the internet; DVDs play on a TV set or a computer screen; TV programs are podcast. Another type of convergence involves media businesses that supply similar things. Cable companies and phone companies provide high-speed internet access; satellite TV shows much of the same programming as cable TV. As companies try to grapple with what the future will bring, they cover their bets by merging and investing in various forms. In so doing, they often converge parts of their business. If, for example, a company has radio, broadcast TV, cable TV, and internet interests, it may use the same staff members to service all the media forms—for example, the same news reporters may gather news for all the entities, even though the presentation of the news is different for each one.

convergence

Because companies cover their bets, they are also eager to adopt new distribution methods that will further the impact of their products and increase their financial bottom lines. The result is that media products are everywhere. Not only are they accessible in your home, but they appear on your grocery cart, in the restaurant or sports bar where you eat (see Exhibit 1.10), in the airport, at the gas station, and in doctors' waiting rooms. Portable equipment is responsible for many lifestyle changes. Children can now watch DVD movies or play video games in the backseat of a car while traveling, eliminating some of the "Are we there yet?" phenomenon. Salespeople can call from the road to set up new appointments. If you enjoy any form of media-oriented work or leisure, you are probably contributing to the bottom line of one of the big conglomerates such as Time Warner, Viacom/CBS, Sony, News Corp./Fox, NBC-Universal, or Disney.

proliferation

The media business is very fast-paced, with media forms changing at a rapid clip. However, despite all the new media, for the most part the older media remain. Movie theaters have been declared dead any number of times, but they still exist (see Exhibit 1.11). They have changed emphasis over the years from grand palaces in the cities to multiplexes in suburban malls, but they have not disappeared. When TV stole radio's comedy and drama programming in the 1950s, many pundits said radio would disappear, but it didn't. Rather, it adapted by changing its programming to disc jockey shows. The audience for broadcast TV keeps shrinking, but it, too, is looking for ways to adapt. The internet is king right now, but invariably something will come along to challenge its dominance.[17]

resilience

1.8 Issues and the Future

One generalization that can be made about the future of the electronic media is that their significance will increase. People are becoming more dependent on electronic media, and will probably use them to a greater extent than at any time in the past.

The internet, in particular, will grow in terms of functions and use. As the information on the internet becomes more reliable, it will further replace the library as the main place to find facts and opinions. The number of functions that the media fulfill may also increase. Perhaps they will become more important for business-to-business communication as air travel becomes more difficult; perhaps they will be used to help solve health care issues.

internet

The criticisms related to media—sex and violence, lack of originality, sensationalism, inaccuracy, bias, stereotyping, overzealousness—probably won't change. They represent human characteristics that have been around for ages. But advertising is likely to undergo many changes in the future, mainly because the old models are not working well. Advertising provides the main financial support for the various media forms, so it will not disappear, but it could become more targeted and less intrusive in order to increase its effectiveness.

changes

The connecting function is the newest of the functions and is still in a state of flux. Some of the equipment that connects people also disconnects them. For example, people are shifting from family phone numbers to individual cell phone numbers. Although this has its advantages in that people only get calls directed at them, it is one more media form (following radios, TV sets, and game consoles) that no longer facilitates family interaction. Many other applications that relate to connectivity may become common—for example, GPS calculations that let people know exactly where you are.

connecting

Individuals have the tools to further democratize the media. Each new recording device seems to be smaller and easier to use. Whether individuals will continue to have the time and inclination to produce programming and website material remains to be seen.

democratization

Convergence and proliferation tend to come and go in cycles. Companies merge and buy out other companies until they become large and unwieldy; then they start breaking themselves down into smaller pieces. The next several chapters of this book look at separate media, giving you information about the history and characteristics of each. You will no doubt notice that media forms have always converged to some extent and that program material has proliferated over the years. The difference today is in degree; everything is faster and more intense, making for exciting challenges. You will also notice that all the media forms have exhibited resilience—a trait that will continue into the future.

continued resilience

1.9 Summary

The electronic media are well used and well known. Those who wish to work in media jobs should learn the intricacies of the field, and those who use the media should understand their effects.

Over the years media forms have proliferated from radio and TV networks and stations to include public broadcasting, cable TV, satellite TV, low-power TV, and corporate TV. Narrowcasting, VCRs, DVDs, video games, remote controls, and DVRs have reduced the power of programming chiefs and put more power into the hands of individuals. The internet and portable devices continue to alter the media scene.

People own many communication devices, but their importance and penetration shift as they compete with each other. People use these devices many hours a day, with internet content being a growth area.

The functions of the electronic media include presenting entertainment, disseminating news and information, aiding commerce, transmitting culture and customs, acting as a watchdog, providing relaxation and companionship, and connecting people to each other. In all areas, there are criticisms of how the media handle their functions.

Recent times have seen a democratization of the electronic media as individuals take control of what they watch and when they watch it. Individuals also contribute to the news and create user-generated content. Convergence is bringing various media forms together as companies work to hedge their bets. Proliferation of media has increased over the years. Through it all the media forms remain resilient.

Suggested Websites

www.broadcastingcable.com (the website for a magazine that contains a great deal of information about electronic media)

www.pewtrusts.org/our_work_category.aspx?id=230 (a site for statistics about electronic media)

www.online-publishers.org/page.php/prmID/421 (the site that shows internet use discussed in this chapter)

www.c-span.org (C-SPAN, a watchdog organization that cablecasts government activities)

www.youtube.com (YouTube, the site that accepts videos from almost anyone)

Notes

1. Douglas Blanks Hindman and Kenneth Wiegand, "The Big Three's Prime-Time Decline: A Technological and Social Context," *Journal of Broadcasting & Electronic Media,* March 2008, 126; and "Ratings Dive in 2006–2007 Season," *Broadcasting and Cable,* May 28, 2007, 3.
2. In gathering statistics about electronic media, it is difficult to find apples and apples. Some figures are based on households and some on individuals; some give percentages, and some give

raw numbers; some figures are for all media and some just for selected electronic media; it is hard to find all of the ones you want for a specific year. Independent sources are certainly the best to use, as companies and advertisers who list statistics often give them a biased slant. Certainly it is possible to use statistics to prove just about anything you want—see Darrel Huff's perennial tongue-in-cheek favorite *How to Lie with Statistics* (New York: Norton, 1954). Nevertheless, statistics do lend interest and credence to the significance of electronic media, and even though they may not be totally accurate, they do show past behavior and future trends. To put the statistical material together for sections 1.3 and 1.4, the author accessed a wide variety of sources during the summer of 2008, the most significant of which were the following: www.census.gov; www.informationweek.com; www.broadcastingcable.com; www.hollywoodreporter.com; www.latimes.com; www.reed-electronics.com; www.wallstreetjournal.com; www.mediaweek.com; www.nielsenmedia.com; www.pewtrusts.org; www.stateofthemedia.org; http://technology360.typepad.com; www.metrics2.com; and http://nabshowdaily.nbmedia.com.

3. "TV Is Dying? Long Live TV!" *Fortune,* February 2, 2007, 43; and "Household Hours of Viewing Per Day," *Broadcasting & Cable,* August 2, 2004, 5.
4. "Grand Theft Auto IV Steals Video Game Record," *Los Angeles Times,* May 8, 2008, C-1; and "Sales Rise Despite Economic Worries," *Los Angeles Times,* February 1, 2008, C-1.
5. "Internet Activity Index," www.online-publishers.org/?pg+activity (accessed November 4, 2004); and "Internet Activity Index," www.online-publishers.org/page.php/prmID/421 (accessed June 2, 2008).
6. The entertain, inform, and persuade functions are well delineated in chapter 4, "The Purposes of the Media," in Edd Applegate, *Print and Broadcast Journalism: A Critical Examination* (Westport, CT: Praeger, 1996). The functions given here are the author's.
7. "Prime Time's New Reality," *Broadcasting & Cable,* August 23, 2004, 8; "Network Fear: The Net as a Copilot," *Los Angeles Times,* March 27, 2007, E-1; and "Something Borrowed, Two New," *Hollywood Reporter,* May 14, 2008, 3.
8. "Crisis Coverage Showed TV Reporting at Its Best," *Hollywood Reporter,* September 13, 2001, 3; "A Lifeline Sent by Airwave," *Los Angeles Times,* September 10, 2005, 1; and "The Impact of Virginia Tech on the News," *Broadcasting & Cable,* April 23, 2007, 14.
9. The study that is referred to in the box is Julia R. Fox, Glory Koloen, and Volkan Sahin, "No Joke: A Comparison of Substance in *The Daily Show with Jon Stewart* and Broadcast Network Television Coverage of the 2004 Presidential Election Campaign," *Journal of Broadcasting & Electronic Media,* June 2007, 213–227.
10. Arvind Diddi and Robert LaRose, "Getting Hooked on News: Uses and Gratifications and the Formation of News Habits among College Students in an Internet Environment," *Journal of Broadcasting & Electronic Media,* June 2006, 193–210; and "Documentary TV," *Television Week,* June 6, 2005, 22.
11. "Strike Up the Brand," *Hollywood Reporter,* April 4–6, 2008, S-10; and "Pay as You Show," *Emmy,* September/October 2007, 76–79.
12. "Latino TV Gets Serious," *Newsweek,* March 20, 2007, 48; and "Turns Out World Really Is Flat," *Hollywood Reporter,* May 2, 2008, 5.
13. "CNN Rolls Out Election Express," *TV Technology,* October 17, 2007, 6; and Joseph Graf and Sean Aday, "Selective Attention to Online Political Information," *Journal of Broadcasting & Electronic Media,* March 2008, 86–100.
14. "Study: Young Kids Immersed in TV," *Broadcasting & Cable,* November 3, 2003, 22.
15. "Counter Intelligence," *Emmy,* March/April 2007, 56–59; and "Staying Connected," *Newsweek,* December 1, 2003, E14.
16. "Why Everybody Is a Reporter," *Broadcasting & Cable,* August 22, 2005, 14; "Cell Phones Change the View of Disaster," *Los Angeles Times,* July 8, 2005, 1; and "Social Media Success Demands Listening," *TV Technology,* May 14, 2008, 36.
17. "Jumping on the Band Wagon," *Emmy,* September/October 2006, 82–85; "The Brave New World of TV," *Broadcasting & Cable,* November 14, 2005, 6; "New Media, New Newsrooms," *Broadcasting & Cable,* January 14, 2008, 22; and "Television Everywhere," *Broadcasting & Cable,* July 24, 2006, 14–15.

Chapter 2

THE INTERNET, PORTABLE DEVICES, AND VIDEO GAMES

The internet has become *the* place to be. It is closely aligned with many forms of entertainment and information, including video games and portable devices (cell phones, PDAs, iPods) discussed in detail in this chapter. But it has also encompassed the more traditional media from the past. Whereas radio, television, and movies were separate for many years, the internet now incorporates material from all of them—as well as from newspapers, magazines, the record business, the telephone industry, and many other media entities.

> If you understand it, it's obsolete.
> **Mosaic Computer, Inc.**

The internet's interactive capabilities have taken media in a new direction. Previous media forms were basically one way—the producers of content distributed to the public, most of whom simply listened to or watched the material passively. With the internet, everyone can be a producer, distributing to anyone who clicks on the site. When people interact with the internet, be it for shopping, gaming, emailing, blogging, or any of the many other activities that the internet has spawned, they are in an active stage that involves thinking and doing. The internet did not start out that way, however. It had humble beginnings and then took hold in ways no one had originally envisioned.

2.1 Origins of the Internet

The internet's roots can be traced to two phenomena in the middle of the 20th century. First was the development of computers. Electromechanical calculating machines existed more than 100 years ago, but the first practical **vacuum tube** computer was developed by the U.S. Army during World War II to calculate ballistic artillery trajectories quickly. The general public learned of computers for the first time in 1952 when UNIVAC (see Exhibit 2.1), a computer manufactured by Remington-Rand, was used for election coverage. With only 5 percent of the vote counted, it predicted that Dwight Eisenhower would defeat Adlai Stevenson for president.

computers

The second phenomenon leading to the internet was the cold war. In 1957, the Soviet Union launched the first communications satellite, *Sputnik.* Realizing the potential of space-based communication for defense, in 1958 President Dwight Eisenhower formed the Advanced Research Projects Agency, commonly called ARPA, and it became part of the Defense Department. Its mission was to manage and direct scientific research in the pursuit of a stronger military.[1]

cold war

ARPA

In 1962, Joseph C. R. Licklider, a psychology professor at the Massachusetts Institute of Technology (MIT), was selected to direct ARPA's behavioral science office. "Lick" put together a team of young researchers to develop his vision of using computers to enhance intelligence. One of these scientists, Bob Taylor, became frustrated that the three computer connections he had in his office at the Pentagon in 1966 (University of California at Berkeley, Strategic Air Command in Santa Barbara, and Lincoln Labs at MIT) could not talk with one another because each had a different computer language and operating protocol. He realized that research and communication could be greatly enhanced if scientists at one location could interact with computers at other locations. He hired a team to make this happen.

Exhibit 2.1
Dr. J. Presper Eckert (center) describes the functions of the UNIVAC 1 computer he helped develop in the 1950s to newsman Walter Cronkite.

(AP/Wide World Photos)

IMPs

The scientists reasoned that if each computer had to learn the language of the other computers, that would require too much computing power—

Exhibit 2.2
The BBN team that developed the first IMP in 1969.

(© Photo courtesy of BBN Technologies)

computers in the 1960s didn't have anywhere near the calculating ability of today's personal computers, in spite of their room-filling size. They decided that each computer would have a smaller computer, called an **Interface Message Processor (IMP),** which would translate data, leaving the larger computer free to do its calculations. In January 1969, the bid to build an IMP went to a small Massachusetts engineering firm, Bolt, Baranek, and Newman (BBN) (see Exhibit 2.2), which met a deadline of just nine months and delivered the first IMP to the University of California, Los Angeles (UCLA). On September 2, Leonard Kleinrock led a team that hooked up a mainframe computer to the IMP and soon had the big and little machines sharing messages (see Exhibit 2.3). Thus was born ARPA's computer network, **ARPANET,** the forerunner of today's internet.

Exhibit 2.3
Len Kleinrock is pointing out that his modern-day watch has more computing power than the first IMP installed at UCLA in 1969. Kleinrock has said that in the early days he thought the internet would be useful, but what he never envisioned was that someday his 97-year-old mother would be using it.

(AP/Wide World Photos)

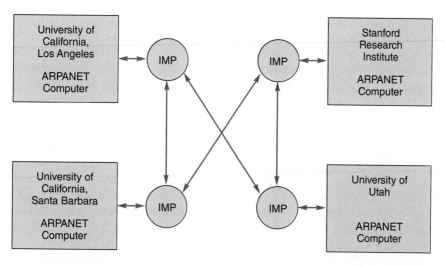

Exhibit 2.4

Diagram of the Interface Message Processors connected to the first four ARPANET computers in December 1969.

At first this network grew slowly. By December 1969, four universities were hooked up (see Exhibit 2.4). In 1984, there were approximately 500 computers on the ARPANET when the National Science Foundation took it over. Since it was no longer part of the Defense Department, people stopped using the name ARPANET and began referring to this interconnected network of computers as the "internet." Through the rest of the 1980s, the number of linked computers grew astronomically. By 1991, more than 300,000 computers were connected, and the internet was no longer the sole domain of research universities. It had permeated corporations and was beginning to enter homes.[2]

ARPANET

other networks

2.2 Standardizing the Internet Design

The idea of networking computers did not belong to ARPA alone. Other computer networks sprang up, such as USENET and TELNET. Each network had its own way of transferring messages and files among its computers. For all the networks to be interconnected into one big network, it was clear that one standard scheme, or protocol, had to be created and adopted for file sharing. The task of creating this single standard went to Vinton Cerf (see Exhibit 2.5), a graduate student in 1969 at UCLA—where the ARPANET launched—who became a professor at Stanford in 1972. He decided there needed to be a computer between each network that functioned in the way IMPs functioned between computers—concerning themselves only with a common language for information sharing rather than with the complex codes of each network.

Cerf called this new scheme **Transmission Control Protocol/Internet Protocol (TCP/IP),** and it is the internet protocol used today. The idea was to enclose each information packet in a **datagram,** something like an envelope that contains a letter (the header on an email

Exhibit 2.5

Vinton Cerf, often referred to as the "Father of the Internet."

(AP/Wide World Photos)

TCP/IP

message functions the same way). The letter itself is not important for delivery, but the envelope is. The datagrams would contain standardized information for delivery so that the packets were sent to the appropriate computers. Those computers would then decode the packets (open the envelopes) to access the content. TCP/IP was used by programmers for other networks (e.g., TELNET), so their networks could interface with the ARPANET. Though there was no directive that all computer networks be interconnected, this seemed to be happening because researchers saw the usefulness of this standard protocol.[3]

Because the idea of interconnecting computers for defense was born during the cold war, it was important that the network be able to survive a nuclear attack. Paul Baran, a researcher at the RAND Corporation, was charged with designing this network in the 1960s. Baran sketched three different types of networks (see Exhibit 2.6). A *centralized* network has all the information flow to and from a central unit, such as the Pentagon, making it too vulnerable because if the central unit goes down, the entire network goes down. A *decentralized* network has a number of different units through which information flows, which makes it less vulnerable, but if any one unit goes down, the part of the network connected to that unit goes down. A **distributed network** has no central unit, making it the least vulnerable because it has multiple paths for information flow. Like a fishing net, each piece is connected to more than one other piece, so if one piece breaks, the information can be routed around that break and move to its destination. This was the network pattern chosen for the internet.

distributed network

This distributed network consists of many **nodes**—the dots on the sketch in Exhibit 2.6. A node functions like a post office. When it receives information, it routes it to the next node, which in turn routes it to the next, and so on, until the information gets where it's going. If any path or node is broken, the information is returned to the previous node, which then reroutes it through a different path or node. Of course, all this happens at the speed of electrons.

packet switching

How does a node know where to route information? Leonard Kleinrock of UCLA had the answer: **packet switching.** On the sender's computer, a message (text, graphic, photo, audio, video) is broken up into little packets of digital ones and zeros. The computer assigns a datagram (header) to each packet, noting its origin, destination, date, time, what it contains, and where it belongs in the overall

Exhibit 2.6

A drawing representing Paul Baran's 1960s sketch of three types of computer networks.

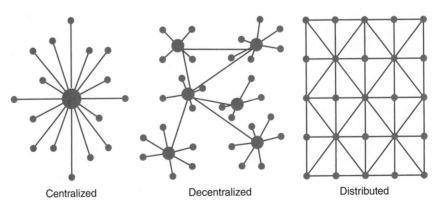

Centralized   Decentralized   Distributed

message. The computer launches the packets into the distributed network. The packets are switched along different paths (e.g., copper wire, coaxial cable, fiber optic, satellite). They arrive at the receiver's computer and are reassembled as the original message. If you have ever watched a webpage load on a slow computer, you have seen how the page "chunks in" piece by piece. Each piece is a packet, arriving at the computer and loading into its respective place. This packet switching on a distributed network became the backbone design of the internet.[4]

2.3 The Beginning of Email

The pioneer credited with the invention of **electronic mail** (**email**) is Ray Tomlinson. While working on the ARPANET for BBN in 1971, he thought it would be a good idea if the scientists could share messages across computers in addition to sharing research data files. On his own time—writing email was not part of anyone's job description—he tweaked an electronic messaging program that allowed users to send messages to others' mailboxes on the same computer. Combining that program with an experimental file transfer program, he was able to send an electronic message to himself between two networked computers sitting in the same room (see Exhibit 2.7).

Tomlinson

Exhibit 2.7

The first email computers in 1971. Ray Tomlinson sent himself a test message from the computer in the back to the one in the front. Although the computers were side by side, this first network email was delivered via the ARPANET, to which both computers were connected. Note: the computers are on the tables by the racks; the machine on the left is a Teletype.

(Photo courtesy of Dan Murphy/BBN)

That first email message is lost to history. Tomlinson doesn't remember it, but guesses it could have been something as simple as the top row of capital keys—QWERTYUIOP (see "Zoom In" box). Nonetheless, the use of electronic mail caught on like wildfire, making it the first "killer application" of the new ARPANET. By the next year, 1972, nearly every network user was regularly sending and receiving electronic messages. In addition to exchanging research information, these scientists engaged in heated debates about the Vietnam War (an early chatroom) and developed an early, computer-based video game they called Space War.[5]

2.4 The World Wide Web

Another important aspect of the internet is the **world wide web (www),** also known simply as "the web." It consists of a vast array of content provided by corporations, nonprofit associations, universities, government agencies, and individuals from all over the planet. It is a distributed information system that links users to much of human knowledge.

 ZOOM IN: **Email Myths**

No one is more surprised about the interest in the history of email than Ray Tomlinson, its creator. To this engineer who graduated from Rensselaer and MIT, creating email was no big deal and certainly nothing that he took any pains to record. In fact, one of his colleagues told him not to tell his boss what he had done because that wasn't part of his job. He was supposed to just work on a file transfer system.

On his website, Tomlinson points out some myths that have arisen about the invention of email. One is that he invented the "at" sign. Actually @ had been around for a long time, but he may have saved it from extinction because it was used so rarely that some were considering removing it from the keyboard.

He also did not press shift-2 to type @. On the keyboard he used, @ was just to the right of "P" where the left bracket and brace are on modern English keyboards.

Although some accounts have said that the first email message was "QWERTYUIOP," Tomlinson points out that what he said is that it was something *like* QWERTYUIOP. It could have been "TESTING 1 2 3 4" or something equally insignificant, but he does think it was uppercase.

Ray Tomlinson, the creator of email.

(© Photo courtesy of BBN Technologies)

The pioneer credited with inventing the web is Tim Berners-Lee (see Exhibit 2.8), a British physicist working at the European Center for Nuclear Research in Geneva, Switzerland. Berners-Lee wanted to accelerate automatic information sharing among scientists at different institutions around the world. He reasoned that research could be swapped almost instantaneously using the internet. To make this happen, he combined three ideas. One was the *TCP/IP* protocol created by Vint Cerf, discussed previously. Another was *hyptertext,* a term first used in the 1960s related to the concept of connecting virtual documents in computers. Third was the *Domain Name System (DNS),* created in the 1980s to translate web and email addresses into standard numbers that computers can read to route information packets through the internet.

Exhibit 2.8
Tim Berners-Lee, often called the "Father of the World Wide Web."

(Nemereofsky/IPOL/Globe Photos)

Putting all this together, Berners-Lee introduced, in 1990, a network computer language called **HyperText Markup Language (HTML).** It made instant information sharing possible by creating **hyperlinks**—words and images that appear highlighted in one document and that allow users to click and be taken instantly to other documents. He envisioned this automatic, instantaneous, global sharing of information as a spider's web spun around the world, so he called his creation the world wide web.

Berners-Lee

hyperlinks

Berners-Lee also wrote the protocol that made possible the transfer of hypertext documents: **HyperText Transfer Protocol (HTTP).** To be available on the web, each document must be assigned a unique **universal (or uniform) resource locator (URL),** also called a web address. Thus was born the string of code letters that make up the addresses you see when you use the web (e.g., http://www.mcgraw-hill.com/index.html).[6]

URLs

2.5 Politicians Boost the Internet

Until the 1990s, the internet was used primarily by government offices and universities. Started by the U.S. Defense Department, it had evolved into a worldwide amorphous entity. It had political support, most notably from Senator (and later Vice President) Al Gore, who had sponsored legislation in the 1980s and early 1990s that funded and expanded the **information superhighway**—a useful metaphor for the grid of cables, fiber optics, and wireless transmitters that allow information to flow among networked computers. As the internet grew, it no longer needed government funding and intervention; it could operate on its own. In 1992, Congress passed legislation that privatized the net, allowing individuals and corporations to profit from it commercially.

privatization

This resulted in a free ownership structure. Most electronic media are owned by companies, such as media giants Time Warner, Disney, and Viacom, as well as smaller private companies. These businesses create and distribute

no owners

content using channels, such as radio and television networks, that they also sometimes own. Unlike these traditional media, however, no one owns the internet. Anyone can be an information provider. You don't need a license, government approval, or a lot of money.[7]

The privatization of the net greatly increased its use among individuals and corporations. By the end of 1993, 3.2 million computers were connected, representing about 10 million users. By the end of 1994, about 10,000 web **servers** were online, of which some 2,000 were commercial.[8]

2.6 Refining the Internet

In the internet's early days, the people using it were fairly technically minded and could wade through complex codes and operational procedures to exchange information. However, once it became private and ordinary people started using it, much attention was devoted to making the internet more user friendly. This involved the development of browsers, internet service providers, and search engines.

2.6a Browsers

Mosaic

Netscape

The World

Like maps in nature parks or shopping malls, **browsers** serve as launching points to guide users to the information they seek. History credits Marc Andreessen (see Exhibit 2.9), a graduate student at the University of Illinois, as the inventor of the first truly viable browser for personal computers, Mosaic. Developed in 1993, Mosaic contributed to the net's explosion that year. Mosaic had an easy-to-use interface, incorporating both text and graphics, as well as powerful search capabilities, especially for its day.

In 1994, Andreessen joined forces with Silicon Graphics founder Jim Clark to develop and market his browser. They renamed the company Netscape Communications and boasted the lion's share of the browser market: as high as 80 percent in 1995, the year the company became public for investors to buy and sell its stock. But when Bill Gates's Microsoft launched its computer operating system Windows 95 in August of that year, it included a free browser called Internet Explorer. Although Explorer initially was considered inferior to Netscape, after a few years and a few generations of improvements it was able to surpass Netscape in market share, and it still holds the lead today. Netscape still exists with a browser called Firefox, which has a small percentage of usage.[9]

2.6b Internet Service Providers

To use a web browser, one must connect to the internet. Various **internet service providers (ISPs)** offer this connection, usually for a fee, though some ISPs are free—receiving their revenue from advertisements. The first ISP, called The World, came to market in 1990. The two major browsers, Microsoft Internet Explorer and Netscape-Firefox, operate as free ISPs. Other popular ISPs have included Yahoo, America Online (AOL), EarthLink, and Verizon.

Exhibit 2.9

Marc Andreessen, creator of Mosaic, the first commercial browser.

(Globe Photos, Inc.)

ISPs sometimes provide subscribers with specially prepared information and services in addition to internet access. For example, some can open up at log-in with news tailored to the client's interests. Some also allow subscribers to place their own information on the net by assigning them server space to create webpages. At first, ISPs charged for internet access by the minute, but in 1995 EarthLink offered the first "all-you-can-eat" service, and now most charge a monthly fee for unlimited (or almost unlimited) access.

services

Prior to actual ISPs, there were text information providers. The main one was CompuServe, started in 1979 as a service that consumers could access from rather modest computers to view large amounts of information (news, sports scores, stock market quotes, weather, etc.) stored within a large CompuServe computer. Early CompuServe and its contemporary services were expensive and complicated, and the standards, operational procedures, and codes necessary to obtain the information made the whole concept inaccessible to most people. In 1997, AOL purchased CompuServe, folding some of its services into AOL's ISP.[10]

CompuServe

2.6c Search Engines

Compared with early CompuServe, searching and retrieving information from the web today is relatively simple. The development of ISPs and browsers naturally led to **search engines**—giant indexes of information. Attempts at search engines date back to the early 1990s when students at several universities set up systems that indexed names of files but not words within the files. The first full-text search engine was WebCrawler, introduced in 1994. A number of improvements followed, such as determining word proximity and being able to search photos and music.

WebCrawler

In 1998 Sergey Brin and Larry Page (see Exhibit 2.10), students at Stanford University, founded Google, the most popular of today's search engines. Not only does Google claim to index the most webpages and register the most hits,

Google

Exhibit 2.10
Google founders Sergey Brin and Larry Page.

(Photo courtesy of Google, Inc. Used with permission.)

but it has led to a new verb: to "google" means to search for something on the web. Google and some of the other search engines use a **raw website return** scheme in which the user enters keywords into a box and the engine scours the web for pages that contain those keywords.[11]

2.7 Internet Growth and Temporary Downfall

e-commerce

As the internet companies of the 1990s grew, merged, and bought each other out, many of them provided browsers, internet service, search engines, email, websites, and a host of other services that were springing up on the internet. They were joined by commercial companies who, because the internet had been privatized, saw it as a means of making money and coined the term **e-commerce** (electronic commerce).

sex

In 1995, amazon.com, founded by Jeff Bezos (see Exhibit 2.11), sold its first book and quickly became the net's biggest retailer. Many cyberstores followed, some existing only on the net and others as offshoots of brick-and-mortar businesses. Pierre Omidyar founded AuctionWeb in 1995 and changed the name to eBay in 1997; the first item sold was a broken laser pointer for $14.83. Omidyar actually called the buyer to make sure he realized the pointer was broken and the buyer said, "I collect broken laser pointers." Realtors, toy stores, stock brokers, airlines, herbal medicine companies, and many others established an internet presence. Many of these did not make money initially, but they felt there was great potential.[12]

Exhibit 2.11
Amazon.com founder Jeff Bezos.

The "product" that made money consistently almost from the beginning was sex—taking in about $1 billion a year. Approximately one-third of all internet usage is visits to porn sites, most of which display a little free material before charging. About 70 percent of visits to these sites are during the 9 A.M. to 5 P.M. workday.[13]

Other web-based activities that came to the fore during the 1990s were **instant messaging (IM)** for live chatting, **bulletin boards** for posting messages, web logs (**blogs**) for chronicling life's events, **chatrooms** for private conversations, and **distance learning** for college coursework.

Because of the rapid development and enormous public acceptance of the internet during the 1990s, the companies involved with it caught the fancy of the stock market. They could sell stock at phenomenal prices, even though they weren't earning any profits. For example, when the auction site eBay went on the stock market in 1998, its stock soared 163 percent the first day of trading, and Pierre Omidyar became an instant billionaire. That same year, Yahoo's stock multiplied sevenfold. This financial euphoria came to a temporary end in the early 2000s, however, when investors started evaluating the dot-com companies more realistically. Many of their stocks plummeted. People who thought they had created a "golden goose" found themselves in hard times. However, a number of young people became very rich during the short period that the stocks were riding high.[14]

stocks soar

stocks plummet

Within a few years, the boom-then-bust wave smoothed out, and public internet corporations could sell stock at reasonable prices. For example, in 2004 Google announced its first public stock offering, and because it was the most used search engine in the world, speculation was that the stock would sell very high, perhaps as much as $160 per share. By the time the sale finally happened in August of that year, reality had set in, and Google shares sold for $85, a respectable price, though not as high as had initially been expected.[15]

2.8 Radio on the Internet

Internet technology improved rapidly during the 1990s. The decade started with crudely typed words but rather quickly moved into attractive layouts that featured graphics. Toward the middle of the decade a few people began experimenting with audio on the net. It is difficult to pinpoint when radio programs first became a part of the internet, but it could have been on April Fools' Day in 1993 when the now defunct Internet Multicasting Service started a talk radio program called "Geek of the Week." It is known that the University of Kansas was a pioneer in internet radio, placing its student station, KJHK-FM, on the net in December 1994. The first commercial station to use the internet, KLIF-AM in Dallas, didn't do so until September 1995.[16]

early programs

That year internet radio stations proliferated rapidly. At first the audio "hiccuped" as the computer **buffered** the stream: the processors and **modems** of the day weren't fast enough to accommodate the data. But as technology and storage improved, both smooth playback and sound quality became acceptable. RealAudio (later called RealPlayer), released in 1995 by Progressive Networks, provided software for **encoding** audio so it could be **compressed** (see Chapter 13) as well as software that consumers could use to **decompress** the audio and hear it through their computer speakers. Shortly thereafter, other companies, such as Microsoft (Windows Media) and Apple (QuickTime) provided similar software. Hobbyists, music fans, college students, and just about anyone who wanted to could create an internet-only radio station. No one needed a license, and the equipment required was minimal and inexpensive.[17]

Real Player

Many over-the-air radio stations simply put their live feed onto the internet in addition to broadcasting it. This meant that people all over the world, not just those in its terrestrial signal area, could hear the station. Because computers can store information, both internet-only and over-the-air stations found they could place program material on the internet that consumers could download and

streaming

on-demand

listen to at their convenience. They began building files of past play-by-play sports coverage, yesterday's DJ shows, updated news clips, public-affairs programs, and so on. The live programming became known as **streaming audio,** and the material stored in files for downloading was **on-demand** programming.[18]

The radio stations, whether over-the-air or internet-only, began building elaborate websites to go with their audio. They encouraged listeners to email requests for music or questions for talk shows. In some instances, they included such elements as live shots of their disc jockeys playing music, photos of the artists whose music was being played, or written statistics to accompany a ball game. Of course, it was possible for the consumer simply to listen to internet radio while doing something else at the computer—word processing, spreadsheets, games.[19]

As internet radio grew, it began to attract advertisers. The ads could be part of the program material or they could be part of the webpage—for example, **banner ad** links for listeners to click through to the advertiser's website. Once internet radio appeared to be capable of bringing in revenue, others wanted a piece of the action.

ads

Groups representing the music industry came forth because most of what was aired on internet radio stations was music. The music licensing groups, such as ASCAP and BMI (see Chapter 11), which set up and collect royalty payments from over-the-air stations and disperse them to composers and publishers of music, added a fee structure for internet stations. In addition, the Recording Industry Association of America (RIAA) convinced Congress, when it was drafting **copyright** bills in 1996 and 1998, that internet stations should also pay fees to performers and record labels, something over-the-air stations don't do. The fees were based, to some extent, on the amount of revenue a station generated, so small stations that didn't earn much had small payments. However, a number of stations that were funded totally out of the owner's pocket shut down because the operator did not want to pay what would amount to several hundred dollars.

royalties

Then, in 2007, the RIAA convinced the government organization that oversees copyright to change how the royalty was calculated and charge stations each time a listener heard a song. This would greatly increase the amount internet stations had to pay—and also greatly increase how much performers and record labels earned. The potential effect on small stations, such as college stations, would be enormous, in most cases representing fee increases of more than 100 percent. So organizations representing internet stations, such as the Intercollegiate Broadcasting System, asked that the new fee structure, and indeed the fees themselves, be rescinded. As of this writing, the copyright board and Congress have this matter under consideration, the RIAA seems willing to make some compromises for small stations, and internet radio stations are still paying under the old rules.[20]

In part because of the copyright squabbles, internet radio has not achieved the success that was predicted in the late 1990s; still, there are thousands of internet stations. Many of them pay a modest monthly fee to services such as

loudcity.net (see Exhibit 2.12) or Live365.com that netcast, organize, and promote internet radio stations, helping listeners find the particular stations that interest them. Internet radio is a viable option for audio listening and an easy, inexpensive way for those who want to create radio to have an outlet.

modest success

2.9 Napster and Its Aftermath

During the 1990s, when it became easy to store audio on a computer, another phenomenon surfaced: music **file sharing.** People in chatrooms began sending each other cuts from CDs to reduce the cost of purchasing music. Shawn Fanning, a 19-year-old college dropout, developed software in 1999 that made this exchange easier, and in 2000 he launched Napster. Anyone could download the software for free and send a request to Napster for a particular song. During its peak month, February 2001, Napster reported 2.8 billion songs downloaded.[21]

Shawn Fanning

The encoding system used for this exchange, **MP3,** had been introduced in 1998 for the audio part of digital video editing. Once people downloaded music onto their computers, they could play it through their speakers, burn it to a CD, or transfer it to newly developed MP3 players. A website, MP3.com, was developed as a place where new artists could release music for people to sample at no cost.[22]

MP3

Teenagers and college students became enamored with Napster and MP3. For a while, "MP3" was the second-most-searched term on the internet, after "sex." Because Napster attracted such immediate attention and heavy use, Fanning was able to obtain financial backing from Silicon Valley companies. But the recording companies did not take well to Napster. Citing copyright violations, the RIAA sued, followed soon by the heavy metal group Metallica and others. In 2001, the courts ordered Napster to shut down, and it filed for bankruptcy. The following year Roxio Inc., a California digital media company,

shutdown

bought the Napster name at auction and relaunched Napster as a legal, for-pay, music downloading service, initially charging $14.95 per month for "all you can eat" music. Several companies, including Grokster and Kazaa, tried to pick up where Napster left off with free file sharing by using a different computer configuration, but the courts eventually ruled that they, too, were violating copyright, and they either closed down or switched to legal pay services.[23]

RIAA suits

In addition, starting in 2003, the RIAA initiated a historic series of lawsuits. On behalf of the major recording labels it represented, the RIAA brought litigation against 261 people, many of them quite young, who had shared music files illegally. Because files carry datagrams (headers) with embedded information when they travel across the net, the RIAA was able to use technology to trace the origins of these illegal file transfers and find the people whose computers had sent them. The cases got so much publicity that they had a chilling effect that led many people to stop downloading music for free.[24]

Also, the record industry, noting a downturn in music sales, changed its tactics from total opposition to downloading to working with legal downloading services, such as Apple's iTunes, to allow music to be purchased over the internet at reasonable rates. They started to realize that the internet was an ideal distribution system for music that could eventually lessen their costs because it did not need brick-and-mortar buildings or an elaborate distribution network of wholesalers.

compromise

Certainly it is still possible to obtain music illegally for free using the internet, but the extent to which that happens has greatly declined since Napster's heyday. The business model for making money through internet music distribution is not totally worked out yet, but the record companies now seem willing to take less money per song than they used to, and music fans seem willing to pay at this lower rate, especially if being sued is the alternative. There has definitely been a decrease in the acrimony that once existed.

2.10 Voice over the Internet

VoIP

Because sound can be sent over the internet, it is possible for people to talk to each other over the internet, and thus was developed **Voice over Internet Protocol (VoIP)**—a telephone system that doesn't operate through traditional telephone company facilities. Rather, VoIP converts voice to digital at the handset and then sends the information over the internet as packets of data, using the packet technology developed by Vinton Cerf and Leonard Kleinrock. A caller connects a microphone, internet phone, or regular landline phone to a computer, and software does the rest. Because of the net's reach, VoIP is cheaper than regular telephone lines.

Vonage and Skype

Two companies are at the forefront of VoIP. One is Vonage, founded by Jeffrey Citron in 2001, and the other is Skype, the 2003 product from Scandinavians Niklas Zennstrom and Janus Friis. For Vonage, callers pay a monthly fee of about $25 for unlimited calling around the world, using their regular phones. Cable systems and ISPs have licensed the technology to offer their customers internet phone service. Skype charges about 2 cents a minute for calls to regular landline phones and cell phones but also has a free service wherein Skype users can call each other if both are talking through headphones connected to their computers. The computer-based version includes other features such as instant messaging and file transfer.[25]

2.11 Video on the Net

Video on the internet developed later than audio, primarily because the files are larger and require more advanced technology. **Streaming video** was mostly experimental in the early 1990s. For the information to get through the internet **pipes**—the physical cables, fiber optics, microwaves, and satellites that make up the internet's delivery grid—and onto the computer monitor, only some of the video information was sent. It did not have as many frames or as high resolution as regular video, so the picture was jerky and fuzzy. The picture also had to be small, and the overall "program" had to be short; otherwise, it would take hours to download. But as high-speed internet connections, usually referred to as **broadband,** became available and other technologies improved, video on the internet became more viable.[26]

broadband

Some of the first material to stream was promotional in nature. TV program websites featured short clips from shows, and movie sites featured trailers. Other early video included news stories because they are short. Networks simply took the individual news features they had on the air, encoded them, and placed them on their websites. People could download these stories and watch them at their leisure. Several sites, such as iFilm and AtomFilms, gathered short independent and student films and made websites specifically for these projects. Sports organizations, such as the National Basketball Association, showed highlights of games. Interviews and music videos also fit into the "short but sweet" category.

early videos

As the years passed, some producers became a little more adventuresome and used the interactive aspects of the internet for their programming. In 1999, Warner showed a cartoon, *The God and the Devil Show,* in which God and Satan interviewed people. Viewers could then send the people to either heaven or hell. Viewers could also be the "judge" for *People's Court,* or play *Wheel of Fortune* or *Jeopardy* on Sony's site. Starting in the late 1990s, some video suppliers offered full programs on the net, primarily on a one-time-only basis. For example, Comedy Central webcast an episode of *South Park,* and CNN showed *Larry King Live.* In 2004, NBC provided coverage of the Summer Olympics on the net, showing events not seen during its evening tape-delayed, televised broadcasts.[27]

later videos

Established networks and individuals started making programs and series specifically for **webcasting,** such as *Waiting for Woody,* a comedy about an unlucky actor who sees hope when his agent arranges a meeting with Woody Allen. It was shown on AtomFilms.com and later picked up by HBO. As technologies improved even more, networks allowed people to download copies of their older programs for free, and movie services set up pay services for people to download movies. For example, both video store Blockbuster (through its Movielink.com site) and Netflix, which distributed movies through mailed DVDs, set up movie downloading sites. Entire cable TV channels, especially independent ones that had trouble getting space on cable TV systems, have moved onto the internet; examples include Employment and Career Channel, Horror Channel, and Horse TV.[28]

copyright

Copyright issues similar to those with music have surfaced with video, but the movie and television industries learned from the experiences of the music industry and adjusted quickly to the realities of the internet. Just as the RIAA represents the music industry, the Motion Picture Association of America (MPAA) represents the film industry and protects the intellectual property of the studios it represents. It, like the RIAA, brought lawsuits against defendants whom it accused of illegally downloading or sharing copyrighted movies.[29]

2.12 Enter YouTube

rapid growth

The video sharing site YouTube was founded in the spring of 2005 by Chad Hurley and Steve Chen (see Exhibit 2.13), who met while they were employees of PayPal. Apparently while at a San Francisco party they hatched the idea for a site that would collect and organize short videos from a wide variety of sources and make them available to the public at large. The service started in April 2005 and grew quickly. Just one month after it started, it was hosting 30,000 viewers a day, and by November it was showing more than 2 million videos per day. In January 2006 the number of videos played per day reached 25 million, and by July it was 100 million. In the fall of 2006, only a year and a half after YouTube was founded, Google bought it for $1.65 billion.[30]

The site includes how-to videos, comedy sketches, trailers for movies, promos for TV programs, and many other types of fare, but many of its most popular videos are from amateur videographers who shoot with consumer cameras and load (and sometimes edit) the material on their computers. From there, they upload to YouTube's vast storage network. YouTube also allows organizations and individuals to set up their own channels within its structure; the Motion Picture Academy and the Public Broadcasting Service are among those who currently have channels.[31]

Many of the videos have enabled their creators to have their 15 minutes (or more) of fame. In the summer of 2006, episodes of what appeared to be a video diary of "Lonelygirl" appeared on YouTube. Hundreds of thousands of people viewed them, thinking they were real. Eventually the creators confessed that this was a fictionalized account. Singers have been signed to recording contracts

Exhibit 2.13
Steve Chen and Chad Hurley, founders of YouTube.

(Photo courtesy of Google, Inc. Used with permission.)

based on their YouTube performances, and people have been hired to write for Comedy Central.[32]

users

YouTube has figured prominently in politics. A widely viewed 2006 video of former Virginia Senator George Allen calling his rival's campaign worker "macaca" was credited with being at least partly responsible for his losing the election. During the 2007 presidential primary contest, CNN and YouTube co-sponsored a debate in which candidates fielded questions (often accompanied by creative video) selected from those submitted by users of YouTube. The candidates also used YouTube to upload both serious and light-moment videos of themselves, hoping to enhance their image with the YouTube crowd.[33]

politics

Copyright issues have surfaced with YouTube also. In its early days, people put up material, such as portions of their favorite movies or TV shows, for which they did not own copyright rights. The organizations that did own the copyrights objected, so YouTube offered to take down any material the organizations cited, no questions asked. This presented a dilemma for networks and movie studios because so many people viewed the clips that they generated extra promotion, even though they were illegal. Unlike the file sharing of music, the YouTube clips did not deprive the creators of income because the entire programs or movies were not shown—just short clips.[34]

copyright

Internet video in general and YouTube in particular are moving targets, dispersing in a multitude of directions and peopled with participants who come up with new ideas at an accelerated rate. But, as with all technologies, there are always new forms around the corner to challenge the existing leaders and push possibilities even further.

2.13 Social Networking

Early 1990s chatrooms and bulletin boards were forms of **social networking** in that they helped people connect with others and share interests. A little later, dating sites hooked together those looking for partners, and classmates.com let people connect with people they had known in high school and college. In the early 2000s, a site called Friendster was set up wherein people invited their friends to join and, in turn, those friends invited their friends. The site was popular for a while, but it suffered from technical difficulties and fake profiles and began losing members.

Friendster

Some of those members went to MySpace, which had actually been started in 1999 but became better known in 2003. Its roots are a little muddy because it received financial and logistical support from another company called eUniverse, and most of the early users were eUniverse employees, but Tom Anderson and Chris DeWolfe are given credit for much of the innovation and success of the site, which built up to 115 million users worldwide. Members post bios, photos, blogs, videos, and other things that strike their fancy, and some TV programmers have taken to producing programs to air on MySpace. In 2005, Rupert Murdoch's News Corporation (parent of Fox Broadcasting) bought MySpace for $580 million.

MySpace

A competitor to MySpace, Facebook was started in 2004 by Mark Zuckerberg while he was a student at Harvard and grew rapidly starting about

Facebook

2007. At first Facebook was solely for college and high school students, but Zuckerberg opened it to everyone and, like MySpace, it encourages all types of member postings. There are also smaller social networks for more focused groups such as churchgoers, dog lovers, or surfers.[35]

2.14 Internet Abuses

predators

Although most uses of internet services are positive or at least harmless, there are serious negative uses. **Predators** use social networking sites, chatrooms, email, and instant messaging to gain the confidence of young people and then arrange to meet them for the purpose of engaging in sexual activities. Adults have also been recruited as victims. In one notorious case, a German man solicited someone willing to be killed and eaten. As if that's not disturbing enough, more than 200 people responded! After selecting his victim and doing the deed, he was caught and sentenced to prison, but the case revealed the dangers that lurk on the net from disturbed people.[36]

However, the internet also helps authorities catch predators and other criminals or potential criminals because they brag of their exploits in blogs and sometimes even post videos of themselves committing crimes. Sites such as MySpace and Facebook are open to police, who can browse profiles of suspects, including people who have committed past crimes such as child molestation. Detectives were able to arrest several alleged graffiti artists because they had photos of their work on their profiles; two boys who firebombed an airplane hangar uploaded video of themselves committing the crime; tip-offs regarding stockpiling of guns and ammunition have helped thwart school shootings.[37]

identity theft

Another internet problem is **identity theft.** When you send an email message into cyberspace, it carries coded information that identifies you, including your username and host computer—something a hacker can steal. Hackers have also been known to break into databases and steal social security numbers, credit card numbers, and other valuable personal information that they can then use to make purchases or sell to others who use it for illicit purposes.

spoofing

Another problem is **spoofing.** A spoofer can send a message using your email address, pretending that the message comes from you when it does not. Spoofers have been known to send a variety of fake emails, including jilted revenge messages from the addresses of ex-boyfriends or ex-girlfriends, solicitations for money or sex or other products or services that the apparent sender never sent, and even computer **viruses** spread via trusted—yet spoofed— email accounts.

spam

Another widespread nuisance is **spam:** unsolicited bulk messages, or junk email. Spam covers the gamut from supposed foreigners-in-exile asking for your bank account numbers to offers for supercheap software to sexual solicitations. A special kind of spam, called **phishing,** occurs when a cybercriminal spoofs or pretends to be a legitimate company, hoping to scam recipients into divulging private information that can be used for identity theft.

Many companies offer software solutions that attempt to **filter** spam by looking for messages that appear to have been sent in bulk (e.g., coming from large servers, being sent to large numbers of receivers, etc.). These solutions take employee time to install and require bandwidth and storage. Information technology people working at companies erect **firewalls** and install additional filters in the ongoing attempt to identify, block, and eradicate spam. Sadly, as fast as programmers devise new ways to fight spam, electronic junk mail peddlers devise new ways to get around spam blocks.[38]

2.15 Portable Devices

The field of portable devices is evolving, with numerous pieces of equipment, such as **laptop** computers, MP3 players, small DVD players, and quite a few inventions under development all falling within the portable realm. The main ones that will be discussed here are cell phones, personal digital assistants, and pods—all of which can interface with the internet through wireless technology (see Exhibit 2.14). They weren't necessarily originally developed with the internet in mind, but the internet has become so popular and powerful that people want to be connected no matter where they are. These devices utilize wireless technologies such as **Bluetooth** or **wi-fi,** both of which were designed for connectivity of portable devices.

wireless

2.15a Cell Phones

The **cellular phone** is an excellent example of a technology that was developed for a different purpose but now includes many features that relate to the internet. Portable wireless phones actually existed as far back as the 1920s, but they were cumbersome and expensive and did not catch on with the general

early development

Exhibit 2.14

A cell phone, a PDA, and an iPod that can all receive video, audio, and phone calls.

public. AT&T's Bell Labs developed the concept for what eventually became the cell phone in 1947. But the Federal Communications Commission (FCC) had no inclination back then to assign spectrum for wireless telephones, so AT&T kept the idea on the shelf for a few decades. In 1981, the FCC authorized spectrum space for cell phones. The first commercial cell phone call is credited to Bob Barnett, an executive at the phone company Ameritech. In 1983, from his car in Chicago, he phoned Alexander Graham Bell's grandson in Berlin. The quality wasn't great, but the commercial viability of cell phones was apparent.

varied uses

Cell phones caught on fairly quickly; by 1993, some 20 million had been sold, and people throughout the world were using them. However, they only used them to make phone calls. It wasn't until the mid 2000s that cell phones were equipped to take pictures, access the internet, enable **text messaging,** play music, and show video. A number of producers have created program material especially for cell phones, taking into account their small screens, but consumers can also access all the print, audio, and video material available on the net.[39]

2.15b Personal Digital Assistants

early uses

The term **personal digital assistant (PDA)** was first used in the early 1990s in relation to the Apple Newton, a handheld computer. At first these devices were primarily a replacement for the paper calendar—a way for people to keep track of what they needed to do each day. They also stored names, addresses, and phone numbers of friends and business associates, and most of them contained a calculator, alarm clock, and similar tools. They were able to synchronize with the owner's larger computers so that the information was the same on the PDA and main computer.

later uses

PDAs still include all these functions, but now they are also used for email and accessing the internet as well as taking photos, listening to audio, and watching video. Some of them also have telephone capabilities, so the line between cell phone and PDA is blurring. The early PDAs, such as Apple Newton and Palm Pilot, operated through the use of a touch screen, but later ones, such as BlackBerry, also have keyboards and thumb wheels to facilitate navigation.[40]

2.15c Pods

iPod

Although there are a number of **pods** on the market, the main one is **iPod** from Apple. The idea for a new form of MP3 player that was hard-drive based and tied to a music delivery system where users could obtain music legally came from Tony Fadell, who pitched it to several companies before approaching Apple. Apple's executives, including CEO Steve Jobs, were enthusiastic about the idea and hired Fadell in early 2001 to head a team to develop it. The project was kept very secret, and when Jobs announced it in October of 2001, people were surprised that Apple would offer a music player, especially one that didn't operate at all like its computers.

But the iPod became very popular, especially after the iTunes Music Store opened on the internet in 2003 (see "Zoom In" box). Jobs had worked hard with music publishers and record companies to convince them that it was in their interest to allow Apple to sell music singles (instead of albums) for the low price of 99 cents each. To appease the music industry, Apple applied a form of copyright protection called **digital rights management (DRM)** onto the music that consumers downloaded from iTunes. It limited the number of times that someone could copy the music from one computer to another, first to three and later to five. The iTunes concept became very popular and helped reduce illegal copying. As time passed, the iPod sported a screen, and Apple added radio and television programs to iTunes. The term **podcasting** was invented to refer to the act of sending material to a pod. Apple also came out with the iPhone, further blurring the line between a cell phone, a PDA, and a pod.[41]

iTunes

 ZOOM IN: **Beethoven and iTunes**

Ludwig van Beethoven.

(Courtesy of Brian Gross)

One of the advantages of basing the music business model on downloading rather than selling CDs in music stores is that music that does not have a wide audience can easily "sit on the shelf." Individual music stores are reluctant to stock music that has a limited audience because they may never sell it, and it takes up room that could be used for hit music that large numbers of people are craving. But when music is stored on a large computer, it can sit there forever and still prove profitable. Part of the reason for this is that the internet is international, so the possible buyers consist of the accumulated people around the world who are interested in the music.

Such has been the case with classical music, which for many years had been on the decline. But in 2007, when all categories of music fell 5 percent, classical music sales grew 22 percent. Downloading from the internet has been of benefit to everyone in the classical music feeding chain, including Beethoven, Bach, Hayden, and others from centuries past. The costs of distribution via internet stores is so much lower than the costs associated with music stores that even classical music that sells only 10 units a year can make a profit, in part because it can be priced rather high. Classical music aficionados are usually fairly well-off and happy to find their music at any price.

2.16 Video Games

Video games are ubiquitous. Certainly they are abundant on the internet and in arcades, and they are also a feature of cell phones, PDAs, pods, DVDs, computers, cable TV systems, and dedicated consoles. Games have become a huge industry that takes in more money than the movie industry domestic box office. The game industry is poised to take advantage of each technological advance that makes computers faster, smaller, and cheaper.

2.16a The Early Days of Video Games

The first popular video games were in amusement arcades, usually near pinball machines. The "Father of Video Games" was Nolan Bushnell, an egotistical but likable entrepreneur who founded Atari in northern California in 1972. He and his main engineer, Al Alcorn, invented *Pong,* an electronic version of table tennis, which they placed in arcades. It became so popular that the main technical problem was that the coin boxes filled up with quarters too quickly. Atari's next successful step was a home version of *Pong,* which sold out almost instantly during the Christmas 1975 season. It was encased in a separate unit that plugged into a TV set, allowing the operator to interact with the set by manipulating the unit's joystick.

Pong

Atari was not the first console video game that connected to a TV set. In 1972 Magnavox had come out with a product called Odyssey, developed by Ralph Baer, that had two controllers and an overlay that stuck to the face of the TV tube. One overlay was a tennis court for a tennis game, and another was a hockey court. Magnavox sued Atari for violating Baer's patent, but they settled amicably. Atari paid Magnavox a small amount, and Magnavox became the first licensee for Atari's products, so everyone won. Atari had the better technology in that *Pong* had sound and the ball speeded up as the player improved.

Odyssey

Bushnell came up with the idea of putting the chip for the game into a cartridge that would slip into the console unit. That way people could buy individual games without having to buy a whole console for each game. But he didn't have enough money to develop this concept, so he sold Atari to Warner Communications. Bushnell's company had a very laid-back culture that pleased his creative workers (see "Zoom In" box), and Warner was a much more straight-laced, formal company, so the cultures clashed. In addition, the first product that Warner and Atari worked on together failed (in part because they didn't get it out in time for Christmas), so in 1978 Warner fired Bushnell from his own company.

Bushnell sells

Over the next several years Atari's products sold well, and for several years in the early 1980s Atari accounted for more than half of Warner's earnings. But Warner never treated the game inventors well, and most of them left.[42]

Meanwhile Atari acquired some competitors, including the American company Midway, which in 1980 brought out a Japanese-developed game it had licensed called *Pac-Man.* (The name in Japan had been the Japanese equivalent of Puck-Man, but the Midway executives were afraid people would change the first letter of Puck-man, so they changed the name to *Pac-Man.*) *Pac-Man* was a round character (in true 8-bit color) with a big mouth that ate everything in his

Pac-Man

ZOOM IN: **Inside Atari**

Nolan Bushnell, founder of Atari, seen here with an Androbot robot.

(Roger Ressmeyer/Corbis)

Nolan Bushnell hired young people for Atari, many of whom were unkempt and undisciplined. He recruited them from some of the seedier parts of northern California during the era when the "flower children" were in full bloom. Bushnell didn't pay much attention to when they got to work, what they wore, whether or not they engaged in personal hygiene, or what they imbibed. But he did care about results in terms of creating video games. So he set deadlines for them to accomplish certain tasks, and if they met the deadlines he threw them a party—something that proved to be an excellent incentive. Atari, rightfully, became known as a party company.

One of the scruffiest employees he hired was a young man named Steve. Steve had a friend, also named Steve, who was never officially an employee of Atari but who hung out at the company with his buddy Steve and kibitzed with other employees. After a while the two Steves decided to form their own company, and they asked Bushnell if he wanted to invest. He declined. In case you haven't already guessed, the first Steve was Steve Jobs and the other was Steve Wozniak, the founders of Apple Computer.

But don't weep for Bushnell because he missed a great financial opportunity. After he got kicked out of his own company, he went on to make his fortune by founding the Chuck E Cheese chain.

path. He became a national personality that was licensed for products such as toys and food. He even had his own song and TV show.

Nintendo

Another competitor was Nintendo, a Japanese playing card company that had been formed in 1889, but which decided in 1980 that it wanted to cash in on the U.S. video game market. It came out with a game called *Donkey Kong* (again the title was the result of a translation problem). It had a "story" to motivate the action: a gorilla runs away from a carpenter and steals the carpenter's girlfriend, so the carpenter chases the gorilla through a factory to rescue the girl. The carpenter's name was Mario, and he, like Pac-Man, caught the fancy of the public. Nintendo gave Mario a family and developed many more games around him that are still among the most popular game franchises today.[43]

2.16b Down and Out—And Up Again

disasters

Back at Atari, things weren't going so well. Having lost its best creative people, Warner, in a 1982 move of desperation, licensed the game console rights for *Pac-Man*. They put it on their Atari 2600 console, but it never worked right. It was hard to control, the characters didn't move well, and it was just plain ugly. Warner had a lot of unsold inventory. Then, in what turned out to be an even bigger mistake, they brought in Steven Spielberg to help them make a game based on his popular 1982 movie, *E.T.: The Extra-Terrestrial*. They couldn't get this game to work either, and legend has it that they bulldozed a hole in the desert to bury all the unused inventory. In 1984 Warner sold Atari at a loss.[44]

The *Pac-Man* and *E.T.* fiascos affected the whole industry, and for several years hardly anyone was playing video games either in arcades or in their homes. In 1986, Nintendo launched a new console, NES, that had better graphics and sound and brought video games to a new level. Accenting its Mario games, Nintendo revived the video game industry. Sega, another Japanese company, came out with a console in 1986, and for a number of years the video game business belonged to the Japanese.[45]

Exhibit 2.15

As evidence that video games are big business, Electronic Arts built this large complex in Los Angeles to take advantage of the talent in the film and TV industry located there.

2.16c Games Take on a Variety of Forms

In the mid to late 1980s, computer manufacturers became interested in games, especially Steve Jobs and Steve Wozniak who founded Apple Computer and who had games in their blood (see previous "Zoom In" box). They made sure the computers they developed had the technological capabilities to handle gaming. One of Apple's employees, Trip Hawkins, left Apple and founded Electronic Arts (EA), a company that created software for games, originally games for computers although later it developed games for consoles also. EA (see Exhibit 2.15) made its mark with sports games and attractive packaging that featured sports stars who endorsed the games. EA had problems with its first football-based game, however. Technology only allowed there to be seven players on each team. Football players nixed the seven-player concept and would not endorse it. Eventually, when technology improved, EA developed a very successful interactive football game, *Madden.*[46]

computers

Husband and wife team Ken and Roberta Williams also developed games in the 1980s. She wrote stories as the basis for games such as *Mystery House* and *King's Quest,* which IBM commissioned for its computers. *King's Quest* was the first third-person perspective game, in which the player could move the characters through the game. Ken developed the technology for the graphics needed for Roberta's stories and in doing so advanced video game possibilities. In the early 1990s, Robyn and Rand Miller released *Myst* for the Macintosh computer. It was a totally absorbing story with advanced graphics that people immersed themselves in for hours, or even days. Another popular game was *Sim City,* in which players could build simulations of people and run their lives, something that its creator, Will Wright, likened to having a dollhouse. The games of the 1970s and 1980s had been "script driven"—the same monster would appear behind the same tree every time the game was played. But now games changed depending on how the player interacted with them.[47]

stories

In the early 1990s, the **first-person shooter** games arrived, led by *Doom.* For these games, the player's eyes were the monitor and players could see their hands and the gun in front of them. When *Doom* came out, its creators, IDS Software, tried to upload it to a server so that people could download it. But so many people were waiting for it and writing notes asking when it was coming that IDS couldn't get it uploaded and had to tell everyone to stay off the site for a while. The first-person shooter games were quite violent and bloody and caught the attention of parents' groups and the government. Congress held hearings, and the games were blamed for a number of social problems, including the 1999 Columbine massacre. The result was that the gaming industry developed a ratings code (such as M for "Mature" and AO for "Adults only") to give consumers information on which to base their purchases.[48]

first-person shooter

Video games' visual and sound effects also improved greatly during the 1990s. *Sherlock Holmes, Consulting Detective,* was the first computer game to incorporate full-motion video. It was released on CD-ROM, as were many games during that decade. Once full-motion video was possible, producers began using recognized actors to star in the games or to serve as body models for animated characters. Sound effects dazzled with futuristic audio as well as wind, rain, and other aural elements to make the players feel as though they were really in their virtual worlds.[49]

visuals and sound

2.16d Console Wars

In the 1990s, Nintendo was riding high as the leader of video consoles. It had used a 1984 Russian-based puzzle game called *Tetris* to launch its new console, Game Boy. (Incidentally, the Russian inventor of the game did not get money for his invention; the Communist Party took it over and sold it around the

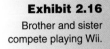

Exhibit 2.16
Brother and sister
compete playing Wii.

Nintendo-Sega   world.) Sega Master System still existed and occasionally gave Nintendo a run
for its money, especially when it came out with a very fast-paced game, *Sonic.*
Consoles were no longer tied to TV sets; they were stand-alone units.

Sony-Microsoft   In 1995 Sony entered the game business with PlayStation. Its technology
was so far ahead of Sega's that Sega dropped out of the video game business. In
2001, Microsoft joined in with Xbox. Skeptics weren't sure that Microsoft,
which previously had dealt only with software, could develop a hardware game
console, but the product proved to be very successful. In the mid-2000s,
Nintendo upped the ante with Wii (see Exhibit 2.16), a video game concept that
involves not just the thumbs but the entire body. In 2008, video game company
Take-Two Interactive brought out *Grand Theft Auto IV* for PlayStation and
Xbox, and it sold a record $500 million in the first week. Currently the three
main video game console companies are Nintendo, Sony, and Microsoft.[50]

2.16e Games for the Internet and Beyond

Meanwhile the internet was growing, and gamers started using it as a
distribution means for their games. This brought about an entirely new game-
playing experience. For example, the internet game *Majestic* infiltrated people's
lives by sending them cryptic emails, menacing phone calls, threatening faxes,
and other "real-life" information related to the game. In addition, people around
the world who didn't know each other could play games with each other. One of
the first games, *Quake,* enabled six people to play each other. These people

formed clans and had meetings in Texas where *Quake* players could actually meet the clan people they had been playing with. As the internet technology improved, games labeled **massively multiplayer online role-playing games (MMORPGs)** came to the fore. *World of Warcraft (WoW),* for example, has more than 11.5 million people playing it worldwide. Like other MMORPGs, it charges players and allows them to customize much of their play.[51]

MMORPG

Video games, which originally targeted young males, have broadened their audience base. There are educational games for young children, games that appeal primarily to women, and games such as Bingo and chess that attract an older demographic. The military uses game technology to train enlistees, many of whom have excellent hand-eye coordination from growing up with games. In arcades, players can don gloves with motion sensors that make a projected virtual world react to their movements. Gamers can enjoy lifelike simulations of flying, turning upside down, and fighting with monsters. Some games are created so players can get into the programming and modify them. Although the gaming industry brings in more money than theatrical movies, moviemakers do not need to fear games because many games are based on movies or TV shows for which the studios receive payment.[52]

user base

2.17 Issues and the Future

Undoubtedly, the internet, portables, and video games will continue to grow and change as technological improvements allow for greater electronic speed and capacity. With each technological advance, creative minds find new ways to use the technology, which then drives the technology even further. One technological area that has been under development and has yet to reach fruition is the one-screen concept (see Chapter 13). Now that there is so much video on the internet, some people want to place that video on the screen of TV sets. Several companies have been working to develop ways to direct internet content to TV screens, and do it with a split screen so that users can watch TV and surf the net at the same time. None of the technologies developed so far have resonated with consumers, but this is an area that may blossom in the near future.[53]

TV sets

At the other end of the spectrum is small-screen reception, such as that on cell phones, PDAs, and pods. There will no doubt be technological and aesthetic advances to enhance the small-screen experience, perhaps through new, as yet not thought of, portable devices.

small screens

Copyright may continue to be an issue, although fortunately the trend appears to be toward reasonable compromises that take into account the economic needs of the individuals and companies creating material and the information and entertainment needs of the consumers.

copyright

Abuses of the internet such as predators, identity theft, spoofing, viruses, spam, and phishing will probably remain with us, but individuals and the government will attempt to lessen their impact. For example, Congress passed

spam

legislation in an attempt to fight spam, most notably the 2003 CAN-SPAM Act. Among other things, this act requires bulk emailers to include in each message a note regarding how the receiver can unsubscribe from the list to stop getting messages from that sender. Although compliance is spotty, the CAN-SPAM act gives consumers methods to cope with spam.[54]

Violence and sexual content of video games is an issue that is not likely to go away. Coupled with it are complaints about the addictive nature of both games and the internet. People who spend 24 hours at a time playing games or surfing the net are obviously neglecting other important aspects of life—if nothing other than sleep.

There is an internal debate going on between those who have content on the internet and those who provide access to the net. It involves a concept called **net neutrality.** Search engines and owners of websites are in favor of net neutrality because they want everything on the internet to have an equal chance of being seen by users. The service providers, such as cable TV and phone companies, oppose net neutrality. They want to manage the traffic on their networks to prevent sluggish operation. Sometimes that means blocking certain sites or messages that are hogging too much bandwidth. Proponents of net neutrality say that blocking material threatens freedom of speech; as proof, they point to Verizon's blocking of text messages from a pro-abortion group and AT&T's blocking of lyrics critical of President George W. Bush. The issue is being debated in Congress and will probably be resolved there, although if history is any indication, what may solve this problem is advances in technology.[55]

The internet, portable devices, and video games have established themselves as viable parts of the modern social fabric. Their futures look healthy and exciting.

net neutrality

2.18 Summary

Technology has been a very important element of the internet and its related forms. The early visionaries associated with ARPA, such as Licklider, Kleinrock, Cerf, and Baran, worked on methods to interlink computers and send information through them by developing IMPs, TCP/IP, a distributed network, and packet switching. Once the internet was formed, the emphasis was on using technology in practical ways. Tomlinson created email and Berners-Lee constructed the world wide web. About the same time, Bushnell was using technology to create the *Pong* video game. When there was a need to make the internet user-friendly, others, such as Andreessen, Brin, and Page, developed browsers, ISPs, and search engines. Technology has also been important in making the internet portable through cell phones, PDAs, and pods. And it was instrumental in providing video games to arcades, consoles, computers, and the internet.

In addition to being a technological marvel, the internet is also an important economic, political, and social phenomenon. The growth of e-commerce through such sites as amazon.com and eBay has changed how many products are marketed. Music, in particular, has changed its business model as a result of Napster and its aftermath. Video games bring in more money than theatrical movies.

The success of the internet is partly due to the fact that Congress privatized it in 1992. Other government actions have been focused on issues such as copyright violation, predators, identity theft, spam, violence in video games, and network neutrality.

Chatrooms, VoIP, MMORPGs, and social networking sites such as MySpace and Facebook have altered how people interact. Older media such as radio and television have had to adapt to changes brought about by sites such as YouTube.

Suggested Websites

http://openmap.bbn.com/~tomlinso/ray/home.html (Ray Tomlinson's site that describes the beginning of email)

www.riaa.org (site of the Recording Industry Association of America that has the latest on music copyright provisions)

www.Live365.com (a site that lists and indexes internet radio stations)

www.vonage.com (a VoIP site)

www.worldofwarcraft.com/index.xml (a massively multiplayer online role-playing game)

Notes

1. "A Brief History of the Net," *Fortune,* October 9, 2000, 34–35.
2. "The Birth of the Internet," *Newsweek,* August 8, 1994, 56–57.
3. "Internet Pioneers: Vint Cerf," http://www.ibiblio.org/pioneers/cerf.html (accessed May 1, 2008).
4. *The Internet: Behind the Web,* Video, History Channel, 2000.
5. "The Evolution of the Internet," *Telecommunications,* June 1997, 39–46.
6. "How the WWW Is Put Together," *Byte,* August 1995, 138; and Barbara K. Kaye and Norman J. Medoff, *The World Wide Web* (Mountain View, CA: Mayfield, 1999).
7. Rijiv C. Shah and Jay P. Kesan, "The Privatization of the Internet's Backbone Network," *Journal of Broadcasting & Electronic Media,* December 2007, 93–109.
8. "The Birth of the Internet," 56–57; and "A Brief History of the Net," 34–35.
9. "The Killer Browser," *Newsweek,* April 21, 2003, E-6–E-11.
10. "A Trip to Cyberspace," *Los Angeles Times Advertising Supplement,* Summer 1995, 8.
11. "The World According to Google," *Newsweek,* December 16, 2002, 47–49.
12. "Once Upon a Time," *Newsweek,* February 21, 2005, E10.
13. "Selling Sex on the Web," *Newsweek,* June 12, 2000, 76J.
14. "Tale of Two Onliners: Stripped vs. Steady," *Hollywood Reporter,* October 19–21, 2001, 1.
15. "Google Inc. Prices Initial Public Offering of Class A Common Stock," http://www.google.com/press/pressrel/ipo.html (accessed April 29, 2008).
16. Kaye and Medoff, *The World Wide Web,* 78–80; and "Radio on the Net," *Electronic Moviemaking,* February 8, 1999, 18.
17. "RealAudio Gives Rise to Online Radio Programs," *Los Angeles Times,* July 8, 1999, 1.

18. "That Dammed Streaming," *Broadcasting & Cable,* May 14, 2001, 42; and Cheryl L. Evans and J. Steven Smethers, "Streaming into the Future: A Delphi Study of Broadcasters' Attitudes Toward Cyber Radio Stations," *Journal of Radio Studies,* Summer 2001, 5–28.

19. "Web Radio Adds Video to Audio," *Broadcasting & Cable,* August 9, 1999, 32.

20. "Web Battles Efforts to Expand Royalties for Music," *Wall Street Journal,* July 25, 2007, B-1; and "Like an LP, Two Sides to This Debate," *Hollywood Reporter,* June 26–July 2, 2007, 6.

21. "The Noisy War Over Napster," *Newsweek,* June 5, 2000, 46–52.

22. "How the Internet Hits Big Music," *Fortune,* May 10, 1999, 96–102.

23. "Grokster Case Fit to Be Tried," *Hollywood Reporter,* June 28–July 4, 2005, 1.

24. "64 Individuals Agree to Settlements in Copyright Infringement Cases," http://www.riaa.com/news/newsletter/092903.asp (accessed November 5, 2004).

25. "Hi! The Net Is Calling," *Newsweek,* January 31, 2005, E4–E5.

26. "The Not Ready for Prime Time Medium," *Broadcasting and Cable,* May 25, 1998, 22–28; and "TV Goes Online—in a Flash," *Broadcasting and Cable,* November 18, 2006, 14.

27. "New Venue for Video on the Internet," *TV Technology,* July 27, 1998, 16; "CNN Streams 'Larry King,' 'Crossfire,'" *Broadcasting and Cable,* May 25, 1998, 28; and "Dilbert Gets a New Cubicle," *Electronic Media,* January 21, 2000, 12.

28. "Crossover Episode," *Hollywood Reporter,* September 2000, Fall TV Season; "CBS, NBC Offer Oldies Online," *Broadcasting & Cable,* February 25, 2008, 31; "TV Channels Move to Web, Think Outside the Cable Box," *Wall Street Journal,* August 10, 2007, B-1; "Serving Up Movie Downloads," *TV Technology,* September 5, 2007, 30; and "A True Union of TV, the Web," *Los Angeles Times,* June 16, 2008, E-1.

29. "Lights! Camera! Legal Action!" http://www.broadcastingcable.com/article/CA478095 .html?display=Breaking+News&referral=SUPP (accessed November 5, 2004).

30. "Google-YouTube May Be Good Fit," *Hollywood Reporter,* October 9, 2006, 7.

31. "AMPAS Sets Up Channel on YouTube," *Hollywood Reporter,* February 22–24, 2008, 8.

32. "'Lonely' at the Top," *Hollywood Reporter,* June 20, 2007, S-2.

33. "How to Run for President, YouTube Style," *Newsweek,* January 7, 2008, 69.

34. "Viacom Sets Off Content Conflict," *Los Angeles Times,* February 3, 2007, C-1.

35. "Your Guide to Social Networking Online," http://www.pbs.org/mediashift/2007/08/ digging_deeperyour_guide_to_so_1.html (accessed February 29, 2008); "The New Wisdom of the Web," *Newsweek,* April 3, 2006, 47; and "TV Takes Step Into 'Afterworld,'" *Los Angeles Times,* August 23, 2007, C-1.

36. "Germans Get a Look at Dark Side of Cyberspace," *Los Angeles Times,* December 31, 2003, A-3.

37. "Walking a New Beat," *Newsweek,* April 24, 2006, 48.

38. Core Competence, Inc., "Phishing and Fraud Prevention Resources," http://hhi.corecom .com/phishing.htm (accessed March 1, 2008).

39. "AT&T's $12 Billion Cellular Dream," *Fortune,* December 12, 1994, 100–12; and "Married to the Mobisode," *Hollywood Reporter,* April 4–10, 2006, 19–20.

40. How Stuff Works, "How PDAs Work," http://communication.howstuffworks.com/pda.htm (accessed March 1, 2008).

41. "Apple's Jobs Unveils Song Service: 'We Believe in the Future of Music,'" *Los Angeles Times,* April 29, 2003, C-1; "The Power of iPod," *Newsweek,* October 23, 2006, 72–74; and "Survival of the Fittest," *Emmy,* January/February, 2006, 50–53.

42. "Atari and the Video Game Explosion," *Fortune,* July 27, 1981, 40–46.

43. *Video Game Invasion: The History of a Global Obsession,"* DVD, Game Show Network, 2004.

44. "Warner Sells Atari to Tramile: Will Report a Loss of $425 Million," *Los Angeles Times,* July 4, 1984, D-1.

45. *Video Game Invasion.*

46. "EA Raises the Ante for Rival," *Los Angeles Times,* February 25, 2008, C-1.

47. "Online Games Get Real," *Newsweek,* February 5, 2001, 62–63.

48. "The Players," *Wall Street Journal,* March 22, 1999, R14.

49. "Games in the Video Age," *AV Video,* April 1996, 75–76.

50. "Consolations for a Console," *Newsweek,* February 12, 2001, 7; "New Consoles to Fuel Vid Games," *Hollywood Reporter,* May 12–14, 2006, 34; "Wii's Unintended Consequences," *Newsweek*, June 2, 2008, E16; and "Just How Grand Is $500 Mil?" *Hollywood Reporter,* May 8, 2008, 1.

51. Dimitri Williams, "Groups and Goblins: The Social and Civic Impact of an Online Game," *Journal of Broadcasting & Electronic Media,* December 2006, 651–670; and "Family Values," *Newsweek,* November 25, 2002, 47–53.

52. "RTS Games Still Going Strong," *Mix,* February 2008, 58; and "Games for the Rest of Us," *Los Angeles Times,* July 21, 2009, C-1.

53. "Microsoft Buying Web-TV," *Broadcasting & Cable,* April 9, 1997, 11.

54. "CAN-SPAM Act of 2003," http://www.spamlaws.com/federal/108s877.html (accessed November 5, 2004).

55. "Net Neutrality Back in Spotlight," *Broadcasting & Cable,* November 12, 2007, 19–23; and "Comcast Limits Download Volume," *Wall Street Journal,* August 30–31, 2008, B5.

Chapter 3

EARLY TELEVISION

Today's television universe consists of hundreds of programming services, some attempting to appeal to a large audience, but most targeting a niche of viewers who have a particular interest in the type of programming the service provides. It is possible to receive television programming through a wide variety of distribution channels—cable television, satellite television, internet downloads, or over the airwaves through antenna systems. If you can't watch a program when it is shown, you can probably find a rerun, record it in any number of ways, or rent or buy it from a local video store.

Such has not always been the case. For many years three networks dominated programming, and if you didn't catch a show when it aired, you were out of luck. This chapter and the next will trace the path that television has taken from its humble beginnings in the 1880s to its dominant position today.

> Many people are disappointed because television has not been put to general use already. They have been reading about it for so long that they are beginning to doubt that it will ever be of much importance. But those who are working with television are confident that someday television will be widely used to entertain people and to inform them.
>
> **"What Will Be the Future of Television?"**
> ***The Junior Review*, February 13, 1939**

3.1 Early Experiments

The first experiments with television used **mechanical scanning** (see Exhibit 3.1), invented in Germany in 1884. This process depended on a wheel that contained tiny holes positioned spirally. A small picture was placed behind the wheel, and each hole scanned one line of the picture as the wheel turned. Even though this device could scan only very small pictures, attempts were made to develop it further. For example, a General Electric employee, Ernst F. W. Alexanderson, began experimental programming during the 1920s using a revolving scanning wheel and a 3-by-4-inch image. One of his "programs," a science fiction thriller of a missile attack on New York, scanned an aerial photograph of New York that moved closer and closer and then disappeared to the sound of an explosion.[1]

Other people developed **electronic scanning,** the system that has since been adopted. One developer was Philo T. Farnsworth (see Exhibit 3.2), who in 1922 astounded his Idaho high school teacher with diagrams for an electronic TV system. He applied for a patent and found himself battling the giant of electronic TV development, RCA. In 1930 Farnsworth, at the age of 24, won his patent and later received royalties from RCA.[2]

The RCA development was headed by Vladimir K. Zworykin (see Exhibit 3.3), who patented an electronic pickup tube called the **iconoscope.** Beginning in the early 1930s, he and other engineers (including Alexanderson) systematically attacked such problems as increased lines of scanning, brightness, and image size. They started with a system that scanned 60 lines using a model of Felix the Cat (see Exhibit 3.4). Gradually they improved this scanning to 441 lines.[3]

In 1939, David Sarnoff, then president of RCA and a strong advocate of TV, decided to display television at the New York World's Fair (see Exhibit 3.5). President Franklin D. Roosevelt appeared on camera and was seen on black-and-white sets with 5-inch tubes. RCA (which owned NBC) by this time had an experimental TV program schedule broadcasting only in New York that consisted of one program a day from its studio, one from a mobile unit traveling the streets of New York, and several assorted films. The studio productions included plays, puppets, and household tips. The mobile unit consisted of two huge buses, one jammed with equipment to be set up in the field and one containing a transmitter. It covered such events as baseball games, airport interviews with dignitaries, and the premiere of *Gone with the Wind*. The films were usually cartoons, travelogues, or government documentaries.[4]

mechanical scanning

Exhibit 3.1

Early mechanical scanning equipment. Through a peephole, J. R. Hefele observes the image re-created through the rotating disc. The scanning disc at the other end of the shaft intervenes between an illuminated transparency and the photoelectric cell. This cell is in the box that is visible just beyond the driving shaft.

(Courtesy of AT&T Co. Phone Center)

Exhibit 3.2

Philo T. Farnsworth in his laboratory (about 1934).

(Smithsonian Institution, Photo No. 69082)

Farnsworth

Zworykin

World's Fair

early color TV controversy

CBS, which existed as a radio network in the 1930s, began experimentation with color television, utilizing a mechanical color wheel of red, blue, and green that transferred color to the images. This color system was not compatible with the system RCA was using. In other words, the sets being manufactured could not receive either color or black-and-white pictures from the CBS mechanical system, and proposed CBS receivers would not be able to pick up existing RCA black-and-white pictures. In 1940, a group led mainly by RCA personnel tried to persuade the Federal Communications Commission (FCC) to allow the operation of the 441-line system. The FCC, however, was not certain this system had adequate technical quality, so it established an industrywide committee of engineers, the **National Television System Committee (NTSC)**, to recommend standards. This committee rejected the 441-line system and in 1941 recommended a 525-line system that became the standard in the United States for many years.

Originally there were to be 13 **very high frequency (VHF)** channels, but channel 1 was eliminated to allow spectrum space for FM radio, so for many years channels 2 through 13 were available for over-the-air broadcast. Twenty-three stations went on the air in 1941 and 1942, 10,000 sets were sold, and commercials were sought. The first commercial was bought by Bulova and consisted of a shot of a Bulova clock with an announcer intoning the time. All of this stopped, however, in 1942 because of World War II. During the war, only six stations remained on the air, and most sets became inoperable because spare parts were not being manufactured for civilian use.[5]

beginning stations

Exhibit 3.3
Vladimir K. Zworykin holding an early model of the iconoscope TV tube.

(Courtesy of RCA)

Exhibit 3.4
Felix the Cat as he appeared on the experimental 60-line, black-and-white TV sets in the late 1920s. This picture was transmitted from New York City all the way to Kansas.

(Courtesy of RCA)

Exhibit 3.5
David Sarnoff dedicating the RCA pavilion at the 1939 New York World's Fair. This dedication marked the first time a news event was covered by television.

(Courtesy of RCA)

3.2 The Emergence of Broadcast Television

Television activities did not resume immediately after the war. The delay was partly due to a shortage of materials. Building a TV station was expensive, and the owners expected that they would operate at a loss until there were enough receivers in the area to make the station attractive to advertisers. In addition, CBS and RCA were still feuding about color. CBS stated that its mechanical color system was so well developed that TV station allocations should not continue until the FCC resolved the color question. RCA promised black-and-white sets on the market by mid-1946 and a color system shortly thereafter that was electronic and compatible with its black-and-white sets. In 1947, the FCC declared that CBS's color system would be a hardship on set owners because they would have to buy new sets. It therefore stated that television should continue as a 525-line black-and-white system.

feud about color

Television emerged as a mass medium in 1948. The number of stations, sets, and audiences all grew more than 4,000 percent within that one year. Advertisers became aware of the medium, and networks began more systematic programming.[6]

enormous 1948 growth

Television networks had existed before 1948 as offshoots of radio networks. As early as 1945, TV networks were organized by NBC, CBS, and ABC. A fourth network, DuMont, existed for a short while. The network programming was broadcast through local stations in cities throughout the country. In the mid-1940s, most cities had only one or two TV stations, and NBC and CBS usually recruited them as **affiliates,** making it difficult for ABC and DuMont to compete. ABC survived because it merged in 1953 with United Paramount Theaters and thus gained an increase in operating funds. DuMont, however, went out of business in 1955.[7]

networks

3.3 The Freeze

Television grew so uncontrollably that in the fall of 1948 the FCC imposed a **freeze** on television station authorizations because stations were beginning to interfere with one another. The proposed six-month freeze lasted until July 1, 1952. During this period, 108 stations were on the air, and no more were authorized to begin operation. What occurred between 1948 and 1952 could be termed an explosive lull. Many cities, including Austin, Texas; Denver, Colorado; and Portland, Oregon, had no television stations. Others had only one or two. New York and Los Angeles were the only cities to boast seven. Although TV networking still could not be considered truly national, the number of sets, audience size, advertising, and programming continued to grow. By 1952, 15 million homes had sets, the largest of which had 20-inch tubes and sold for about $350. Television advertising revenues reached $324 million.

freeze

When the FCC ended the freeze in 1952, it established a station allocation table that assigned specific channels to specific areas of the country. Between 1948 and 1952, the FCC realized that the 12 VHF channels being used would not be sufficient to meet demand. The FCC engineers added 70 stations in the

allocation tables

ultra-high frequency (UHF) band, for a total of 82 channels. UHF was at a much higher frequency than VHF, and very little was known about the technical characteristics of UHF at that time. The FCC engineers thought that by increasing the power and tower height of UHF stations, they would be equal in coverage to VHF stations. In addition, the FCC reserved 242 channels (80 VHF and 162 UHF) for noncommercial educational television.[8]

early cable TV

Another thing that happened during the freeze was the introduction of **cable television.** The only way that people could receive TV if they were not within the broadcast path of one of the 108 stations on the air was to put up an antenna where the signal could be received and run that signal through a wire to where they wanted it (see "Zoom In" box). As a result, neighbors in remote and mountainous areas built rudimentary cable TV systems in order to provide television reception for themselves (see Exhibit 3.6).[9]

Within a short time, small "mom-and-pop" companies took over the stringing of cable as a moneymaking venture, charging the people who wanted to receive television an initial installation fee and monthly fees compatible with modest profits. If there were no local signals available, these

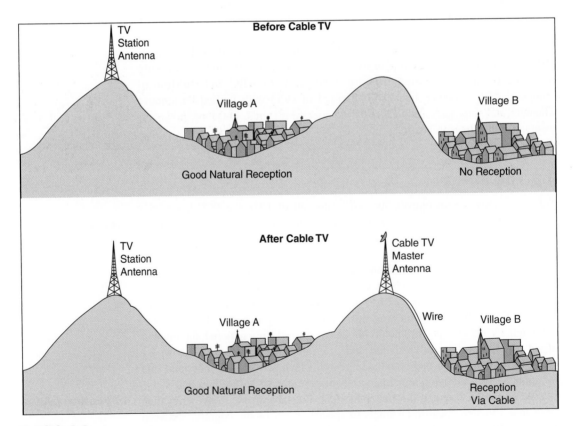

Exhibit 3.6

This diagram shows the structure of very early cable TV when its purpose was to bring TV signals to areas that could not receive them.

ZOOM IN: Where Did Cable TV Begin?

There are several stories about how cable TV actually began. One is that a man who sold TV sets in a little appliance shop in a hilly area of Pennsylvania started cable TV in 1948. He noticed that he was selling sets only to people who lived on one side of town. After investigating, he discovered that the people on the other side of town could not obtain adequate reception, so he placed an antenna at the top of a hill, intercepted TV signals, and ran the signals through a cable down the hill to the side of town with poor reception. When people on that side of town bought a TV set from him, he would hook their home to the cable.

Another story is that a "ham" radio operator in Oregon was experimenting with TV signals because of his interest in the field and because his wife wanted "pictures with her radio." He placed an antenna on an eight-story building and ran cables from there to his apartment across the street. Neighbors became excited about the idea, so he wired their homes, too. The initial cable subscribers helped pay for the cost of the equipment, and he charged newcomers $100 for a hookup.

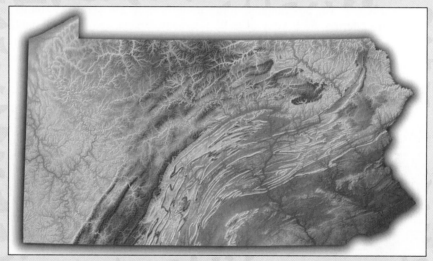

Pennsylvania, with its many hills and mountains, had a need for the retransmission of early cable TV.

(Courtesy of Brian Gross)

companies carried the three network signals by importing them from stations in nearby communities, a practice that became known as **distant signal importation.** With 65,000 subscribers and an annual revenue of $10 million in 1953, cable TV was only a minor operation. Most broadcasters were unconcerned about this business that was growing on the fringes of their signal contour.[10]

distant signals

The lift of the freeze led to an enormous rush to obtain broadcast stations. Within six months, 600 applications were received and 175 new stations were authorized. By 1954, 377 stations were broadcasting, and TV could be considered truly national.[11]

rush for stations

3.4 Early TV Programming

sports

During the 1948–49 season, 30 percent of sponsored evening programs on broadcast TV were sports related. Wrestlers, both men and women, competed to outdo one another in costumes, hairdos, and mannerisms. The large number of TV sets in bars spurred the sports emphasis. During the 1949–50 season, however, sports comprised less than 5 percent of evening programming, and children's programming was tops, indicating that TV had moved to the home.

Milton Berle

In TV cities, movie attendance, radio listening, sports event attendance, and restaurant dining were all down—especially on Tuesday night, which was Milton Berle night. People with TV sets stayed home and often invited their friends (or allowed their friends to invite themselves) to watch this ex-vaudevillian's show, which started in 1948 and included outrageous costumes, slapstick comedy, and a host of guest stars (see Exhibit 3.7). "Uncle Miltie" on his *Texaco Star Theater* became a national phenomenon and was the reason many people bought their first television set.[12]

radio's influence

The programming forms used by early TV were very similar to those used during the heyday of radio—comedy, drama, soap operas, public affairs, and children's programs (see Chapter 5). Some of them were reconstructed radio shows with radio stars such as Groucho Marx (see Exhibit 3.8), Jack Benny, and

Exhibit 3.7

Milton Berle in one of his outlandish costumes. His *Texaco Star Theater* was TV's first big hit.

(Courtesy of NBC)

Exhibit 3.8

Groucho Marx (left), shown here with assistant film editor Bruce Bilson, was the host of *You Bet Your Life,* which started on radio in 1947 and moved to TV in 1950. The format was a game show, but the real entertainment came from the spontaneous comments and freewheeling interviews Groucho conducted with the contestants.

(Courtesy of Bruce Bilson)

ventriloquist Edgar Bergen (who didn't last long because people could see his lips move).

In 1951, *I Love Lucy* (see Exhibit 3.9) began as a maverick of the TV world because it was filmed earlier, whereas other shows were aired live. There was a stigma against film at the time, partly because it added to the cost and partly because the TV networks inherited the live tradition from radio and assumed that all shows should be produced live. The film aspect was particularly useful and dramatic when Lucille Ball became pregnant, and the story line dealt with Lucy's pregnancy. The episode involving the birth of Lucy's baby was filmed ahead of time, and Lucille's real baby was born the same day the filmed episode aired—to an audience that comprised 68.8 percent of the American public. Eventually, the filming more than paid for itself because the program became the first international hit. Copies of the film were made, dubbed into numerous languages, and sold overseas.[13]

Exhibit 3.9

Lucy tries to make a hit playing the saxophone. The *I Love Lucy* series, starring Lucille Ball, captured America's heart and can still be seen in reruns.

(Photo/Viacom, Hlwd.)

Television newscasts of the 1950s developed slowly. Networks found it easy to obtain news and voices, but pictures were another matter. At first they contracted with the companies that supplied the newsreels then shown in movie theaters. This did not exactly fit television's bill because much of it was shot for in-depth stories, not news of the day. Networks hired their own film crews, but limited budgets and bulky film equipment meant that camera operators could attend only planned events, such as press conferences and ribbon-cuttings. The 15-minute newscasts tended to be reports on events that had been filmed earlier.[14]

I Love Lucy

Interview-type news shows were a further development. *Meet the Press* (which is still on the air) began its long run of probing interviews with prominent people. *See It Now* started in 1951 as a news documentary series featuring Edward R. Murrow. A historic feature of the first program was showing for the first time both the Atlantic and Pacific oceans live on TV. At this point, television could be proud of its achievements. Sets had increased in size and quality, TV programming had increased in hours and variety, the public was fascinated with the new medium, and advertisers were providing increasingly strong financial backing.[15]

newscasts

public affairs

3.5 Blacklisting

To this fledgling industry came some of the country's best-known talent. They came from radio, Broadway, and film—all of which were experiencing downturns as television was burgeoning. Unfortunately, some of these people became caught in the anti-Communist **blacklisting** mania of the 1950s led by Senator Joseph R. McCarthy. A 215-page publication called *Red Channels: The Report of Communist Influence in Radio and Television* gave information about 151 people, many of whom were among the top names in show business, that suggested they had Communist ties. Some of these charges, such as associating people with "leftist" organizations to which they had never belonged, were

Red Channels

proved to be false. Other allegations were true, but were "leftist" only by definitions of the perpetrators of *Red Channels*. These included such "wrongdoings" as signing a cablegram of congratulations to the Moscow Art Theater on its 50th birthday.

Although many network and advertising executives did not believe these people were Communists or in any way un-American, they were unwilling to hire them, in part because of the controversy involved and in part because sponsors received phone calls that threatened to boycott their products if programs employed these people. Some well-established writers, for the better part of a decade, found that all the scripts they wrote were "not quite right," and certain actors were told they were "not exactly the type for the part." Many of these people did not even know they were on one of the "lists" because these were circulated clandestinely among executives.

McCarthy's downfall

In time, the blacklist situation eased. Ironically, broadcasting was influential in exposing the excesses of the Communist witch hunt, which had spread beyond the entertainment industry. Edward R. Murrow prepared several programs on Senator McCarthy, who had alarmed the country by saying that he had a list of hundreds of Communists in the State Department. The Murrow telecasts helped expose this claim as false. In 1954, television covered hearings in which McCarthy took on the army. As the nation watched, McCarthy and his aides harassed and bullied witnesses. Public resentment built against McCarthy, and the Senate voted 67 to 22 to censure him.[16]

3.6 The Live Era

Programming of the 1950s was predominantly live. *I Love Lucy* continued to be filmed, and several other programs jumped on the film bandwagon as foreign countries began developing broadcasting systems. Americans could envision the reuse of their products in other countries and, hence, the possibility of recouping film costs. Reruns in the United States were as yet not thought of, although some programs were **kinescoped,** so they could be shown at various times. These

kinescopes

kinescopes were low-quality, grainy film representations of the video picture. Most of the popular network series of the day, however, originated in New York and were telecast as they were being shot (see "Zoom In" box). Local stations produced some of their own programming, such as children's shows, that were also live.

"golden age"

The programming of the early 1950s is often referred to as the "golden age of television." This is mainly because of the live dramas produced during this period. One of the most outstanding plays was Rod Serling's *Requiem for a Heavyweight* (see Exhibit 3.10), the psychological study of a broken-down fighter. Another was Paddy Chayefsky's *Marty,* the heartwarming study of a short, stocky, small-town butcher who develops a sensitive romantic relationship with a homely schoolteacher.[17]

Weaver

In addition to conventional drama, innovative formats were tried, many of them the brainchild of Sylvester L. "Pat" Weaver, president of NBC from 1953 to 1955. One of his ideas was the spectacular, a show that was not part of the

ZOOM IN: **The Challenges of Live Programming**

Live programming, especially dramas, created problems for writers, actors, and technicians. Costume changes needed to be essentially nonexistent. The number of story locales was governed by the number of sets that could fit into the studio. These sets were arranged in a circle on the periphery of the studio so the cameras could have easy access to each new scene. Timing was sometimes an immense problem. In radio, scripts could be quite accurately timed by the number of pages, but television programs contained much action, the time of which often fluctuated widely in rehearsals. One writing solution was to plan a search scene near the end of the play. The actor could find what he or she was looking for right away if the program was running long or could search the room for as long as necessary if the program appeared to be moving too quickly.

Because the programs were live, they were usually performed twice—once for the Eastern and Central Time zones and a second time for the Mountain and Pacific Time zones. During the interval between the two broadcasts, the cast and crew would go out to eat and drink. The second performances were usually a little more "lively" than the first ones.

A scene from *Marty,* a 1953 drama starring Rod Steiger and Nancy Marchand that was originally aired live and later made into a movie.

(Culver Pictures)

regular schedule. One outstanding spectacular was *Peter Pan,* which was viewed by some 165 million Americans. Weaver was also involved in developing the *Today* and *Tonight* shows. *Today* was first hosted by Dave Garroway (see Exhibit 3.11), who had a chimpanzee as a sidekick. *Tonight* originally starred Steve Allen, and from 1962 to 1992 was hosted by Johnny Carson. Weaver believed that programs should be controlled by the network rather than by advertising agencies. Most of the early TV and golden era radio program content had been produced by the **advertising agencies** (see Chapter 5). Weaver developed a **magazine concept** whereby advertisers bought insertions in programs, whose content was supervised and produced by the networks.[18]

While commercial television was prospering, educational television programming was struggling. Although the FCC had reserved channels, **Ford Foundation**

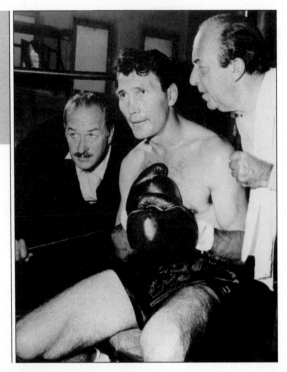

Exhibit 3.11

Dave Garroway with chimps J. Fred Muggs *(left)* and Phoebe B. Beebe. Muggs was discovered in classic show-business style by one of *Today's* producers, who spotted him in an elevator.

(Courtesy of NBC)

activating the stations was difficult because of the huge sums of money needed. The stations were not allowed to engage in advertising, and the organizations interested in establishing stations—universities, school districts, nonprofit community organizations—were not known for deep pockets.

Fortunately, the Ford Foundation stepped in and provided money for facilities and programming. Two innovative programming concepts it funded were the Chicago TV College and the Midwest Program on Airborne Television Instruction (MPATI). The Chicago project (see Exhibit 3.12) was a fully accredited set of televised courses that enabled students to earn two-year college degrees through a combination of at-home viewing and on-campus class attendance. MPATI used an airplane that circled two states to broadcast programs to schoolchildren. The Ford Foundation also helped establish a **bicycle network** system for distributing programming, the National Educational Television and Radio Center, which operated out of Ann Arbor, Michigan. This helped stations acquire programming, but because tapes were mailed to stations in a round-robin fashion, nothing timely could be exchanged. Despite this outside help, educational programming of the 1950s and 1960s was, in a word, dull. It was produced on a shoestring budget and consisted primarily of talking heads discussing issues and information.[19]

Exhibit 3.12

Professor Harvey M. Karlen conducting a lesson on national government for Chicago TV College.

(Courtesy of Great Plains National Instructional Television Library)

3.7 Color TV Approval

Color TV underwent a number of changes and did not become widespread for several decades. CBS continued to advocate its mechanical system while RCA favored an electronic one. In 1950, in a reversal of its 1947 decision, the FCC accepted the CBS system, believing it provided higher quality color pictures. RCA, however, continued to fight the CBS system by refusing to program in color and by gaining allies among other set manufacturing companies and TV stations. CBS ran into difficulty manufacturing its color sets, and a general state of confusion concerning color reigned in the TV industry.

CBS system accepted

To help solve the problem, the National Television System Committee (NTSC), the same committee of engineers that had decided on the 525-line system, volunteered to study the situation. Because this committee included more members who favored the RCA system than the CBS system, very few people were surprised when it recommended the compatible system. To its credit, however, the RCA system had improved, in part in response to suggestions from the NTSC. The FCC took the NTSC recommendation and in 1953 sanctioned RCA's electronic compatible system. At the time, even CBS supported the adoption.

RCA system accepted

For a long time, however, RCA-NBC was the only company actively promoting color. NBC constructed new color facilities and began programming in color, but both CBS and ABC dragged their heels, and most local stations did not have the capital needed to convert to color equipment. Consumers were reluctant to purchase color TV sets because they cost twice as much as black-and-white sets, and the limited color programming did not merit this investment.

slow growth

But as more color sets were sold, the prices fell, and programming in color increased, causing even more sets to be sold. Not until the late 1960s were all networks and most stations producing color programs.[20]

3.8 Broadcast-Cable Clashes

During the 1950s and 1960s, cable TV became more sophisticated. Wires placed on telephone poles replaced the house-to-house wire loops augmented here and there by a tree. In addition to importing signals into areas where there was no television, cable companies began importing distant signals into areas where there were a limited number of stations. For example, if a small town had one TV station, the cable system would import the signals from two TV stations in a large city several hundred miles away.

distant signal harm

This importing of distant signals caused the first squabbles between broadcasters and cable TV systems. Existing TV stations in an area found that the size of their audience shrank because people watched the imported signals. Sometimes an imported signal was playing the same program that was showing on the local station. For example, a local station might be showing a rerun of *I Love Lucy* and find that the imported station was showing the same rerun, splitting the show's audience in half. Because of the smaller audience, local stations could not sell their ads for as high a price as before the importation.

FCC nonintervention

In the late 1950s, some stations in areas affected by cable TV appealed to the FCC for help. The FCC, however, maintained a policy of nonintervention in cable TV matters during the early years, hoping the courts would settle the problems between the cable operators and broadcasters. But the situation only became more confused as court cases piled up.[21]

3.9 Prerecorded Programming

The days of live broadcast programming, other than news and special events, began to disappear in the mid-1950s for several reasons. One was the introduction of videotape in 1956. The expense of the equipment prevented it from taking hold quickly, but once its foot was in the door, videotape revolutionized TV production. Programs could now be performed at convenient times for later airings. As the equipment became more sophisticated, a taping could be stopped for costume and scene changes. As the equipment became even more sophisticated, mistakes could be corrected through editing.[22]

videotape

film

The live era, however, began to yield to film even before tape took hold. Film companies were originally antagonistic toward TV because it stole much of the audience that attended movie theaters. Some film production companies would not allow their stars to appear on TV and would not even allow TV sets to appear as props in movies.[23]

The 1953 merger of United Paramount Theaters and ABC opened the door for Hollywood film companies and the New York TV establishment to begin a dialogue. The first result of this dialogue was a one-hour weekly series, *Disneyland,* produced for ABC by the Walt Disney Studios. This was a big hit, and soon several other major film companies were producing film series for TV.

Among the early filmed TV series, westerns predominated. By 1959, 32 western series were on prime-time TV. The one with the greatest longevity was *Gunsmoke* (see Exhibit 3.13), which revolved around Dodge City's Matt Dillon, Chester Goode, Doc, and Miss Kitty. The TV boom continued in the late 1950s—more TV sets, more viewers, more stations, more advertising dollars. To this rising euphoria came a dark hour.[24]

Exhibit 3.13

Gunsmoke, the longest-running western. It started on CBS in 1955 and lasted until 1975. Shown here are two of the principals, Dennis Weaver and James Arness.

(Globe Photos, Inc.)

3.10 The Quiz Scandals

Quiz programs on which contestants won minimal amounts of money or company-donated merchandise existed on both radio and television. In 1955, however, a new idea emerged in the form of *The $64,000 Question* (see Exhibit 3.14). If contestants triumphed over challengers for a number of weeks, they could win huge cash prizes. Sales of Revlon products, the company that produced and sponsored *The $64,000 Question,* zoomed to such heights that some were sold out nationwide. The sales success and high audience ratings spawned many imitators. Contestants locked in soundproof booths pondered, perspired, and caught the fancy of the nation.

westerns

The $64,000 Question

Exhibit 3.14

Master of ceremonies of *The $64,000 Question* was Hal March *(right).* Gino Prato, an Italian-born New York shoemaker, mops his brow while listening to a four-part question about opera.

(From United Press International)

**Charles Van
Doren**

Then, in 1958, a contestant from *Twenty-One* described situations that indicated the program was rigged. The networks and advertising agencies denied the charges, as did Charles Van Doren, a Columbia professor who was the most famous of the *Twenty-One* winners. A House of Representatives subcommittee conducted hearings, and in 1959 Van Doren testified. He read a long statement describing how he had been persuaded in the name of entertainment to accept help with answers in order to defeat a current champion who was becoming unpopular. Van Doren was also coached on methods of building suspense, and when he did win, he became a national hero and a leader of intellectual life. He asked to be released from the show and finally was allowed to lose after months on the program. He initially lied, he said, so that he would not betray the people who had faith in him.

**effects of
scandals**

In retrospect the **quiz scandals**' negative effect on TV was short-lived. The medium was simply too pervasive a force to be permanently afflicted by such an incident. Congress did amend the Communications Act to make it unlawful to give help to a contestant, but for the most part the networks rectified the errors by canceling the quiz shows and reinstating a higher percentage of public-service programs.

Networks also took charge of their programming to a much greater extent. All three networks decreed that from then on most program content would be decided, controlled, and scheduled by networks, which would then sell time to advertisers. This was a further extension of Weaver's "magazine concept." Beginning in 1960, most program suppliers contracted with the networks rather than with advertising agencies. This made life more profitable for the TV networks, too, because they established profit participation plans with the suppliers.[25]

3.11 UHF and Cable TV Problems

UHF problem

When the FCC ended its freeze and established stations in the UHF band, it intended that these stations would be equal with VHF stations. In reality, they became second-class stations. UHF's weaker signal was supposed to be compensated for by higher towers, but this did not work in practice. People did not have sets that could receive UHF, so to tune in UHF stations they needed to buy converters. Many were unwilling to do this, and UHF found itself in a vicious circle. To persuade people to buy UHF converters, UHF had to offer interesting programming material; to finance interesting program material, UHF stations had to prove to advertisers that they had an adequate audience. The FCC tried to help the fledgling UHF stations in several ways—by passing regulations to encourage networks to buy UHF stations, by making some markets all UHF and some all VHF, by requiring UHF receivers in all sets. But nothing seemed to work. The thing that helped the most was when cable TV systems started putting UHF stations on their systems, enabling UHF channels to have signals that were as strong as VHF channels.[26]

The cable TV industry, however, had a number of its own obstacles to overcome without worrying about UHF. In 1963, a court ruled that the FCC could refuse to authorize additional facilities to the Carter Mountain cable company in the Rocky Mountains, because the cable company might damage the well-being of the local broadcast stations. This gave the FCC power to restrict cable to protect broadcasters. As a result, the FCC abandoned its hands-off policy regarding cable TV and began initiating rulings that generally favored broadcasters.

In 1963, the FCC issued a notice that covered two main areas: (1) All cable systems would be required to carry the signal of any TV station within approximately 60 miles of its system; and (2) no duplication of program material from distant signals would be permitted 15 days before or 15 days after a similar local broadcast. The rule of local carriage, which became known as **must-carry,** caused little or no problem at the time because cable owners were glad to carry the signals of local stations. However, must-carry would raise many issues in the future—issues that still plague the broadcast-cable relationship. **FCC ruling**

The 30-day provision did cause bitter protest from cable operators because it limited their rights with regard to what they could show on their distant imported stations. This rule, known as **syndicated exclusivity,** meant that if a local station was going to show an *I Love Lucy* rerun on June 15, a cable TV operator would have to black out that rerun on a distant imported station for the whole month of June. The cable TV industry marshaled its forces and succeeded in having the 30-day provision reduced to only one day. This, of course, angered the broadcasters. The debate and actions related to syndicated exclusivity lasted many years and only abated when cable systems had enough specific cable programming that they no longer felt the need to import distant stations. **syndicated exclusivity**

Another report issued by the FCC in 1966 put more restrictions on cable service. This order came when about 600 cable systems were starting construction and 1,200 had applications pending. All these systems were required to prove to the FCC that their existence would not harm any existing or proposed broadcast station. By not increasing its staff to handle this load, the FCC was, in essence, freezing the growth of cable. This made broadcast station owners very happy. But the effect of this ruling on cable operators was not what the FCC had envisioned. Cable companies that were unable to expand were sold to large corporations that could withstand the unprofitability of the freeze period. **Multiple-system operators (MSOs)**—companies that owned a number of different cable systems in different locations—became quite prevalent by the 1970s.[27] **cable "freeze"**

Another regulatory issue that caused consternation between broadcasters and cable operators during the 1970s was the payment of copyright fees. When cable systems transmitted broadcast signals, they did not pay any fees to those who owned the copyright to the material. In some instances, networks or stations created the material, so they owned the copyright. In other cases, the networks or stations purchased program material from film companies or independent producers and paid copyright fees. The stations, networks, film companies, and independent producers believed the cable companies should pay copyright fees for the retransmission of material because these retransmission rights were not included in the broadcast TV package. Cable operators thought they were **copyright** exempt from paying fees for retransmission rights because they were merely extending coverage. In 1976, a new copyright law was passed, and one of its provisions was that cable TV systems were required to pay **compulsory license** fees to a newly created government body called the Copyright Royalty Tribunal. This body would then distribute the money to copyright owners. The amount of this compulsory license fee was 0.7 percent of the cable operator's revenue from basic monthly subscriptions, and the copyright law gave the tribunal authority to adjust the rate for inflation. Although both cable operators and copyright holders seemed happy with this plan at the time, it would prove contentious in the future.[28]

3.12 Early Cable TV Programming

Despite its legal problems, the cable TV business did make a few forays into programming, mostly during the 1960s and 1970s. When cable TV first began it was simply a **common carrier**; that is, cable companies picked up signals and brought them into homes for an installation charge and regular monthly fees and did not provide any programming themselves. The very early systems had only three channels for the three broadcast networks. As television and its resulting

12 channels

technology grew, cable systems provided as many as 12 channels of programming. Different cable systems placed varying programming on these 12 channels, but this usually consisted of converting all the local VHF and UHF stations to VHF space on the dial, thus making the UHF signal equal to that of VHF. If there weren't many local stations, then the cable system would bring in stations from nearby communities.

beginning "programming"

Gradually, however, some cable facilities began to undertake their own programming. The most common "programming" involved unsophisticated weather information. Cable TV operators would place a thermometer, barometer, and other calculating devices on a disc and have a TV camera take a picture as the disc slowly rotated. This would then be shown on a vacant channel, so people in the area could check local weather conditions. Some systems had news of sorts. This might involve a camera focused on bulletins coming in over a wire-service machine or on 3-by-5-inch cards with local news items typed on them (see Exhibit 3.15). It was a simple, inexpensive, one-camera type of local origination. Some cable systems had more complex local programming, usually in the form of local news programs, high school sports events, city council meetings, and talk shows on issues important to the community.[29]

In October 1969, the FCC issued a rule that required all cable TV systems with 3,500 or more subscribers to begin local programming. This was a burden on many cable systems that did not have the funds to build studios, buy equipment, and hire crews. As a compromise, the FCC changed the order so that systems that did not want to produce programming themselves could simply make equipment and channel time available to members of the community who wished to produce programs.

public access

This brought about a different type of local programming known as **public access.** Individuals or groups used the equipment provided by the cable operator

Exhibit 3.15
A simple cable TV system whereby local announcements are placed on the TV screen one after the other as the drum rotates.

(Courtesy of TeleCable of Overland Park, Kansas)

to produce programs without the operator's input or sanction, and those programs were then cablecast over one of the system's channels. This differed from **local origination,** which was what programming planned and produced by the cable system came to be called. In most areas, public access was not a huge success, and the equipment provided by the cable company remained largely unused for lack of interest. In other areas, particularly where cable systems showed an active interest in local programming, some exciting and innovative projects were undertaken. Sometimes, unfortunately, the people using public access time were would-be stars who used cable for vanity purposes or people from fringe groups who promoted various causes or even lewd modes of behavior. Such individuals and groups gave public access an unsavory reputation.[30]

Another form of local programming instituted by some cable systems was the showing of movies. A cable company would use one of its channels to show **movies** movies that only subscribers who paid an extra fee could receive. These movies were leased from film companies and shown without any commercial interruptions. Regulations prevented cable systems from showing the well-known films that broadcasters wanted to show during prime time, but because the small cable systems had only a limited number of subscribers willing to pay for the movies, they could not afford to lease blockbuster movies anyway. The channels for movies were not a huge success, but they did bring extra income to some cable systems.

Throughout the 1960s and 1970s, promises were made, broken, and remade concerning the potential services and programs that would be available through cable, such as home security systems and home shopping services. For the most part, however, early cable was a medium to bring broadcast signals to areas with poor reception.

3.13 Reflections of Upheaval

news

Broadcast television journalism gathered force and prestige during the 1960s—the decade of civil rights revolts, the election of John F. Kennedy, assassinations, the Vietnam War, and student unrest. The networks encouraged documentaries and, in 1963, increased their nightly news from 15 to 30 minutes, thereby assigning increased importance to their news departments. Anchoring on camera for NBC were Chet Huntley and David Brinkley (see Exhibit 3.16), and CBS had Walter Cronkite, the person who became known as the most trusted man in America.

documentaries

The quiz scandals had helped precipitate a rise in documentaries. To atone for their sins, networks increased their investigative fare. Documentaries were now easier to produce because technical advances allowed film and sound to be synchronized without an umbilical cord between two pieces of equipment, and wireless microphones enabled speakers to wander freely without having to stay within range of a mike cord. News-gathering flexibility became even greater when portable video cameras were developed at the end of the 1960s. Technological advances also made it possible for TV to show the 1969 moon landing, one of the most watched events ever (see Exhibit 3.17).

civil rights

Many documentaries reported on racial problems. *Crisis: Behind a Presidential Commitment* chronicled the events surrounding Governor George Wallace's attempted barring of the schoolhouse door to prevent blacks from attending the

Exhibit 3.16

A Huntley-Brinkley newscast. Chet Huntley broadcast from New York, while David Brinkley, seen on the television screen, broadcast from Washington, D.C.

(Courtesy of NBC)

Exhibit 3.17

The 1969 moon landing was one of the most watched events on TV.

University of Alabama; *The Children Are Watching* dealt with the feelings of a six-year-old black child attending the first integrated school in New Orleans. Television reacted to the civil rights movement in another way—it began hiring African Americans. Both radio and TV had been lily-white, but TV started hiring African Americans in the 1960s. Scriptwriters began including stories about blacks, and Diahann Carroll (see Exhibit 3.18) in *Julia* became the first black TV heroine.

Television was credited, through the **Great Debates** between John F. Kennedy and Richard M. Nixon, with having a primary influence on the 1960 presidential election results. Kennedy and Nixon met for the first debate at the studios of Chicago's CBS station. Kennedy, tanned from campaigning in California, refused the offer of makeup. Nixon, although he was recovering from a brief illness, did likewise. Some of Nixon's aides, concerned about how he looked on TV, applied Lazy-Shave, to create a clean-shaven look. Some people believe that Kennedy's apparent victory in the first debate had little to do with what he said. People who heard the program on radio felt Nixon held his own, but those watching TV could see a confident, attentive Kennedy and a haggard, weary-looking Nixon whose perspiration streaked the Lazy-Shave. Three more debates were held, and Nixon's makeup and demeanor were well handled, but the small margin of the Kennedy victory is often attributed to the undecided vote that swung to Kennedy during the first debate.[31]

Three years later, television devoted itself to the coverage of the assassination of John F. Kennedy (see Exhibit 3.19). From Friday, November 22, to Monday, November 25, 1963, there were times when 90 percent of the American people were watching television. One New York critic wrote, "This was not viewing. This was total involvement." From shortly after the shots were fired in Dallas until President Kennedy was laid to rest in Arlington Cemetery, television kept the vigil, including the first "live murder" ever seen on TV as Jack Ruby shot alleged assassin Lee Harvey Oswald. Many praised television for its controlled, almost flawless coverage. Some thought TV would

Exhibit 3.18

Diahann Carroll *(left)* was the first African American heroine in a TV series.

(Supplied by NBC/Globe Photos, Inc.)

Great Debates

Exhibit 3.19

The Kennedy funeral. This off-monitor shot shows JFK's son saluting the flag covering his father's caisson.

(Courtesy of Broadcasting magazine)

Exhibit 3.20

Walter Cronkite visiting Vietnam.

(Reprinted with the permission of Broadcasting & Cable, 1987–1994 by Cahners Publishing Company)

have made it impossible for Lee Harvey Oswald to receive a fair trial and that it was the presence of the media that enabled the Oswald shooting to occur.[32]

Television brought war to the American dinner table for the first time in history. The networks established correspondents in Saigon as the troop buildup began in Vietnam during the mid-1960s. Reports of the war appeared almost nightly on the evening news programs. In 1968, amid rising controversy over the war, Walter Cronkite decided to travel to Vietnam and returned believing that the United States would have to accept a stalemate in that country (see Exhibit 3.20). Much of the controversy surrounding the war originated on the country's campuses, where students became increasingly dissident. This, too, was covered by the media, as was the 1968 Democratic convention in Chicago, where youths outside the convention hall protested the nomination of Hubert Humphrey.

Kennedy assassination

Vietnam

The media became embroiled in the controversy. On the one hand, they were accused of inciting the riot conditions because the demonstrators seemed to be trying to attract media coverage. On the other hand, many people inside the convention hall learned of the protest by seeing it on a TV monitor and might not otherwise have known of this show of discontent.[33]

3.14 A Vast Wasteland?

Minow's speech

In 1961, Newton Minow, Kennedy's appointee as chairman of the Federal Communications Commission, spoke before the annual convention of the National Association of Broadcasters. During his speech, Minow, who had seemed favorable toward broadcasting, startled his audience with the following words:

> I invite you to sit down in front of your television set when your station goes on the air and stay there without a book, magazine, newspaper, profit-and-loss sheet, or rating book to distract you—and keep your eyes glued to that set until the station signs off. I can assure you that you will observe a vast wasteland.[34]

The term **vast wasteland** caught on as a metaphor for television programming (see "Zoom In" box). The executives were not happy with Minow's phrase.

violence and antiviolence

During the early 1960s, the dominant fare on TV was violence. The original versions of such programs as *The Untouchables, Route 66,* and *The Roaring 20s* all featured murders, jailbreaks, robberies, kidnappings, and blackmail. Saturday morning children's programming was also replete with violent cartoons. A surge against violence, aided by Minow's challenge, spurred a shift to shows about doctors, such as *Dr. Kildare,* and more comedies, such as *The Dick Van Dyke*

ZOOM IN: **Still a Wasteland?**

Here are some other excerpts from Newton Minow's 1961 speech. How do you think they apply (or don't apply) to television today?

When television is good, nothing—not the theater, not the magazines or newspapers—nothing is better. But when television is bad, nothing is worse.

You will see a procession of game shows, violence, sadism, murder, western badmen, western good men, private eyes, gangsters, more violence and cartoons. And, endlessly, commercials—many screaming, cajoling, and offending.

It is not enough to cater to the nation's whims—you must also serve the nation's needs. If some of you persist in a relentless search for the highest rating and the lowest common denominator, you may well lose your audience.

I am unalterably opposed to government censorship. There will be no suppression of programming that does not meet with bureaucratic tastes.

We need imagination in programming, not sterility; creativity, not imitation; experimentation, not conformity; excellence, not mediocrity.

The power of instantaneous sight and sound is without precedent in mankind's history. This is an awesome power. It has limitless capabilities for good—and for evil. And it carries with it awesome responsibilities—responsibilities which you and I cannot escape.

Newton Minow.

(Reuters/TimePix)

Show, Gilligan's Island, The Beverly Hillbillies, and *Hogan's Heroes* (see Exhibit 3.21).

During the 1960s, old Hollywood movies established themselves on TV (see Chapter 6). In 1961, *Saturday Night at the Movies* began a prime-time movie trend that, by 1968, saw movies every night of the week. This rapidly depleted Hollywood's supply of old films, and some of the low-quality films that made it onto the airwaves enhanced the vast wasteland theory. In 1966, NBC made a deal with Universal to provide movies that were specially made for television and would not appear in theaters first. This concept caught on, and by 1969, all three networks had **made-fors** on a regular basis.[35]

old movies

Also during the 1960s, the number of local stations that were on the air grew. Most cities had more than three stations, and because there were only three networks, some of the stations, referred to as **independents,** did not have

local station fare

Exhibit 3.21

Hogan's Heroes, a comedy set in a World War II prisoner-of-war camp, was on CBS from 1965 to 1971.

(Courtesy of Bruce Bilson)

a network to feed them programming. Although some of these stations produced their own programming, most of them bought the bulk of it from **syndicators**—companies that had started businesses to provide programming for these stations. Syndicated programs were sent to stations on tape, and they could air them whenever they wanted. By this time the U.S. television industry had discovered the value of the rerun, so much of what the syndication companies provided was reruns of programs that had been on one of the networks. They also provided packages of older movies, often giving them a theme—such as horror movies or Cary Grant movies. This type of programming fare gave Minow further fuel for his "vast wasteland" comment.[36]

3.15 The Public Broadcasting Act of 1967

Meanwhile, educational broadcasting was undergoing its own transformation. The Ford Foundation, feeling that it was shouldering too much of the support for educational television, cut back on funding. A number of organizations and councils appeared and disappeared, trying to solve the financial problems, but the result was a lack of focus and political infighting.

Carnegie Commission

The Carnegie Foundation finally came to the rescue by setting up the Carnegie Commission on Educational Television. This group of highly respected citizens spent two years studying the technical, organizational, financial, and programming aspects of educational television and in 1967 published its report. The commission changed the term *educational television* to *public television* to overcome the pedantic image the stations had acquired. It also recommended that "a well-financed and well-directed system, substantially larger and far more pervasive and effective than that which now exists in the United States, be brought into being if the full needs of the American public are to be served."[37]

Public Broadcasting Act

Most of the Carnegie Commission's many recommendations were incorporated into the Public Broadcasting Act of 1967. There were two major changes between the Carnegie recommendation and congressional passage of the act, however. First was the addition of radio. The Carnegie Commission addressed itself only to TV, but the radio interests that led to National Public Radio were included by Congress (see Chapter 5). Second, the Carnegie group recommended that **public broadcasting** be given permanent funding, perhaps through a tax on TV sets, but Congress opted for one year's funding of $9 million with additional funding to be voted on later.

The Public Broadcasting Act of 1967 provided for the establishment and funding of the Corporation for Public Broadcasting (CPB) to supply national leadership for public broadcasting and to make sure that it would have maximum protection from outside interference. Government money was given to CPB, which in turn gave money to stations and networks. In this way public broadcasters were protected from government influence. The CPB's 15-member board (later changed to 9) was appointed by the president with the consent of the Senate. The main duties of the board were to help new stations get on the air, to obtain grants from federal and private sources, to provide grants to stations for programming, and to establish an interconnection system for public broadcasting stations.

CPB

The CPB was specifically forbidden from owning or operating the interconnection system. The corporation, therefore, created the Public Broadcasting Service (PBS), an agency to schedule, promote, and distribute programming over a wired network (the use of wires disappeared in 1978 when PBS became the first network to distribute programming totally by satellite). PBS had a governing board consisting of station executives. This service was not to produce programs, but rather to obtain them from such sources as public TV stations, independent producers, and foreign countries. Hence a three-tier operation was established: (1) The stations produced the programs; (2) PBS scheduled and distributed the programs; and (3) CPB provided funds and guidance for the activities.

PBS

public TV programs

The advent of the Corporation for Public Broadcasting and its accompanying funding allowed public television to embark on innovative programming of high quality. The first series to arouse interest was the successful children's series *Sesame Street,* first produced in 1969 by a newly created and newly funded organization, the Children's Television Workshop (CTW). This series helped strengthen PBS as a network because it was in demand throughout the country.

Other early PBS series that met with sustained popularity were *Mister Rogers' Neighborhood* (see Exhibit 3.22), a children's program produced in Pittsburgh; *The French Chef,* Julia Child's cooking show produced in Boston; *Black Journal,* a public affairs series dealing with news and issues of importance to African Americans produced in New York; and *Civilisation,* a British import on the development of Western culture.[38]

Exhibit 3.22

Fred Rogers *(right)* with Mr. McFeely, one of the many characters who visited regularly on *Mister Rogers' Neighborhood.*

(Courtesy of Family Communications, Inc.)

3.16 Government Actions

Nixon-era controversies

The organization for public broadcasting set up by the Public Broadcasting Act of 1967 looked good on paper, but it had problems in operation. A major controversy surfaced during the Nixon administration when the CPB, which had been formed to insulate public broadcasting from government, became somewhat of an arm of the government. The conflict centered on programs that were critical of the government, primarily nationally aired documentaries. The administration carried its ire into the budgeting area, and Richard Nixon vetoed a bill that would have given public broadcasting $64 million for 1972–73. With this action, the chairperson and president of CPB resigned, and Nixon appointed new people to the posts who were more in line with his way of thinking. With CPB now more in tune with the administration, a schism developed between CPB and PBS over who should control the programming decisions for public broadcasting. PBS wanted to continue the documentaries while CPB pressed for an emphasis on cultural programming that would be less political in nature. The controversy continued but gradually abated as Watergate consumed the administration and Nixon was forced to resign.[39]

fin-syn

In 1970, the FCC adopted a rule barring networks from acquiring a financial interest in independently produced programs and from engaging in domestic syndication of these programs or of their own network-produced programs. This rule, called **financial interest–domestic syndication (fin-syn),** was adopted because the FCC (and the Justice Department) thought the networks, which at the time had about 90 percent of the prime-time audience, were too powerful in the programming business. The rule stated that networks could produce only 3.5 hours of the 22 prime-time hours available each week.

The bulk of the shows during the next three decades were produced by independent production companies. For example, *The Mary Tyler Moore Show* and its **spin-offs** *Rhoda* and *Phyllis* were produced by MTM, headed by Grant Tinker. *All in the Family* (see Exhibit 3.23), the first situation comedy to deal with previously taboo subjects such as politics, ethnicity, and sex, came from Tandum-TAT, headed by Norman Lear and "Bud" Yorkin. After these programs aired on the networks, the rights reverted to the production companies, which gained the profits from syndication. Fin-syn lasted until 1995, when the networks were able to prove their power was no longer dominant, given all the other programming sources that had arisen. The abolishment of fin-syn enabled the networks to produce more of their own product and to profit from it financially. Obviously, the independent Hollywood production companies that had supplied the programming were not happy with this situation.[40]

Exhibit 3.23

The cast of *All in the Family* on the set of their Queens home.

(Photo/Viacom, Hlwd.)

In 1971, the FCC established the **prime-time access rule (PTAR).** The purpose was to break the networks' monopoly on programming between 7:00 and 11:00 P.M. so that some of this time could be programmed by stations to meet community needs. The rule stated that networks would be allowed to program only three hours a night on stations in the top 50 markets, leaving the other hour to their local affiliates. During this hour, there were to be no old network programs or old movies; all programs had to be new. However, the rule was modified so that prime-time access became 7:30 to 8:00 P.M., Monday through Saturday. Although prime-time access was established to allow stations to broadcast local programming of high quality, this did not happen to any significant degree. The stations filled the time with the cheapest thing they could find—game shows and remakes of old formats, although some developed local **magazine shows.** As the years progressed, the various entities involved argued about the efficacy of the rule, and in 1995 the FCC abolished it.[41]

prime-time access

Family hour began in 1975 as an attempt to curtail sex and violence before 9:00 P.M., when children are most likely to be watching. The family-hour concept originated with the code of the National Association of Broadcasters (NAB), an industry organization. Some believed the code came about as a result of subtle pressure from the FCC and, hence, represented an abhorrent attempt by the commission to regulate program content. The family-hour idea was widely opposed by writers, producers, and directors, who took the concept to court. Eventually courts decided that the FCC had overstepped its powers and that the family-hour restrictions should be removed from the NAB code.[42]

family hour

3.17 Corporate Video

Another form of television known as **corporate video** grew up in the 1960s and 1970s. It involved the use of television equipment by corporations, educational institutions, the government, and nonprofits. Pre-1960s video equipment was so bulky and expensive that only TV stations and networks could afford and house it. Black-and-white cameras weighed about 50 pounds and cost about $30,000. Videotape recorders stood 6 feet high, used 2-inch-wide tape reels, and cost several hundred thousand dollars. During the late 1960s, lower quality, smaller black-and-white cameras and recorders that used 1-inch tape were developed. Organizations purchased this equipment and used it primarily to train personnel and to communicate with employees (see Exhibit 3.24).

equipment

Educational institutions purchased this equipment also. Colleges used cameras and other equipment to teach students to become broadcasters. Schools also used the equipment to enhance learning. For example, a biology professor taped close-ups of dissections and then showed the tapes in class, giving all the students a "front-row view" of the dissection process. Basketball coaches used a camera to tape a game and then watched the tape with the team to discuss strategies that did and did not work (see Exhibit 3.25). At one time many corporations and educational institutions had **closed-circuit TV (CCTV).** All the rooms of a building or complex of buildings were wired so that video material from one location could be sent to all rooms. Companies used CCTV

educational uses

Exhibit 3.24

In the 1970s, General
Telephone employees
used this studio to tape
a training program that
instructed telephone
operators in safety
procedures.

*(Courtesy of General
Telephone Company)*

when the president wanted to speak to all the employees, and schools used them
to distribute instructional material to various rooms on campus.

ITFS   In 1971 the FCC authorized **instructional television fixed service (ITFS),**
a special form of over-the-air broadcasting. Schools obtained licenses to
transmit in the 2,500 MHz range, which is higher than broadcast frequencies so

Exhibit 3.25

An early semiportable
TV unit being used to
tape a college
basketball game so the
team can assess its
strengths and
weaknesses.

*(Courtesy of Orange Coast
College)*

cannot be received by regular TV sets. Some school districts used ITFS to interconnect all the grade schools in a district, thereby allowing, say, a Spanish teacher to teach all third graders in the district at the same time. Universities used ITFS to transmit lectures to locations away from the main campus where students could watch the material and gain course credit. This concept of **distance learning** still exists, but the internet has replaced ITFS, and many schools have rented or sold their frequencies to business that are using them for other wireless applications.

In the late 1970s, companies started using satellites for **teleconferencing.** This form of communications involves leasing satellite time and using it to transmit video information from place to place (see Exhibit 3.26). Teleconferencing usually has an interactive telephone or internet element so that viewers can interact with the speakers, asking questions and offering comments.

Organizations continue to use video equipment in a variety of ways. Congressional representatives and other government officials who used to hand out printed news releases now make **video news releases (VNRs)** for broadcast and cable TV outlets. Many companies make videos for their consumers—how to operate a new oven, how to assemble a toy airplane. Hospitals use HDTV cameras to record surgical operations, so the procedures can be shown to interns and medical students.[43]

Exhibit 3.26

Teachers at a local school participating in a satellite teleconference on new teaching techniques.

(© Cynthia Roesinger/Photo Researchers, Inc.)

teleconferencing

continuing uses

3.18 HBO's Influence

The stage was set in 1975 for a dramatic change in cable programming when Home Box Office (HBO) began distributing movies and special events via satellite. Home Box Office was actually formed in 1972 by Time, Inc., as a movie/special pay service for Time's cable system in New York. The company decided to expand this service to other cable systems and to set up a traditional **microwave** link to a cable system in Wilkes-Barre, Pennsylvania. During the next several years, HBO expanded its microwave system to include about 14 cable companies, but this was not a successful venture, and it was not profitable for Time.

In 1975 Time decided to show the Ali-Frazier heavyweight championship fight from Manila by satellite transmission on two of its cable systems. The

early HBO

HBO on satellite

Exhibit 3.27

A live in-concert taping by HBO of singer Diana Ross before a regular nightclub audience at Caesar's Palace in Las Vegas. This was one of HBO's early programs.

(Courtesy of Home Box Office)

marketing issues

HBO success

Turner's superstation

experiment was very successful, and HBO decided to distribute all of its programming by satellite (see Exhibit 3.27). As soon as HBO sent its signals to a satellite (which was then a relatively new technology), they could be received throughout the country by any cable system that was willing and able to buy a receiving dish.

HBO then began marketing its service to cable systems nationwide, which was no easy chore. The original receiving dishes were 10 meters in diameter and cost almost $150,000, a stiff price for cable systems, many of which were just managing to break even. But satellite technology moved quickly enough that by 1977 dishes in the range of 3 to 4.5 meters sold for less than $10,000. Another problem that HBO encountered involved the rules that had been established by broadcasters that prevented cable from bidding on programming, such as movies and sports events, that conventional broadcasters wanted to show. HBO and several cable TV system owners took these rules to court, and in 1977 the courts set aside the rules, allowing HBO to develop as it wanted. Another problem HBO had when it first started marketing its service was its financial relationship with cable systems. At first, it offered the cable system owners 10 percent of the amount they collected by charging subscribers extra for HBO programming. Approximately 40 percent of the fee was to go to HBO and 50 percent to the program producers. When cable owners complained about their percentage, HBO raised it so that the systems retained about 60 percent of the money and HBO and the program producers split the other 40 percent.

With receiving dishes manageable in both cost and size, with appealing programming, and with financial remuneration at a high level, cable systems began subscribing to HBO in droves. Likewise, HBO became very popular with individual cable subscribers who were willing to pay extra to receive commercial-free movies. By October 1977, Time announced that HBO turned its first profit.

Soon after HBO started distributing programming via satellite, Ted Turner, who owned a low-rated UHF station in Atlanta, Georgia, decided to put his station's signal on the same satellite as HBO. This meant that cable operators that bought a receiving dish for HBO could also place Turner's station on one of their channels. This created what was referred to as a **superstation** because it could be seen nationwide. Cable operators paid a dime a month per subscriber for the superstation signal, but they did not charge the subscriber as they did for the HBO pay service. The economic rationale was that the extra program service would entice more subscribers. The charge to the cable companies did not cover the superstation's costs, but the station could charge a higher rate for its advertisements once the superstation had a bigger audience.

With two successful program services on the satellite, the floodgates opened and cable TV took on a new complexion—one that led the television industry into its modern era.[44]

3.19 Summary

Although broadcast TV and cable TV started at about the same time during the 1940s, they have very different early histories. Broadcast TV was the queen bee that envisioned cable as a minor nuisance.

Broadcast TV was the product of experiments with mechanical and electronic scanning dating back to the 1880s, but it did not become truly operational until after World War II, with 1948 being its explosive year. A freeze on stations lasted from 1948 to 1952, during which time programming became popular—especially that of Milton Berle and Lucille Ball. The 1950s brought blacklisting and the Edward R. Murrow program that helped squelch it, the "golden age of television" with its live programs, a color TV standard, the videotape recorder, and the quiz scandals. News and documentaries increased during the 1960s, and TV journalists covered many crucial stories related to civil rights, the Kennedy election and assassination, and Vietnam. Commercial TV in general was criticized for its violent and inane content. Educational television, which had been dull, picked up momentum after passage of the Public Broadcasting Act of 1967. In the 1970s, the FCC initiated fin-syn, PTAR, and family hour, all of which affected programming for several decades but were eventually rescinded.

Cable TV started in the 1940s but did not experience the explosive growth in that decade that broadcast TV did. For several decades it was mainly a retransmission service, partly because the FCC generally favored broadcasters when it came to matters such as distant signal importation, must-carry, syndicated exclusivity, and copyright. Early programming consisted mainly of weather stations and news on printed cards. Eventually public access and local origination started. The turning point for cable was HBO's success with its pay service.

By the end of the 1970s, both broadcast and cable TV had changed enormously from their early days. In the next chapter you will learn about the continued relationship between the two and also see how they reacted, separately and jointly, to new television media forms that surfaced during ensuing decades.

Suggested Websites

www.philotfarnsworth.com (the official family site for TV inventor Philo T. Farnsworth)

www.davidsarnoff.org (the David Sarnoff library, which contains articles about his life and about TV and radio in general)

www.chicagotelevision.com/dumont.htm (a site devoted to the history of the DuMont Network)

www.americanrhetoric.com/speeches/newtonminow.htm (complete text and audio of Newton Minow's speech)

www.museum.tv (site for the Museum of Broadcast Communications, which contains a wealth of information about the history of TV)

Notes

1. These early mechanical systems are described in A. A. Dinsdale, *First Principles of Television* (New York: Wiley, 1932) and in the first volume of the authoritative three-volume history of broadcasting by Erik Barnouw published in New York by Oxford University Press. The titles and dates are *A Tower in Babel: A History of Broadcasting in the United States to 1933* (1966); *The Golden Web: A History of Broadcasting in the United States 1933–1953* (1968); and *The Image Empire: A History of Broadcasting in the United States from 1953* (1970). Barnouw condensed the television material into a one-volume book, *Tube of Plenty: The Development of American Television* (1975).
2. For more on Farnsworth, see Donald G. Godfrey, *Philo T. Farnsworth: The Father of Television* (Salt Lake City: University of Utah Press, 2001); and Stephen F. Hofer, "Philo Farnsworth: Television Pioneer," *Journal of Broadcasting,* Spring 1979, 153–66.
3. Barnouw, *A Tower in Babel,* 66, 154, 210. For Zworykin's own view of the electronics of TV, see Vladimir Zworykin, *The Electronics of Image Transmission in Color and Monochrome* (New York: Wiley, 1954).
4. Barnouw, *The Golden Web,* 126.
5. Ibid., 126–30.
6. Ibid., 242–44.
7. "The Death of the DuMont Network," *Emmy,* August 1990, 96–103.
8. Robert K. Avery and Robert Pepper; "An Institutional History of Public Broadcasting," *Journal of Communication,* Summer 1980, 126–38.
9. David L. Jaffe, "CATV: History and Law," *Educational Broadcasting,* July/August 1974, 15–16; and Thomas F. Baldwin and D. Stevens McVoy, *Cable Communication* (Englewood Cliffs, NJ: Prentice-Hall, 1983), 8–10.
10. Jaffe, "CATV," 34.
11. Barnouw, *The Golden Web,* 295.
12. Barnouw, *The Golden Web,* 285–90.
13. "For the Love of Lucy," *Emmy,* July/August 2006, 60–63.
14. Mike Conway, "A Guest in Our Living Room: The Television Newscaster before the Rise of the Dominant Anchor," *Journal of Broadcasting & Electronic Media,* September 2007, 457–78.
15. Matthew C. Ehrlich, "Radio Prototype: Edward R. Murrow and Fred Friendly's *Hear It Now,*" *Journal of Broadcasting & Electronic Media,* September 2007, 438–56.
16. *Red Channels: The Report on Communists in Radio and Television* (New York: Counterattack, 1950); and "Hollywood Blacklist," *Emmy,* Summer 1981, 30–32.
17. Barnouw, *The Image Empire,* 22–23.
18. "Pat Weaver: Visionary or Dilettante?" *Emmy,* Fall 1979, 32–34.
19. Clifford G. Erickson and Hyman M. Chausow, *Chicago's TV College: Final Report of a Three-Year Experiment* (Chicago: Chicago City Junior College, 1960); and Mary Howard Smith, *Midwest Program on Airborne Television Instruction: Using Television in the Classroom* (New York: McGraw-Hill, 1960).
20. Daniel E. Garvey, "Introducing Color Television: The Audience and Programming Problems," *Journal of Broadcasting,* Fall 1980, 515–26; and "Color-TV Gaining Momentum," *Hollywood Reporter,* October 2, 1963, 1.
21. "Cable: The First Forty Years," *Broadcasting,* November 21, 1988, 40.
22. "The Videotape Recorder Turns 50," *TV Technology,* April 12, 2006, 26–27.
23. Ken Auletta, *Three Blind Mice: How the TV Networks Lost Their Way* (New York: Random House, 1991), 76.
24. "The End of the Trail," *Emmy,* September–October 1984, 50–54.
25. "Dress Rehearsals Complete with Answers?" *U.S. News,* October 19, 1959, 60–62; and Susan L. Brinson, "Epilogue to the Quiz Show Scandal: A Case Study of the FCC and Corporate Favoritism," *Journal of Broadcasting & Electronic Media,* June 2003, 276–88.
26. "The First 50 Years of Broadcasting," *Broadcasting,* May 25, 1981, 94.
27. Jaffe, "CATV," 35.

28. "Righting Copyright," *Time,* November 1, 1976, 92; and Margaret B. Carlson, "Where MGM, the NCAA, and Jerry Falwell Fight for Cash," *Fortune,* January 23, 1984, 171.

29. Ron Merrell, "Origination Compounds Interest with Quality Control," *Video Systems,* November/December 1975, 15–18; and Sloan Commission on Cable Communications, *On the Cable: The Television of Abundance* (New York: McGraw-Hill, 1972).

30. Richard C. Kletter, *Cable Television: Making Public Access Effective* (Santa Monica, CA: Rand Corporation, 1973); and Charles Tate, *Cable Television in the Cities: Community Control, Public Access, and Minority Ownership* (Washington, DC: The Urban Institute, 1972).

31. P. M. Stern, "Debates in Retrospect," *New Republic,* November 21, 1960, 18–19.

32. "Covering the Tragedy: President Kennedy's Assassination," *Time,* November 29, 1963, 84; "Did Press Pressure Kill Oswald?" *U.S. News,* April 6, 1964, 78–79; and "President's Rites Viewed throughout the World," *Science Newsletter,* December 7, 1963, 355.

33. Michael J. Arlen, *Living Room War* (New York: Viking Press, 1969); Oscar Patterson, "An Analysis of Television Coverage of the Vietnam War," *Journal of Broadcasting,* Fall 1984, 397–404; and Edward Fouhy, "Looking Back at 'The Living Room War,'" *RTNDA Communicator,* March 1987, 12–13.

34. Barnouw, *The Image Empire,* 197.

35. A large number of highly pictorial works have been published that deal with TV programming through the years. These include Irving Settl, *A Pictorial History of Television* (New York: Frederick Ungar, 1983); Alex McNeil, *Total Television: A Comprehensive Guide to Programming from 1948 to the Present* (New York: Viking Penguin, 1991); "The Twenty Top Shows of the Decade," *TV Guide,* December 9, 1989, 20–27; and "The Unforgettables," *Emmy,* July/August 2006, 60–63.

36. *1990 INTV Census* (Washington, DC: Association of Independent Television Stations, 1990).

37. The Public Broadcasting Act of 1967, Public Law 90-129, 90th Congress (November 7, 1967).

38. Robert M. Pepper, *The Formation of the Public Broadcasting Service* (New York: Arno Press, 1979); and "In the Beginning, Before PBS," *TV Technology,* April 11, 2007, 46–47.

39. "Twenty Tumultuous Years of CPB," *Broadcasting,* May 11, 1987, 60–75.

40. "Fin-Syn," *Broadcasting and Cable,* January 24, 2000, 30.

41. "FCC Gives Prime-Time Access Rule the Ax," *Los Angeles Times,* July 27, 1995, D-1.

42. "'Family Hour' OK but Not by Coercion," *Daily Variety,* November 5, 1976, 1.

43. Larry G. Goodwin and Thomas Koehring, *Closed-Circuit Television Production Techniques* (Indianapolis, IN: Howard W. Sams, 1970); John Barwich and Stewart Kranz, *Profiles in Video* (White Plains, NY: Knowledge Industry Publications, 1975); "Around the Nation with ITFS," *EITV,* June, 1985, 28–29; "Airwave Licenses Benefit Schools," *Los Angeles Times,* May 20, 2008, C-1; "Teleconferencing Industry: Full-Speed Ahead Growth," *AV Video,* April 1991, 20–24; and "FCC: VNRs Need IDs," *Broadcasting & Cable,* April 18, 2005, 17.

44. *HBO Landmarks* (New York: Home Box Office, n.d.), 1; and Patrick Parsons, "The Evolution of the Cable-Satellite Distribution System," *Journal of Broadcasting & Electronic Media,* March 2003, 1–17.

Chapter 4

MODERN TELEVISION

Between 1940 and 1980, broadcast television became a mature industry. It encompassed three competing networks and more than 1,000 local TV stations. Many stations broadcast programming fed to them by the networks; others produced or bought their

> New media have meant new values. Since the dawn of history, each new medium has tended to undermine an old monopoly, shift the definitions of goodness and greatness, and alter the climate of men's lives.
>
> **Eric Barnouw, *A Tower in Babel*, 1966**

own programming. The fortunes of these companies fluctuated, and the Federal Communications Commission (FCC) issued a few rulings that cut down on the income of some television entities, but in general, being in the broadcast television business in 1980 was akin to having a license to print money.

Up until the end of the 1970s, cable television was a fledgling industry. Although it had started in the 1940s, it was primarily an adjunct of the broadcast business, bringing TV station signals to homes in hilly areas that could not receive TV signals over the air. It was, however, poised to make a giant leap.

Neither broadcast nor cable TV could envision the variety of television forms that would arise in the 1980s and beyond—satellite TV, VCRs, DVDs, phone company ventures, DVRs, HDTV, the internet, portable devices, video games. The last three were discussed in Chapter 2; this chapter will cover the later years of broadcast TV, the growth of cable TV, and the rise of other television technologies.

4.1 Growth of Cable TV Programming Services

As mentioned in Chapter 3, Time, Inc., put its Home Box Office signal on satellite in 1975 and, after several years of struggling, garnered a fairly large audience for its commercial-free movies. Shortly after Time became successful with HBO, Viacom launched a competing **pay cable** service, Showtime (see Exhibit 4.1). Other pay cable services started in rapid succession: The Movie Channel, a 24-hour-a-day service from Warner-Amex; Spotlight, a Times-Mirror movie service; Bravo (see Exhibit 4.2) and The Entertainment Channel, both cultural programming services; The Disney Channel, a family-oriented pay service featuring Disney products; and Playboy, "adult" programming that included R-rated movies, skits, and specials.[1]

pay cable

Exhibit 4.1
Jason Robards starring in Eugene O'Neill's *Hughie,* one of Showtime's cablecasts of a Broadway production.

(Courtesy of Showtime)

Other new satellite-delivered programming services were launched as what became known as **basic cable.** Some of these were supported by advertising, some were supported by the institutions that programmed them, and some were supported by small amounts of money that cable systems paid to the programmers. Most **narrowcast,** meaning they had a specific niche audience in mind. Among them were

ESPN, all-sports programming started by Getty Oil

Black Entertainment Television, programming about and for African Americans

Exhibit 4.2
A scene from *The Greek Passion* performed by the Indiana University Opera Company and taped by Bravo. This was the first opera to be taped for cable television.

(Courtesy of Bravo Networks)

Exhibit 4.3
An early wrestling match on USA.

(Courtesy of USA Network)

basic services

CBN, Christian-oriented family programming

Nickelodeon, a noncommercial children's service

CNN, 24 hours a day of news started by Ted Turner

USA, a network with a variety of programming (see Exhibit 4.3)

C-SPAN, public service-oriented programming, including live coverage of the House of Representatives

ARTS, an ABC-owned cultural service

CBS Cable, an advertising-supported cultural service owned by CBS

MTV, the creator of music videos

Satellite News Channel, a Westinghouse-ABC joint venture of 24-hour news, established to compete with CNN

Daytime, a service geared toward women

Cable Health Network, programming about physical and mental health

The number of **superstations** also grew and were considered part of basic cable. As mentioned in Chapter 3, Ted Turner's WTBS was first, but soon after WGN in Chicago and WWOR in New York also went on satellite.[2]

local programming

Local programming took on an entirely new dimension. The older cable systems that only had 12-channel capacity usually allocated one channel to local programming. The newer systems that took advantage of improved technology to provide 20, then 54, then more than 100 channels usually promised an entire complement of local channels. At least one of these channels was reserved for **local origination**—programming that the cable system itself initiated. A number of others were some combination of **public access.** Some systems had **leased**

access channels for businesses or individuals interested in buying time on a cable channel to present their messages. Not all the local programming planned by cable systems and local groups materialized, but the quality of access programming improved greatly from the early days (see Chapter 3) when it consisted mainly of vanity TV.[3]

Interactive cable was highly touted. Promises were made that cable would perk the coffee, help the kids with homework, do the shopping, protect the home, and teach the handicapped. In some areas of the country, a fair amount of interactive cable was undertaken. The pioneer and most publicized system was Qube, which Warner-Amex operated in Columbus, Ohio, starting in 1977. The basic element of Qube was a small box with response buttons that enabled Qube subscribers to send an electronic signal to a bank of computers at the cable company that could then analyze the responses. An announcer's voice or a written message on the screen asked audience members to make a decision about some question, such as who was most likely to be a presidential candidate, what play a quarterback should call, or whether a city should proceed with a development plan. Audience members made their selection by pressing the appropriate button in multiple-choice fashion. A computer analyzed the responses and printed the percentage of each response on the participant's screen.[4]

Qube

Home security was another interactive area that cable entered. Various burglar, fire, and medical alarm devices connected to the cable system. A computer in a central monitoring station sent a signal to each participating household about every 10 seconds to see if everything was in order. A signal was sent back to the cable company monitoring station, which then notified the police if any of the doors, windows, smoke detectors, or other devices hooked to the system were not as they should be. These interactive services were largely unprofitable, but cable TV seemed like such a gold mine that companies were willing to experiment.[5]

security

4.2 Cable TV's Gold Rush

Cable programming and services were not the only part of the cable business that grew at a phenomenal rate during the late 1970s and early 1980s. The number of homes subscribing to cable TV more than doubled between 1980 and 1985 (see Exhibit 4.4), and the amount of money cable companies brought in skyrocketed. In 1979, pay revenues grew 85 percent over 1978 figures. The industry predicted that this peak would subside to about 50 percent growth, but in 1980 pay revenues grew another 95.5 percent. Between 1975 and 1980, cable TV profits grew 641 percent. With figures such as these, it is no surprise that cable experienced a veritable gold rush.[6]

revenues

One way this gold rush manifested itself was in franchising. In most areas, the local city council became the agency that issued cable **franchises** and stipulated how the cable system was to conduct business. Local governments assumed this responsibility in part because the national government would not intervene in cable TV matters during its early years and in part because cable

franchising

Exhibit 4.4

Growth of cable TV.

Year	Number of Systems	Number of Subscribers	Percent of Homes	Pay Cable Subscribers	Number of Cable Networks
1955	400	150,000	.5	—	—
1960	640	650,000	1.4	—	—
1965	1,325	1,275,000	2.5	—	—
1970	2,490	4,500,000	7.6	—	—
1975	3,506	9,800,000	12.0	469,000	—
1980	4,225	16,000,000	20.0	9,144,000	28
1985	6,844	32,000,000	43.0	30,596,000	56
1990	9,575	54,280,000	58.9	40,100,000	79
1995	11,218	58,834,440	62.5	43,730,000	128
2000	10,845	66,054,000	71.3	49,200,000	224

TV companies had to tear up city streets to lay cable, and that took the approval of the city government. Competing applicants for a cable franchise would present their plans for operation, including such items as the method of hookup (for example, telephone poles or underground cable), the fees to be charged to the customer for installation and for regular monthly service, and the percentage of profit that the company was willing to give the city for the privilege of holding the franchise. Based on this information, the council awarded the franchise to the company it believed was most qualified. During the early years of cable, this was a sleepy process; usually only one or two companies applied for a franchise in a given area. Areas with good reception did not even bother with franchising because no companies applied.

However, with the cable TV boom, cities that could not have given away cable franchises in earlier years because of their clear reception of broadcast TV signals suddenly became prime targets for cable and its added programming services. Most city governments were not accustomed to dealing with such matters. Gradually, through the use of consultants, city councils established lists of minimal requirements that they wanted from the cable companies. Cable companies, in their fervor to obtain franchises, usually went well beyond what the cities required. They, too, hired consultants, who contacted city leaders to learn the political structure and needs of the city and to decide how the company should write its franchise proposal to ensure the best possible chance of winning the contract. Only one company could receive a franchise for a particular area; therefore, cries of corruption sometimes accompanied these procedures.[7]

The development of local programming became an important part of the franchising process. Cities usually requested that cable companies set aside a certain number of channels for access by community groups. These were often referred to as **PEG** (public, education, government) access channels. To

PEG channels

program these PEG channels, cable companies promised to provide equipment and sometimes personnel.

Cable companies began promising cheaper rates, shorter time to lay cable, more channels, more equipment, and generally more and better everything as franchising competition became intense. Sometimes the winning cable company was unable to meet all the requirements stipulated in the bid, especially concerning the speed with which the system was to be built. The laying of cable, however, did move forward at a rapid rate.

As the cable companies promised more, they realized they might not recover their investment for about a decade. This hastened a process that was already prevalent within the cable industry—the takeover of small mom-and-pop cable operations by large **multiple-system operators (MSOs)** and then the consolidation of these MSOs with other large companies. Major companies emerged in the cable industry, partly because they wanted to be part of the gold rush and partly because only large companies had the resources to withstand the expenses of the franchising process and the other start-up costs of laying cable, marketing, and programming.[8]

large systems

Other groups also began to stake their claims in cable's gold rush. Advertisers who saw cable reaching close to 30 percent penetration of the nation's households in the early 1980s began placing ads on cable's programming channels, both local and national. Members of the various unions and guilds that operate in the broadcasting industry were not involved in cable programming when it first began because the cable companies did not recognize the unions. After several long, bitter strikes during the early 1980s, however, the unions won the right to be recognized and to receive residuals from cable TV. Thus, the same people who worked in broadcast programming began working in cable programming.[9]

other participants

Perhaps the biggest winners in the cable gold rush were the equipment manufacturers that supplied the materials needed to build the cable systems. Suppliers of the converters that enable a regular TV set to receive the multitude of cable channels, the satellite dishes, and the cable itself found their order desks piled high. Space on a satellite became a precious commodity as more companies wanted to launch national programming services.[10]

In retrospect, all of this was too much too fast, and by the mid-1980s the bloom was off the rose. A number of cable programming services merged or went out of business, advertisers became more cautious, and MSOs sold some of their systems. All of this was a necessary correction to the hype that existed in the cable world, and once it was over, cable remained a viable business.[11]

4.3 The Beginnings of Satellite TV

Satellite TV would eventually become a competitor for cable TV, but it got off to a slow start. When satellites were first launched in 1962, the technology was very complicated—well beyond what typical homeowners could handle. But the technology developed and reception dishes became more user-friendly, so in 1979 the FCC issued a ruling that no one needed a license to have a **TV receive-only (TVRO)** satellite dish. In other words, individual households were allowed to own satellite dishes that received (but did not send) programming.

TVRO dishes

One outcome of this 1979 ruling was that Satellite Television Corporation (STC) informed the FCC it wanted to develop a new programming service that would go directly from satellites to homes. The satellites would have higher

STC

power than the ones being used by cable TV systems to receive cable TV networks, so the consumer dishes could be smaller and cheaper than the dishes the cable systems needed. The FCC agreed with the basic idea and invited all interested parties to apply for licenses for this service, which was then called **direct broadcast satellite (DBS).** In 1982, the FCC approved eight DBS applications and expected that the services would be operational by 1985.[12]

satellite failures

The FCC mandated that the companies prepare "due diligence" reports in 1984 to make sure the services were on track. With great fanfare, market researchers proclaimed that 50 percent of people would subscribe to DBS, but the companies ran into many technical and financial problems as they dealt with the realities of launching high-power satellites. Several of the original DBS applicants flunked their "due diligence" tests, and others didn't even bother to fill out the forms. In a surprising 1984 move, STC, the company that had pioneered the DBS concept, announced it was pulling out of the business after having invested five years and $140 million gearing up. Prospects for DBS dimmed, and no services were on the air by 1985.[13]

backyard satellite

Another result of the 1979 FCC ruling that no one needed a license to own a TVRO was that individuals, particularly those in rural areas where TV reception was poor, bought the large dishes used by cable TV systems and set them up in their backyards. This became known as **backyard satellite** reception. The dishes cost about $3,000, but the people who bought them could then receive all the cable programming for free. This caused a minor stir among cable TV program suppliers, but the number of backyard dishes was so small that the cable providers did not aggressively pursue the issue.[14]

4.4 Home Video

Videotape recorders for TV stations were introduced in 1956 (see Chapter 3), but they were much too bulky and expensive for consumers. The first **videocassette recorder (VCR)** specifically designed for the home-consumer market was the Sony **Betamax,** introduced in 1975. This deck used half-inch tapes that could record for one hour. Obviously it was **analog** because **digital** recording technology was not yet available. It sold for $1,300, but the price was actually $2,300 because the recorder could only be purchased with a new Sony color TV set.[15]

Betamax

VHS

The price went down and the features went up as Sony encountered competition from Matsushita, which in 1976 introduced the half-inch **video home system (VHS)** and marketed it through its Japan Victor Company (JVC). This JVC recorder (see Exhibit 4.5) had the advantage of recording up to two hours. People who bought these VCRs used them primarily for recording feature films off the air, so the two-hour format enabled them to record an entire feature on one cassette. In the subsequent format war, both Sony and Matsushita introduced new features, such as extended recording time and timers that would record programs while the owner was away. In the end, VHS beat Beta as the format of choice for consumers (see "Zoom In" box).[16]

Hollywood at first fought the onslaught of VCRs, fearing that people would record movies and rent or buy them from **video stores** that were springing up

Exhibit 4.5
An early VHS
videocassette recorder.

*(Courtesy of JVC Industries
Company)*

 ZOOM IN: **Whatever Happened to Betamax?**

The Sony Betamax format had everything going for it. It offered higher technical quality than the VHS format, in part because the tape wrapped around the video head more completely, allowing more information to be stored. Beta images did not suffer from as much smearing and degradation as VHS when they were dubbed from one tape to another. In addition, Betamax came out before VHS, giving it what is often referred to as the "first mover" advantage—the first of anything gets most of the hype, and those that follow have more trouble obtaining market share. It also had a well-known brand behind it—Sony.

So what happened? Matsushita allowed other companies to license its VHS format. Companies such as Sharp, Magnavox, and RCA began manufacturing and distributing VHS machines. This took some sales away from Matsushita's JVC company, but Matsushita received license fees from the other companies. The combined promotion and advertising from all the companies placed the VHS format firmly in the minds of consumers.

Sony chose to go another route. It wanted to keep all the Betamax sales for itself, so it did not license its technology. It also counted on the Sony brand name to carry the day and did not invest heavily in promotion. Eventually it did engage in limited licensing and more promotion, but it was too late. VHS had captured the market.

A Sony Betamax
videocassette
recorder.

*(Courtesy of Sony
Corporation of
America)*

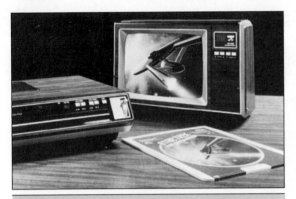

Exhibit 4.6
RCA's stylus videodisc player.

(Courtesy of RCA Corporation)

(e.g., Blockbuster) instead of going to the theater. In a landmark lawsuit in 1976, Universal and Disney brought suit against Sony, claiming that any device that could copy their program material violated copyright laws and should not be manufactured. The case dragged through the courts for years, but in 1984 the Supreme Court ruled that home VCRs do not, in or of themselves, violate copyright laws. This decision was fortunate for Sony and for consumers because by 1984 more than 13 million recorders had been sold.[17]

In the end, video did not destroy Hollywood—it saved Hollywood. People bought and rented movies on tape in droves. By the early 1990s, movie studios were earning more from video sales and rentals than from theatrical box office sales—a business model that continues today. Theatrical release generates the buzz; home video generates the profit.[18]

copyright suit

videodiscs

Analog **videodiscs** were also available in the 1980s. In 1978, MCA and a number of other companies brought to market the **laser disc** ($700 for a player), which used a laser beam that read information embedded in a plastic disc. RCA followed in 1981, introducing a **capacitance disc** with a $500 player (see Exhibit 4.6); it had a diamond stylus that moved over grooves similar to those of a phonograph record. Both videodisc systems were intended to market movies to viewers. They never became a dominant consumer medium, however, because they suffered from high costs, large size, and the inability to record.[19]

4.5 Broadcast TV in the 1980s

The 1980s were not financially good times for over-the-air commercial television. Stations and networks were faced with competition from cable TV, which splintered the audience and the advertising dollar. Profits were down almost universally, and some stations even went bankrupt. This downturn in broadcasting, combined with a national phenomenon of company takeovers, led to major changes in the management and ownership of all three networks.

ABC

The first network affected was ABC, which was purchased in 1985 for $3.5 billion by Capital Cities Communications, a company one-third the size of ABC. Although the takeover was termed "friendly," it was the first time in history that a major network had been purchased, and it was referred to as "the minnow swallowing the whale." CapCities kept ABC for 10 years and then sold it to Disney for $18.5 billion, a tidy profit for a minnow.[20]

NBC

Less than a year after Capital Cities bought ABC, a second major network, NBC, was purchased as part of an overall takeover of its parent company, RCA. The purchaser was General Electric, a company that had been involved in the formation of NBC radio and then eased out of ownership by RCA (see Chapter 5). This, again, was called a "friendly takeover." In 2004, NBC spread its wings further and merged with Universal, garnering the network a strong production facility.[21]

Things were not so "friendly" in the CBS executive suites. In 1985, Ted **CBS** Turner attempted an unfriendly coup that CBS thwarted only by buying back much of its own stock, an act that greatly damaged CBS's financial footing. In 1986, CBS founder, William Paley, who had retired, came back to CBS and formed an alliance with Lawrence Tisch, who also was nibbling at the edges of CBS in what some thought was a takeover bid. Tisch was appointed chief operating officer, so his "takeover" did not involve any formal exchange of stock. In 1995, Tisch arranged to sell CBS to Westinghouse for $5.4 billion, placing another pioneer radio company back in the network business (see Chapter 5). In 1997, Westinghouse changed its name to CBS and then, in 1999, merged with Viacom, a major TV distribution company.[22]

A new network, Fox, appeared on the scene in 1987. Rupert Murdoch, an **Fox** Australian media baron who had recently obtained U.S. citizenship, purchased 20th Century Fox and six TV stations owned by Metromedia. He then built a network using his six stations and other independent stations throughout the **LPTV** country. Fox increased the number of programs offered to its affiliates slowly, but a number of its shows, such as *The Simpsons* (see Exhibit 4.7), quickly became hits, and during the 1990s Fox established itself as a major network force.[23]

Beginning in 1980, the FCC also started accepting applications for a new type of over-the-air TV called **low-power TV (LPTV).** These 10-watt VHF and 1,000-watt UHF stations can be put on the air for as little as $50,000. They are sandwiched between regularly operating stations and cover only a 12- to 15-mile radius. The stations must prove that they do not interfere with the signal of any full-power stations. The FCC established LPTV to enable groups not involved in TV ownership to have a voice in the community, hoping that women and minorities, in particular, would apply. Many did, but many large companies also applied. The FCC received more than 5,000 applications by April 1981, when it decided to impose a freeze to sort things out. Eventually the FCC began giving out allocations, and at present more than 2,700 low-power stations are on the air. Some provide movies, foreign-language programs, music videos, or religious programs. Others make modest profits from local programs, such as high school band concerts and farm reports. Still others are affiliated with major broadcast or cable networks.[24]

Exhibit 4.7

The Simpsons. This cartoon family of self-proclaimed losers gained great popularity in its prime-time spot on the Fox network and earned themselves a star on the Hollywood Walk of Fame. The real person in the center is Nancy Cartwright, the voice of Bart Simpson.

(Photo by Fitzroy Barrett/Globe Photos, Inc.)

4.6 Broadcast and Cable Legal Issues

must-carry

As cable grew, the animosities between broadcasters and cablers increased. **Must-carry,** which originally caused little anxiety for either broadcasters or cablecasters (see Chapter 3), became controversial. It was a rule that required cable systems to carry all local TV stations, and before all the cable networks were established that was no problem. But cable owners started to want must-carries outlawed so that they could use their channels to carry the more profitable satellite services rather than local stations. They claimed that because they had to carry little-watched religious, educational, and ethnic broadcast stations, they did not have channel space for public access, C-SPAN, and other cable programming services. Local broadcast stations wanted to be carried on the systems and even thought cable systems should pay them because they were of great value to cable. During the 1980s, must-carries were outlawed and reinstated a number of times, depending on which forces lobbied the hardest or won the court cases.

retransmission

In 1992, a new law provided that broadcasters could choose either must-carry or **retransmission consent.** If they chose must-carry, they were guaranteed a spot on the cable system. If they chose retransmission consent, they could ask the cable system to pay them for the privilege of carrying their signal. If the cable system paid, the broadcasters would be richer, but the cable system had the option of not carrying the station if it opted for retransmission consent. This led to a certain amount of acrimony in which popular stations pulled their signals from cable systems that refused to pay. However, in many instances, the broadcasters and cable systems found a way to compromise. Instead of asking for money for retransmission of their signals, the stations asked for channel space on the cable system, so they could program additional material. For example, ABC and Hearst, which owned ESPN, agreed to let cable systems carry all ABC and Hearst-owned stations in return for a channel on which they could place a new service, ESPN2.[25]

copyright

Copyright issues resurfaced (see Chapter 3) in 1983 when the Copyright Royalty Tribunal, acting under its right to adjust fees, ordered cable systems to pay 3.75 percent (a hefty increase from the original 0.7 percent) of their gross receipts for their imported distant signals. In 1993, Congress abolished the Copyright Royalty Tribunal, replacing it with ad hoc arbitration panels chosen by the Librarian of Congress. These arbitration panels have leaned toward even higher rates, but many cable systems, now that they have so many cable networks, have stopped distant signal importation.[26]

deregulation

The 1980s, under the presidency of Ronald Reagan, were years of **deregulation,** wherein the government reduced rules and regulations for many types of businesses. Broadcasting and cable were no exception. For example, in 1984, the number of years that a broadcast station could hold a license before having to renew it with the government was lengthened from three years to five, and the number of TV stations one entity could own was increased from 7 to 12, provided the 12 stations were not in markets that collectively contained more than 25 percent of the nation's TV homes.[27]

For cable TV, Congress passed the Cable Communications Policy Act of 1984, delineating the role of cable systems and local governments. The main positive change for cable operators was that cities with at least three broadcast stations could no longer regulate the rates that cable systems charged their customers for basic service. Instead, these charges were determined by the marketplace. However, the cable industry did not behave nobly under deregulation. Because cable had monopolistic power—there was generally only one cable system in any particular area—many cable systems raised basic cable rates as much as 50 percent over seven years. Cable was also laissez-faire about customer relations, making customers wait on the phone for long periods. This led to **reregulation,** culminating in a 1992 law giving the FCC authority to determine reasonable rates for basic services and to set service standards. The FCC rolled back rates and mandated that cable operators provide a 24-hour-a-day toll-free number and a convenient customer service location. Congress also ordered cable networks to sell their programming to anyone who wanted it, not just cable systems. This was done to help and encourage other types of technologies, so they could compete with cable.[28]

reregulation

4.7 Growth of Hispanic Television

As the demographics of the nation changed to include more Spanish-speaking people, Spanish-language programming increased, both in broadcast TV and cable TV. The two largest program distributors are Univision and Telemundo.

Univision traces its roots back to 1955 and the establishment of the first U.S. Spanish-language TV station, KCOR, in San Antonio, Texas. The station was a money-loser and in 1961 was sold to people who had ties to what became Mexico's main TV force, Televisa. The new owners improved the station's fortunes, mainly by airing Televisa's programming. In 1963, the group signed on KMEX-TV in Los Angeles and called itself Spanish International Network (SIN). For more than 20 years SIN was shown on Spanish-language TV stations and as a cable TV network (see "Zoom In" box).

Univision

In 1986, the owners sold the network to a partnership of Televisa and Hallmark Cards. This group changed the name to Univision and started producing some programs specifically for Hispanic Americans, such as *TV Mujer,* a **magazine show** aimed at Latinas living in the United States. In 2007, Univision was sold to a group of equity investors, most of whom were not of Hispanic descent. The relationship between Televisa and Univision deteriorated over the years, with Televisa threatening to stop supplying Univision with its popular **telenovelas.** Televisa felt it was not being properly compensated and, in 2009, after a nasty court battle, won a handsome financial payment from Univision. The agreement assured that Televisa would continue to supply Univision with telenovelas.[29]

Telemundo was launched as TV station, WKAQ, in Puerto Rico in 1954. During the 1970s and 1980s, the station produced many popular Puerto Rican telenovelas. In 1987, a Spanish-language Los Angeles station purchased WKAQ and several other stations and set up the Telemundo Group. The group went

Telemundo

ZOOM IN: SIN Pays

In the early 1980s, Spanish International Network (SIN) decided to become a cable TV network. In those days the newly formed basic cable networks were trying to come up with a business model through which they could make money. In the broadcast world, networks paid their affiliated stations to play their programs, and the networks collected much of the advertising revenue raised by the programs. There wasn't much advertising revenue for cable networks in the early days, however, so they decided to charge the cable systems for their programming. The cable systems (the companies that laid the cable and connected it to people's TV sets) did have a revenue stream—they charged the consumers a monthly fee. The cable networks were trying to get some of that fee and were experimenting with methods of doing so. Mostly the networks would jockey with the systems, offering their programming for 15 cents a subscriber per month, 10 cents, 6 cents—until they found what the market value was.

Spanish International Network tried charging cable systems, too, but did not have much luck. The systems, many of which had only 12 channels, could choose from many offerings, and Spanish-language programming was not high on their list. Spanish-speaking people at the time were not considered a prosperous demographic, and many cable systems were located in areas that had few, if any, Hispanics. The Spanish International Network reasoned that if it could entice cable systems to carry it, this larger potential audience would allow it to raise its advertising rates. Therefore, it decided to pay the cable systems to carry it and came up with the clever slogan "SIN Pays." The buttons the network made featuring this slogan were in great demand at cable TV conventions, and the network succeeded in getting picked up by many cable systems—some of which did not even have Spanish-speaking viewers.

Exhibit 4.8

Azteca América is a new service started in 2001. Seen here is the cast of its entertainment gossip show *Ventaneando*—(left to right) Daniel Bisogno, Monica Garza, Pati Chapoy, Aurora Valle, and Pedro Sola.

(Courtesy of Azeteca América)

through a number of management changes and programming failures—including trying to let the Hollywood community produce its telenovelas. In 2001, NBC purchased Telemundo and has successfully achieved its aim of narrowing the ratings gap between Telemundo and Univision. Telemundo has always produced most of its own programming, and NBC has supported this policy financially and philosophically.[30]

Although Univision still leads in the ratings, both networks are successfully ensconced in cable TV and Spanish-language broadcasting. This success has encouraged others to enter the Spanish-language TV business (see Exhibit 4.8). Some of the newer services are English-language cable networks that translate their material into Spanish. Others feature specialized forms of Spanish-language TV (see Exhibit 4.9). There is also a Hispanic-oriented cable channel, Sí TV, that broadcasts in English but features material of interest to Hispanic Americans.[31]

Selected Spanish-Language TV Services
Azteca América—a fast-growing network owned and programmed by the Mexican broadcaster TV Azteca
Hispanic Information and Telecommunications Network (HITN)—a service owned by a private nonprofit organization that programs noncommercial educational fare
V-me—a cultural channel that airs on PBS station secondary digital channels
LATV—a network based in Los Angeles that programs comedies and other material aimed at 18- to 34-year-olds
GOL-TV—a network devoted totally to football (soccer)
¡SOPRESA!—a general programming service offered through cable TV, broadband, and mobile platforms
NY1 Noticias—a news channel that serves Time Warner's New York cable systems
Mun2—pronounced "mundos," it is owned by Telemundo/NBC and features music videos
Galavisión—a small network owned by Univision that programs mostly material that has run on Univision
CNN en Español—24-hour-a-day news similar to what is aired on CNN's English service
MTV Tr3s—a fusion of Latino and American music and lifestyle programming that appeals to youth
Discovery en Español—a Spanish-language version of the Discovery Channel

Exhibit 4.9

A partial list of Spanish-language television services.

4.8 Satellite TV Revived

Satellite TV was essentially resurrected from the dead during the 1990s. After STC backed out of the direct broadcast satellite business in 1984, interest in the field remained dormant for quite a few years. In the early 1990s Rupert Murdoch of Fox, whose company News Corp. was involved in satellite businesses in other parts of the world, became interested in U.S. satellite-to-home opportunities and formed an alliance with NBC, Cablevision Systems, and Hughes Communications for a system to be called Sky Cable. The Sky Cable alliance dissolved quickly, the victim of clashing ideas and egos. Hughes, however, plodded along with the idea and eventually joined forces with USSB, a company owned by Stanley Hubbard that was one of the few original DBS applicants still interested in developing the service. In 1993, Hughes successfully launched a high-power satellite, and in 1994 Hughes and USSB jointly offered a 150-channel digital satellite service that became known as DirecTV. The widespread consumer acceptance of DirecTV surprised even the most avid satellite TV supporters. Within six months, more than 500,000 households had paid $700 each for the receiver dish and the decoder box that would enable their television sets to display programming coming directly from high-powered satellites.[32]

Murdoch's interest

DirecTV

Why did Hughes/USSB succeed when all the 1980s ventures had failed? A number of factors were involved. Hughes, a satellite-building company, was

Exhibit 4.10

DirecTV's 18-inch reception dish developed by Sony and RCA.

(Courtesy of Hughes Communication, Inc.)

able to perfect a high-power satellite that could deliver to homes. Also, with the help of RCA and Sony, it developed small, reliable 18-inch dishes that could be mounted on the sides of houses (see Exhibit 4.10). Digital technology had improved greatly during the decade, and by offering digital signals, DirecTV could feature high-quality picture and sound. Digital improvements also allowed DirecTV to offer more than 100 channels, including ones for major sports events and top-rated movies. In 1992, Congress passed a law ordering cable networks to sell their programming to anyone who wanted it, not just cable systems. This law cleared the path for USSB and Hughes to obtain well-known, popular programming such as that on ESPN, MTV, and C-SPAN. Perhaps most important of all, people were angry with their cable companies for raising rates, being slow to fix technical glitches, and not answering their phones. They were willing to spend $700 to quit their cable services and try something new.

backyard satellite

The success of DirecTV was also a factor in killing off most of the backyard satellite reception. The companies with programming on satellites had become more upset that the backyard dish owners were receiving the programming for free. Many services began **scrambling** their signals. This brought forth a surge in illegal decoders, but as scrambling improved and antipiracy laws loomed, these decoders became problematic. Many people who had the large dishes opted for the smaller dishes that DirecTV offered. They were more cost-effective and convenient and did not carry the stigma of illegality.[33]

While DirecTV was becoming successful, Rupert Murdoch still lingered in the wings. In 1995, he formed an alliance with MCI to establish a satellite TV service. This proved to be too expensive for Murdoch, so he looked for an additional partner and found one in Charlie Ergen, who owned another would-be DBS company, EchoStar. Again egos and finances destroyed the relationship, and eventually Ergen established his own service, Dish Network, which has competed quite effectively with DirecTV. Murdoch did eventually get his satellite TV system in 2003, when he purchased DirecTV for $6.6 billion. He didn't keep it long, however, selling it in 2006 to cable TV entrepreneur John Malone.[34]

Dish

local programming

One drawback of satellite TV services of the 1990s was that they lacked permission to deliver local stations to their subscribers. Although they tried a number of times to obtain this permission through Congress, the more powerful cable TV business lobbied successfully to keep them at bay. However, in 1999, Congress passed a bill allowing satellite services to retransmit local stations into their local areas, often referred to as **local-into-local.** Being able to deliver these stations has given satellite TV a valuable asset.[35]

Another drawback for satellite TV is the difficulty of providing interactive services, such as internet and telephone, because satellite broadcasting is only one-way—consumers cannot broadcast back from their satellite dishes. DirecTV and Dish are forming alliances with companies providing interactive services and working on technological solutions for the interactive problem, but they have yet to reach a widespread acceptable solution.[36]

interactive services

4.9 The Telecommunications Act of 1996

The main legislation that governed the broadcasting industries for many years was the **Communications Act of 1934.** Although written largely for radio (see Chapter 5), it served broadcasting well for decades, mainly because the American system of government allows laws to be interpreted by the courts and regulatory bodies such as the FCC. These interpretations lead to precedents that can accommodate changing times. In addition, the act was amended to take into account alterations such as those related to using television for electioneering (see Chapter 11). But by the 1990s it was definitely showing its age because of the advent of many new technologies such as cable and satellite TV.

Congress passed a sweeping new law, the **Telecommunications Act of 1996.** It did not replace the Communications Act of 1934, but it dealt with newer technologies and ramifications that related to older technologies. For the TV broadcasters, the Telecommunications Act of 1996 was part deregulatory and part regulatory. Following the deregulatory line, it lengthened the five-year license renewal period determined in 1984 to eight years, eliminated the 12-station numerical cap, and raised the coverage area from 25 percent to 35 percent.[37]

new law

But the amount of sex and violence on television bothered members of Congress, so they included regulatory provisions related to that subject. The main one was that TV sets had to be equipped with a **V-chip,** a device that consumers could use to block programs with violence, sex, or other undesirable material. The television industry was required to develop a ratings system to earmark these programs and then encode the programs electronically so the chip could identify them and block them from coming into households where they were not wanted. The industry provided an age-based rating system, but Congress and citizens groups also wanted content-based information, so the final result was a tiered rating system (see Exhibit 4.11).[38]

V-chip

The Act's provisions directly related to cable were mainly deregulatory, once again allowing cable systems to set their own rates for basic services. This time, however, cable systems faced growing competition from satellite TV and were intent on keeping their customers happy, so most of them made only modest rate increases.

deregulation

Perhaps the most significant provisions of the Telecommunications Act, stemming from the lawmakers' desire to foster competition within the telecommunications industry, were those that encouraged the intermingling of the broadcast, cable, online, and telephone businesses. It took a while for this to take hold, in part because of the technical complexities involved but also because of the different types of cultures within the various industries—especially the telephone business.

intermingling

Exhibit 4.11

The complete
ratings system, with
both age-based and
content criteria.

TV Program Ratings

TV-Y—for all children

TV-Y7—directed toward older children
TV-Y7-V—mild violence
TV-Y7-L—mild coarse language
TV-Y7-FV—fantasy violence

TV-G—for general audience

TV-PG—parental guidance suggested
TV-PG-V—moderate violence
TV-PG-L—mild coarse language
TV-PG-S—mild sexual situations
TV-PG-D—mild suggestive dialogue

TV-14—probably unsuitable for children under 14
TV-14-V—strong violence
TV-14-L—moderate coarse language
TV-14-S—moderate sexual situations
TV-14-D—moderate suggestive dialogue

TV-MA—mature audiences only
TV-MA-V—extreme violence
TV-MA-L—strong coarse language
TV-MA-S—strong sexual situations

4.10 Telephone Company Entry

Bell

Telephones, of course, were not a new technology. History credits Alexander Graham Bell (see Exhibit 4.12) with the invention of the telephone in 1876. Bell and two of his friends started the Bell Telephone Company to lease phones and charge for short-distance phone calls. For long-distance calls, they formed a

Exhibit 4.12

Alexander Graham Bell
demonstrating an early
telephone.

(AP/Wide World Photos)

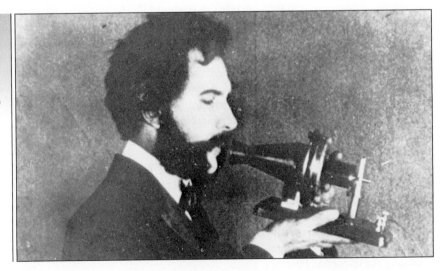

subsidiary, American Telephone and Telegraph Company (AT&T), which eventually became the parent organization, owning both long-distance and local phone companies. AT&T operated as a monopoly, regulated by the government. The telephone business grew into a major national and international industry, but it remained primarily a voice service connecting one individual with another. Improvements were made in switching systems and phone installation, but because AT&T was a monopoly making a guaranteed profit, it had little incentive to add new services for the customer.[39]

During the 1970s, the Justice Department took a hard look at AT&T's monopoly. Several companies wanted to sell telephones competitively to consumers, and some businesses, such as MCI and Sprint, wanted to compete in long-distance service. This was controversial because AT&T's long-distance service had been subsidizing its local phone service in order to keep local rates low. Other companies could offer lower rates than AT&T because they did not need to subsidize local service.

Addressing these issues, the Justice Department in 1980 issued a consent decree that broke up AT&T. "Ma Bell" had to divest itself of its 22 local phone companies. These were spun off into seven new "baby Bells," called **regional Bell operating companies (RBOCs),** which served seven sections of the country. The consent decree also allowed independent companies to compete in long-distance service, giving customers a choice, and it permitted independent companies to manufacture telephones for customers to buy.[40]

breakup of AT&T

With strong competition, the phone companies began offering a new array of services. They started talking about **POTS (plain old telephone service)** and **PANS (pretty amazing new service).** Some features, such as the hold button and answering machines, had been part of phone technology previously, but the **postdivestiture** period of the 1980s brought many new options. These included redial, speed dial, call waiting, conference calling, and caller ID. Cordless phones evolved, fax machines came to the fore, and—for better or worse—voice-mail systems replaced the company telephone operator.[41]

new services

The Telecommunications Act of 1996 further encouraged competition in the phone business and cleared the way for telephone companies to provide other services, including television. The first approach to expanding services was for companies to merge. AT&T, for example, merged with what was then a giant cable TV company, TCI, in 1998 with the idea of becoming the one provider for telephone, cable TV, and the internet. It was an idea whose time would eventually come, but the cultures of the staid phone company and the more flamboyant cable company clashed, and the merged company was not able to make money with this model, so AT&T sold the cable business in 2002.[42]

initial forays

Eventually AT&T was bought, not by a cable company but by one of the "baby bells," SBC, which took the AT&T name. Many of the RBOCs merged over the years, with several of them coming together in 2000 to form a new company, Verizon. Although there still are many regional and local phone companies, the two that dominate the landline and cell phone businesses are AT&T and Verizon.[43]

AT&T and Verizon

Both these companies have decided to provide phone, internet, and television service on their own without assistance from companies that have already been in those businesses. Verizon has a fiber optic system called FiOs that it is rolling out with notable success in various parts of the country. AT&T is not as far along but hopes to be able to use new technology and spectrum it recently acquired to offer a multitude of services. Of course, the cable TV companies are not sitting still. Most of them have gone beyond their basic business of offering TV channels and are offering phone services and high-speed internet.[44]

4.11 Digital Videodiscs

Analog videodiscs never hit it big with consumers, but the same cannot be said for the **digital videodisc** (**DVD**), which was introduced in 1997. It could hold 4.7 gigabytes (133 minutes of video) and had technical quality far superior to that of analog tape. DVD started out standing for digital videodisc, but as it progressed, the DVD designation was dubbed **digital versatile disc** because the discs were also used to hold music, games, text, and other data.[45]

standardization

DVD started its life looking as though it were going to encounter the same traumas as the previous analog discs. After a number of years of development, there were two competing formats—one supported by Sony and Philips and the other supported by Toshiba and Time Warner. Remembering the Betamax-VHS battle, the two sides met to try to establish one format. IBM helped forge a compromise because it could see the possibilities for using DVDs for computer storage. Companies then developed DVDs that could record as well as play back, something that had never reached successful levels with analog discs, and that added to the appeal and practicality of the DVD.[46]

Hollywood reaction

Working out compatibility problems did not end DVD's woes. For about a year, Hollywood repeated its earlier fear of videotape and refused to place material on discs. Industry moguls were worried about copyright because the discs could be so easily copied, so manufacturers came up with better copyright protection. There were some strong believers among the studio personnel, particularly Warren Lieberfarb, then DVD president at Warner Bros. He pitched his hunch that people would buy DVDs, and Warner started selling discs of its movies. Lieberfarb was right.[47]

DVD growth

Consumers took to this new format more than they had previous disc formats, and soon there was such a buzz about DVD that the rest of the movie studios signed on. Between 1997 and 2001, 20 million households bought DVD players, either stand-alone devices that connected to their TV sets or DVD drives built into computers. (Later, portable DVD players came to market.) By 2002, some movies were selling more DVD copies than VHS copies, a trend that has continued and accelerated to the point that video stores now carry very few VHS tapes. In 2002, a new company, Netflix, began a mail-based DVD rental system that successfully stole some of the business from video stores. The main video store chain, Blockbuster, countered by creating a mail order website of its own and by essentially eliminating the late fees charged by its stores.[48]

Movie studios started enhancing the extra material that they had originally put on some of the older analog discs. In addition to interviews with directors

and other background information, they now create such features as games related to the movie, multiple endings, original directors' cuts, and documentaries on the "making of." Other companies offer self-help DVDs, and TV producers put entire seasons of their series on DVD (see Exhibit 4.13), creating an added source of income.[49]

The newest saga of the disc business involves high-definition DVDs. At first it appeared there would once again be a format war between Toshiba, which had developed a format called **HD DVD,** and Sony, whose development was called **Blu-ray.** But by 2008 the major studios had endorsed Blu-ray, mainly because it had six times the capacity of the rival format, and Toshiba stopped development of HD DVD.[50]

Blu-ray

4.12 Enter TiVo

TiVo is a brand name, but it has become the generic word to describe **digital video recorders (DVRs).** They use high-capacity microchips to store many hours of video on solid-state hard drives. They convert the signal coming into the TV set from a cable, satellite, or antenna into digital form, and then compress and store the information until someone plays it back, at which time the signal is decompressed for the TV set. DVRs also enhance viewing. You can pause a program at any time and come back and pick up where you left off, and you can fast-forward through commercials. They display an accurate schedule of everything that is going to be on TV and suggest what you might want to record based on what you have recorded in the past. You can use a search feature to find shows that have particular actors or are of a particular genre.[51]

features

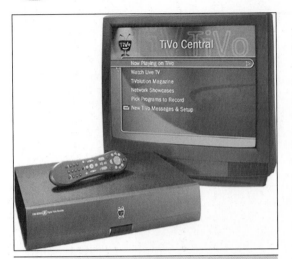

Exhibit 4.14

A first-generation digital video recorder.

reality TV

Exhibit 4.15

The original *Survivor* took many twists and turns before Richard Hatch was voted the winner of the $1 million prize. Here the Tagi tribe, which included Hatch, arrives on the island.

Several companies introduced DVRs in 1999, but it was TiVo, with its clever graphics and user-friendly interface, that caught on (see Exhibit 4.14). TiVo, however, could not figure a way to make money. The problem was that the technology was easy to duplicate, and cable and satellite companies started including their own DVR products in their service packages. TiVo tried licensing its product to others, only to find that the companies canceled the licenses and went with their own product. In 2008, TiVo won a patent lawsuit against Dish and received $74 million, which may help it protect its technology and get on a profitable track.[52]

4.13 Programming Changes

Programming has changed greatly during the modern era of television. If nothing else, there is more of it. Broadcasters, cable systems, satellite TV operators, and even telephone companies all have many more channels that they did prior to the 1980s. Videodiscs have increased their content to include TV series, Spanish-language networks flourish, and people can store programming on their DVRs for future or repeat viewing. Granted, much of the same programming appears on all the distribution sources—a movie shown on NBC might also make its rounds on USA cable, DirecTV, FiOs, DVD, Telemundo (translated into Spanish), and many a TiVo.

In addition to quantity increases, there have also been content and structure changes. One major change is the addition of a new genre—**reality TV.** The genre came to the fore in the 1990s, featuring actual events or reenactments of events, some of which were bizarre. Networks and stations liked these programs because they garnered fairly large audiences and were inexpensive to produce. The reality genre experienced a big jump in popularity in the summer of 2000 when CBS broadcast *Survivor,* the trials and tribulations of specially selected ordinary citizens as they overcame challenges of living in the South Seas and competed for a $1 million prize by voting each other off the island (see Exhibit 4.15).

The series was such a success that many clones and variations appeared in the following years, including more *Survivor* series and NBC's *The Apprentice,* starring Donald Trump, who made "You're fired" a well-known phrase as contestants aspiring to a job with his firm were eliminated. High-stakes game shows, which had been sparse in prime time since the scandals of the 1950s, also made a comeback, particularly with the 1999 ABC hit *Who Wants to Be a Millionaire?* hosted originally by Regis Philbin (see Exhibit 4.16). Fox's singing competition show, *American Idol,* which started in 2002, became a ratings leader week after week and year after year.[53]

Two other new forms of programming, **home shopping** and **infomercials,** started on both broadcast and cable facilities. As part of deregulation, the FCC lifted restrictions on how much advertising a station could program. Home shopping programs were essentially total advertisements, showing and extolling the virtues of products that people could buy over the telephone. Infomercials, often in the form of talk shows or information programs, devoted half-hour time slots to the virtues of products such as weight-control programs and baldness treatments. Stations particularly liked these infomercials because the producers paid the stations to air them.[54]

Exhibit 4.16

Host Regis Philbin on the set of *Who Wants to Be a Millionaire?*

(© ABC Photography Archives)

advertising formats

old genres

pay services

The old **genres** also continued, with comedy hits such as *The Cosby Show, Friends,* and *Everybody Hates Chris* and dramas such as *Law and Order, 24,* and *The West Wing* (see Exhibit 4.17). But some of the most notable programs were coming from a new source, HBO. Having decided to produce original programming, HBO started besting the broadcast networks in major Emmy categories, especially with *The Sopranos* (see Exhibit 4.18), an ongoing dramatic series about a New Jersey mafia family, and *Sex and the City* (see Exhibit 4.19), the candid portrayal of four women friends and their interactions concerning men.[55]

New programming concepts related to program accessibility took hold, mainly in the cable and satellite arena. The first was **pay-per-view (PPV).** It enabled viewers who paid an extra one-time-only sum to see special events such as boxing matches and first-run movies. At first, most PPV plans asked subscribers to phone the cable company to order the event, which would then be sent only to those homes requesting and paying for it. As time progressed, special interactive equipment was developed that allowed the viewer to push a button on a remote-control-like device that enabled the special programming to be delivered to the TV set. PPV was not an instant success, but those

Exhibit 4.17

Aaron Sorkin (center), the creative force behind *The West Wing,* and cast members accept an Emmy at the 2002 telecast.

(Photo courtesy of the Academy of Television Arts & Sciences)

Exhibit 4.18

James Gandolfini accepts a best actor Emmy for his portrayal of Tony Soprano in HBO's *The Sopranos.* This series was the first cable network offering to top the commercial networks for the Emmy for best drama.

(Courtesy of the Academy of Television Arts & Sciences)

UPN, WB, and CW

Exhibit 4.19

The cast and principals of *Sex and the City* receiving an Emmy for best comedy.

(Courtesy of the Academy of Television Arts & Sciences)

companies that stayed with the business experienced a slow but steady increase in subscribers and revenues.

As technologies improved, PPV has been taken a further step to **near-video-on-demand (NVOD).** One movie is shown on several channels, each of which starts it at a different time—usually 15 minutes apart. A viewer who wants to see the movie but misses the 8:00 P.M. starting time on Channel 57 can catch it on Channel 58 starting at 8:15. NVOD is similar to PPV in that the movie is sent to all paying subscribers at the same time; it is just sent more often. Even further up the technological chain is **video-on-demand (VOD).** Individual viewers can watch programs precisely when they want to. VOD uses a large **server** at a cable system that contains an enormous amount of information—movies, TV programs, video games, and so on. When a consumer asks for a particular movie (or anything else), it can be downloaded on one of the cable system channels into a digital box on top of the customer's TV set. The movie will also remain in the cable server so other customers can request and receive it. The consumer can then play the movie, stop it, rewind, and fast-forward at will.[56]

4.14 New and Revised Networks

Structural changes in networks occurred in both cable and broadcast TV. In the broadcast area, Fox's success encouraged the 1995 launch of two other networks, United Paramount Network (UPN) and the WB Network. The ownership of these networks underwent some convoluted changes. Paramount merged with Viacom, which then merged with CBS, meaning that Viacom/CBS owned two networks, CBS and UPN. A 1940s rule stipulated that a company could own only one network, but in 2001 the FCC relaxed that restriction allowing both UPN and CBS to be owned by one company. Warner underwent ownership changes in 2000 when America Online bought it in a highly hyped deal that created AOL Time Warner. This deal did not live up to expectations; Warner finally exerted dominance over the internet company and in 2003 removed AOL from the letterhead, once again becoming Time Warner. In the final analysis, neither UPN nor WB garnered a large audience, so in 2006 Warner and CBS merged the two networks into CW, "C" for CBS and "W" for Warner.[57]

The multitude of cable TV programming networks (which eventually also became satellite TV services) experienced a great deal of merging and ownership change (see "Zoom In" box).

ZOOM IN: **Two Coups**

Two highly touted cable network failures warmed the hearts of cable practitioners because the networks had been started by broadcasters who exhibited a high-and-mighty attitude. One was the CBS-owned cultural service, CBS Cable. With its deep pockets, CBS planned to produce commercially viable original cultural television that was a cut above what any cable companies could afford. It also used those deep pockets to mount a very aggressive marketing campaign. When most cable networks were trying to convince cable systems to carry them by holding receptions with celery and carrot sticks, CBS Cable was offering shrimp and lobster. But neither the cable systems nor the advertisers bought into CBS's plans. The system owners resented what they saw as a broadcaster's attempted encroachment into the cable business, and very few signed up to cablecast the network. Advertisers followed suit. CBS hung on for several years, then stopped programming in 1983 after losing $50 million.

Another well-publicized 1983 coup occurred when Ted Turner's Cable News Network slew the broadcast giants ABC and Westinghouse. Turner, a well-respected maverick in the cable TV business, had gone out on a limb in 1980 to form a 24-hour TV news service—something that was very expensive and innovative. His concept was starting to catch on when, in 1982, ABC and Westinghouse decided to combine their considerable news resources and take over with their own 24-hour channel, Satellite News Channel. Turner, working from an inferior but more flexible position, outmaneuvered the broadcasting executives and managed to buy them out in October 1983. This meant less competition for CNN, which was then able to proceed on less tenuous financial footing because it did not have to compete for advertisers with Satellite News Channel.

Ted Turner, seen here with then-wife Jane Fonda.

Exhibit 4.20 shows what happened to the networks listed in section 4.1. Some of the basic cable channels, in addition to changing ownership, had to change their stripes to keep current. For example, MTV started out programming only music videos. As time went on, this programming drew less attention, so MTV changed to a schedule that included some nonmusic programs aimed at the same audience that appreciated music videos (see Exhibit 4.21).[58]   **cable channels**

Not as many new cable networks have been launched since the early 1980s, but some have been introduced. In 1991 Court TV (now called truTV) became prominent with live, complete coverage of the William Kennedy Smith rape trial and then the O.J. Simpson murder trial in 1995. Several comedy channels that started in 1991 merged into Comedy Central, which made a name for itself with an irreverent cartoon show, *South Park* (see Exhibit 4.22).[59]

Many of the more recent channels have come from well-established companies that operate other channels. These companies have the infrastructure

Exhibit 4.20

Here is what happened to the cable networks mentioned earlier in section 4.1, "Growth of Cable TV Programming Services."

Showtime—merged with The Movie Channel

The Movie Channel—merged with Showtime

Spotlight—went out of business

Bravo—converted from pay to basic

Entertainment Channel—gave its programming to ARTS, which became A&E

Disney—converted from pay to basic

Playboy—changed how hard-core its programming was based on viewer response

ESPN—changed ownership from Getty to ABC

Black Entertainment Television—was purchased by Viacom

CBN—changed from religious programming to family-oriented programs

Nickelodeon—accepted commercials and changed ownership from Warner to Viacom

CNN—changed ownership when Warner bought Ted Turner's networks

USA—brought in a consortium of owners

C-SPAN—added Senate coverage in addition to House of Representatives

ARTS—took programming of the Entertainment Channel and became A&E

CBS Cable—went out of business

MTV—changed from music videos to include other types of programming

Satellite News Channel—was bought out by CNN

Daytime—merged with Cable Health Network to form Lifetime

Cable Health Network—merged with Daytime to form Lifetime

Exhibit 4.21

During the 1992 presidential campaign, MTV encouraged its viewers to become participants in the political process. To this end, MTV invited Democratic candidate Bill Clinton to participate in a program in which he answered questions from 18- to 24-year-olds.

(AP/Wide World Photos)

Exhibit 4.22
These four foul-mouthed third graders became a big hit on the Comedy Central animated show *South Park.*

(AP/Wide World Photos)

to develop and sell channels that will have only small numbers of interested viewers. For example, a do-it-yourself channel that programs primarily information about remodeling homes was started by the same company that owns Home and Garden Television and the Food Channel. Noggin, an educational channel, comes from Viacom, which also has Nickelodeon. Sometimes new channels are just vehicles for repackaged programs from another channel. For example, Discovery has reorganized the material it has cablecast over the years into channels dealing with such subjects as science, health, and children—each of which is a separate channel.[60]

access

Activity on the local public access and local origination channels has slowed greatly from its hyped period of the early 1980s. Cable systems that had promised truckloads of production equipment to local organizations tried their best to delay these costly obligations. Local programming departments that had included five or six employees dwindled. The only local programming for some systems consisted of messages typed on the screen about local news and events. They had, in essence, reverted to a sophisticated version of the early 3-by-5-inch cards (see Chapter 3). Some systems joined to form regional cable networks that programmed primarily sports. Local programming has not disappeared, but it does not fill the multitude of channels promised in many of the original franchise agreements.[61]

interactive services

Interactive services took a nosedive. Warner-Amex killed its highly touted Qube system by 1984, mainly because it was not profitable. In the 1990s, Time Warner experimented with interactive technology by installing a system in Florida called Full Service Network (FSN). Because of Warner's involvement, some jokingly referred to FSN as "Qube Squared." After two years of operation (1995–1997), Time Warner pulled the plug on FSN, attributing the closure to high costs and technical problems. Interactive services have been taken over by the internet, which handles them much more effectively. Cable systems, by offering internet service as part of their package, are fulfilling their interactive functions somewhat vicariously.[62]

4.15 Digital TV and HDTV

DTV and HDTV

In 2009, the United States switched from analog TV to **digital television (DTV).** Digital TV had been a long time coming and brought about a certain amount of confusion. Part of the problem lay in confusion between digital TV and **high-definition television (HDTV).** The two are intertwined but not the same. Digital TV uses technology similar to computers, with signals coded as either off (0) or on (1). High-definition TV has more lines on the TV screen—about 1,000 as compared to the 525 lines of the traditional **standard-definition (SD)** analog TV—and also has a wider screen (see Chapter 13). HDTV can be analog or digital, and digital TV can be standard or high definition, but the big changes that have come are mostly related to digital HDTV.

HDTV history

HDTV was originally developed by the Japanese in the 1970s as an analog system. At one point during the 1980s, it was almost adopted in the United States and other parts of the world, but the plan fell apart. In 1990, the American company General Instruments proposed an all-digital high-definition system that had the technical advantages inherent in digital technology, such as clearer sound and no loss of quality when tapes are copied. In 1993, a consortium of companies called the Grand Alliance joined to further develop this digital system, which in 1997 was adopted by the FCC to be the American system. In 1998, the FCC gave broadcast stations new frequencies for digital HDTV transmission—channels that were to be developed while the stations continued to broadcast on their old analog channels. All stations were supposed to have their digital channels up and running by May 2002, but because of technical problems and because it costs up to $4 million to convert to digital, many stations fell behind schedule. After several postponements of the conversion to digital, it finally took place in 2009.[63]

more channels

Consumers with analog sets have had to buy new sets, buy conversion boxes for their old sets, or be attached to cable TV systems that convert digital signals to analog. One of the controversies that has arisen related to the conversion is the old issue of must-carry. The frequencies given to broadcast stations for their digital signal were originally intended to carry just one HDTV digital channel. But as technology improved, stations found they could use the frequencies to broadcast one or two additional standard definition signals, called **secondary channels.** A number of stations are doing just that, with programming they provide themselves or purchase from other sources, such as continuous local weather, Spanish-language programming, and movies. The stations want the cable systems to carry all of their channels, but the FCC has ruled, at least for the present, that the cable systems must carry only the **primary channel** that is high definition. The FCC has also ruled that cable systems need to carry both digital and analog versions of primary channels for three years. Cable systems are lobbying to carry only the digital versions.[64]

The FCC has taken back the analog frequencies that the TV stations originally had and **auctioned** them off. Most of them have gone to cell phone companies, particularly AT&T and Verizon, with some going to satellite TV companies. Eventually these companies may use them to enhance wireless services, such as access to the internet through portable devices.[65]

auction

One of the casualties of the conversion to digital may be low-power TV. The LPTV stations were not required to convert to digital, so they can still operate as analog stations. However, they must find room for themselves in the new spectrum where there is far less vacant space. Some of them may need to cease operation.[66]

The adoption of high-definition TV, combined with large screens and **surround sound** systems, leads to a better television viewing experience. No doubt it will also lead to many structural changes within the industry as different production and distribution companies vie for their piece of the pie. But change is nothing new for this dynamic industry.

4.16 Issues and the Future

With so many forms of distribution trying to offer the same services, consumers should profit from the competition—theoretically. However, there is an ever-increasing number of ways that consumers can spend their money on video-oriented services. Before the 1980s, most people didn't pay for television at all; they got it for free using antennas on their roofs. Now it is not uncommon for people to pay for cable TV, satellite TV, pay-per-view, extra HDTV channels, DVDs, TiVo, internet service, several types of phone service, and on and on. There is no letup in gadgets and services people can purchase.[67]

number of
services

From the industry point of view, each service that comes along must constantly look over its shoulder to see what might replace or alter it. For example, Blu-ray may be short-lived as the high-definition DVD format because **flash memory** chips stand ready to hold movie content on a much smaller, more versatile medium. Satellite TV may lose out to phone companies and cable TV if it does not solve the problem of inability to offer interactive services. The future of the phone companies' offerings may depend on what they are able to do with the new frequencies they received as the result of the TV stations' giving up their analog frequencies. TiVo may fold if it can't make a profit. Broadcasters have new horizons in terms of their secondary channels, but just how well those will be accepted is yet to be determined.

competition

The trend within the cable industry has been toward consolidation. Small MSOs joke about being listed among the top 25 cable system owners, not because they have grown but because of all the mergers and buyouts among the large MSOs. As companies merge and buy each other out, the total number of companies involved in cable and satellite TV shrinks, but the companies get

consolidation

larger. Many of the cable companies that own distribution systems also own networks, so they have an incentive to run their own networks on their systems. This may not always serve the viewers' best interests, especially if the systems are trying to curtail access by networks owned by another company. For example, a Time Warner cable system might be more likely to give consumers incentives to subscribe to HBO, which it owns, than to Showtime, which is owned by Viacom, even though some of those consumers would prefer Showtime.[68]

government actions

The government can always make life exciting for media entities. As an example, some congressional forces want cable to offer **à la carte** services. In other words, consumers would be able to order and pay for only the channels they want to watch instead of having to buy the packages of channels that the cable systems offer. Of course, the cable networks and systems are opposed to this because they would sell less. The issue of sex and violence on TV is not likely to go away. Politicians try to "clean up the airwaves," but they are hamstrung by the **First Amendment** and often only campaign against sex and violence in election years when they see the issue as one that may garner them votes. Broadcasters feel that they unfairly lose their audience to cable networks because, at present, those networks are not highly regulated and can program edgier material. Politicians' main response to this is to try to impose stricter provisions on cable and satellite TV. Although people indicate they want sex and violence reduced on broadcast television, they do not fully utilize devices such as the V-chip that could help in this regard.[69]

One thing is certain. Issues related to television will not evaporate. There are so many movements and countermovements that there is always room for new clashes and new solutions. All of this makes for an interesting business.

4.17 Summary

Television has seen many changes during the last three decades. The early 1980s were dominated by the three broadcast networks, NBC, CBS, and ABC, but now they are just three of many channels offering a wide variety of programming. All three networks had ownership or management changes in the 1980s, and Fox started in 1987. Low-power TV and deregulation were also aspects of broadcasting during the 1980s and beyond. Spanish-language TV, headed by Univision and Telemundo, grew significantly. Although the Telecommunications Act of 1996 further deregulated broadcast TV, it also required broadcasters to establish a code to work in conjunction with the V-chip. Reality TV became a major force in the 2000s, and UPN and WB networks formed and then merged to establish CW.

Cable experienced amazing growth in the early 1980s. Innumerable pay cable and basic cable networks were established, along with local programming and interactive services such as Qube and home security. Franchising activities were hectic, and MSOs took over many mom-and-pop systems. Advertisers, unions, and equipment manufactures got in on the gold rush. Cable's success led to arguments with broadcasters over must-carry. Cable, like broadcasting, experienced deregulation, but it was also reregulated because of price gouging and poor customer service. New forms of programming such as PPV, NVOD, and VOD arose. New networks continue to be established, but at a much slower pace than previously.

Satellite TV was slow to start in the 1980s, with even the pioneer company, STC, pulling out. It was revived in the 1990s, when Hughes and USSB formed DirecTV, and EchoStar established Dish. Winning the ability to deliver local stations helped its progress; now it needs to find ways to be interactive.

Home video started with competition between Betamax and VHS tape systems. Universal and Disney brought a copyright suit against Sony, but Sony won, thus legitimizing the videocassette recorder. Analog discs never caught on, but DVDs became popular in the 1990s, making tapes essentially obsolete. Personal digital recorders (commonly known as TiVos) are used by many to record programs.

For most of its history, the phone company stuck to the telephone business, with AT&T being a government-regulated monopoly. After AT&T was broken up in 1980, more competition and more services surfaced. A provision of the Telecommunications Act of 1996 allowed phone companies to enter the TV business, something Verizon is undertaking successfully.

All the current media forms are subject to exciting challenges because of the switch to digital TV and HDTV.

Suggested Websites

www.directv.com/DTVAPP/index.jsp (home page for satellite service DirecTV)

www.fox.com/home.htm (the site for the Fox broadcasting network)

www.hbo.com (cable network Home Box Office's main site)

www.tivo.com (information about the digital video recorder, TiVo)

www.univision.com/portal.jhtml (Spanish-language network Univision's website)

Notes

1. "Viacom Becomes Second Satellite Pay Cable Network," *Broadcasting,* October 31, 1977, 64; and "Disney Previews Pay-TV Channel," *Daily Variety,* April 13, 1983, 1.
2. "Basic-Cable Programming: New Land of Opportunity," *Emmy,* Summer 1980, 26–30; and Peter W. Bernstein, "The Race to Feed Cable TV's Maw," *Fortune,* May 4, 1981, 308–18.
3. Don Kowet, "They'll Play Bach Backwards, Run for Queen of Holland," *TV Guide,* May 31, 1980, 15–18; and Ann M. Morrison, "Part-Time Stars of Cable TV," *Fortune,* November 30, 1981, 181–84.
4. "The Two-Way Tube," *Newsweek,* July 3, 1978, 64.
5. "Home Security Is a Cable TV, Industry Bets," *Wall Street Journal,* September 15, 1981, 25.
6. Census Bureau, *Statistical Abstracts of the United States, 2000* (Washington, DC: Department of Commerce, 2000), 567–74; *Cable Television Developments* (Washington, DC: National Cable Television Association, 1995), 1–7; *Broadcasting & Cable Yearbook, 2000* (New Providence, NJ: R. R. Bowker, 2000), xxx; and "Cable Revenues Gain Faster Than Profits, Survey Finds," *Broadcasting,* November 30, 1981, 52.

7. "Cities Issue Guidelines for Cable Franchising," *Broadcasting,* March 9, 1981, 148; and Pat Carson, "Dirty Tricks," *Panorama,* May 1981, 57–59.

8. "Entertainment Analysts Find Cable Mom-Pop Days Are Gone; Big Bucks Rule the Day," *Broadcasting,* June 15, 1981, 46.

9. "Cable Television Is Attracting More Ads; Sharply Focused Programs Are One Lure," *Wall Street Journal,* March 31, 1981, 46; and "Scribes Back at Typewriters," *Daily Variety,* July 16, 1981, 1.

10. "All You Ever Wanted to Know About Buying an Earth Station," *TVC,* May 15, 1979, 16–19.

11. "Cable's Lost Promise," *Newsweek,* October 15, 1984, 103–5; and "Reshaping the Industries," *Los Angeles Times,* February 28, 1996, D-1.

12. "Lofty Bid for First DBS System," *Broadcasting,* December 22, 1980, 23.

13. "Another Nail in the DBS Coffin: Comsat Bows Out," *Broadcasting,* December 3, 1984, 36.

14. "Backyard Satellite Dishes Spread but Stir Fight with Pay-TV Firms," *Wall Street Journal,* April 2, 1982, 25.

15. "The Video Tape Recorder: Crown Prince of Home Video Devices," *Feedback,* Winter 1981, 1–5.

16. "High Tech: The New Videocassettes," *Emmy,* Summer 1980, 40–44.

17. "Hollywood Loses to Betamax," *Daily Variety,* January 18, 1984, 1; and "3rd-Quarter VCR U.S. Population Pegged at 13 Mil," *Daily Variety,* October 25, 1984, 1.

18. "Home Front," *Emmy,* June 1995, 44–46.

19. "Take the Videodisc—Please," *TV Guide,* December 26, 1981, 12–14; and "RCA Gives Up on Videodisc System," *Los Angeles Times,* April 5, 1984, IV-1.

20. "Capcities/ABC," *Broadcasting,* March 25, 1985, 31; and "Disney to Buy Cap Cities/ABC for $19 Billion, Vault to No. 1," *Los Angeles Times,* August 1, 1995, A-1.

21. "General Electric Will Buy RCA for $6.28 Billion," *Los Angeles Times,* December 12, 1985, A-1; and "Feds Bless NBC-U Wedding Plan," *Daily Variety,* April 21, 2004, 4.

22. "CBS Gleam in Ted Turner's Eye," *Broadcasting,* March 4, 1985, 35; "Tisch Does What CBS Feared in Turner," *Wall Street Journal,* November 20, 1987, 6; and "Group W-CBS Deal OK'd," *Electronic Media,* November 27, 1995, 2.

23. "The Fox Files," *Hollywood Reporter,* April 15, 2002, S1–S2.

24. "As the Town Turns: Sit Back, Grab a Beer, See Some Grass Grow," *Wall Street Journal,* June 5, 1998, A-1; and "Low-Power TV Speaks Foreign Languages," *Broadcasting & Cable,* December 13, 1991, 96.

25. Michael G. Vita and John P. Wiegand, "Must-Carry Regulations for Cable Television Systems: An Economic Policy Analysis," *Journal of Broadcasting & Electronic Media,* Winter 1994, 1–19; "Court: Must-Carry Is Constitutional," *Electronic Media,* December 18, 1995, 6; and "Looking Back at Retransmission," *Electronic Media,* March 4, 2002, 1.

26. "Congress Abolishes Copyright Royalty Tribunal," *Broadcasting & Cable,* November 29, 1993, 18; and "Cable Price Hike," *Broadcasting & Cable,* October 23, 2000, 29.

27. "FCC Strikes the Flag on TV Ownership Rules," *Broadcasting,* August 13, 1984, 35.

28. "Free at Last: Cable Gets Its Bill," *Broadcasting,* October 15, 1984, 38; "Cable Rates Up 50% Since Deregulation," *Daily Variety,* September 5, 1991, 1; and "FCC Reins in Cable TV Rates," *Christian Science Monitor,* February 24, 1994, 3.

29. "Univision Sale Facing Vote Today," *Hollywood Reporter,* September 27, 2006, 6; and "Televisa, Univision Settle," *Los Angeles Times,* January 23, 2009, C-1.

30. "NBC to Acquire Telemundo Network for $1.98 Billion," *Los Angeles Times,* October 12, 2001, C-1.

31. "Hispanic TV Takes Off in the U.S.," *Wall Street Journal,* September 7, 2000, B-1; and "Sí TV Celebrates a Year of Speaking English," *Broadcasting & Cable,* February 21, 2005, 30.

32. "Hubbard Broadcasting: Into DBS from Day One," *Broadcasting & Cable,* December 6, 1993, 30–68; and "Hughes Sees Payoff from DBS Gamble," *Aviation Week & Space Technology,* May 1, 1995, 58–59.

33. "All But Unanimous: VideoCipher-Plus," *Broadcasting & Cable,* February 17, 1992, 12.

34. "It Could Have Been Worse," *Broadcasting & Cable,* November 5, 2001, 19; "Rupe Finally Bags His Bird," *Daily Variety,* April 10, 2003, 1; and "Can John Malone Remake His Cable Empire in the Sky?" *Broadcasting & Cable,* January 1, 2007, 18–19.

35. "DirecTV Makes Local Deals," *Broadcasting & Cable,* December 13, 1999, 16.
36. "DirecTV Adds Internet and Phone Network," http://www.broadcastingcable.com/index.asp?layout=articlePrint&articleID=CA6468980 (accessed March 27, 2008).
37. "New Law of the Land," *Broadcasting & Cable,* February 5, 1996, 8.
38. "Finally, Ratings Agreement," *Electronic Media,* July 14, 1997, 1.
39. Some of the materials that give the history of the telephone are J. Brooks, *Telephone: The First Hundred Years* (New York: Harper and Row, 1975); John R. Pierce, *Signals: The Telephone and Beyond* (San Francisco: Freeman, 1981); and "30 Years," *Telecommunications,* June 1997, 25.
40. "Ma Bell's Big Breakup," *Newsweek,* January 18, 1982, 58–59; and Robert Britt Horwitz, "For Whom the Bell Tolls: Causes and Consequences of the AT&T Divestiture," *Critical Studies in Mass Communication,* June 1986, 119–53.
41. "The Exploding Fax Universe," *Computer Telephony,* March 1995, 123; "Cutting the Cord," *Wall Street Journal,* November 9, 1990, R-1; and "Small Scale Telephony," *Byte,* May 1995, 125.
42. "Ma Bell Finally Gets Cable," *Hollywood Reporter,* June 25, 1998, 1.
43. "FCC Clears AT&T Merger," *Los Angeles Times,* December 30, 2006, C-1.
44. "FiOs Joins the Millionaires Club," *Hollywood Reporter,* January 29, 2008, 10.
45. "Fast Forward," *Hollywood Reporter,* July 7, 1998, S-15.
46. "Agreement Reached on a New Format for Video," *Los Angeles Times,* September 16, 1995, D-1.
47. "Thou Shalt Buy DVD," *Fortune,* November 8, 1999, 201–8.
48. "DVD Is Crowned Sell-Through King," *Hollywood Reporter,* January 9, 2002, 1.
49. "Fox Gives Viewers a Taste with TV-DVD Starter Sets," *Hollywood Reporter,* March 17, 2005, 55.
50. "Par Completes Blu-ray's Sweep," *Hollywood Reporter,* February 21, 2008, 19.
51. "TiVo Digital Video Recorder," *Los Angeles Times,* October 11, 2001, T-3.
52. "Comcast, TiVo Record a Deal," *Los Angeles Times,* March 16, 2005, C-1; and "Big Court Ruling for TiVo Is a Pause That Refreshes," *Hollywood Reporter,* February 1–3, 2008, 7.
53. "Taking 'Survivor' Lessons," *Broadcasting & Cable,* August 21, 2000, 12; and "What's So Real About Reality TV?" *Emmy,* October 2000, 47–49.
54. "Infomercial Programs Go for the Big Time," *Broadcasting & Cable,* October 12, 1992, 45.
55. "Why the Sopranos Sing," *Newsweek,* April 2, 2001, 48–55; and "Sex Education," *Emmy,* June 2001, 104–7.
56. "Pay-Per-View Seems a Sure Thing Despite Marketing, Technical Obstacles," *Wall Street Journal,* February 6, 1989, A9A; "Near Video on Demand Has Arrived," *Daily Variety,* May 13, 1994, 1; and "The VOD Roll-Out," *IC,* May 2000, 18–24.
57. "UPN, Time Warner Join Network Fray," *TV Technology,* March 1995, 9; "Back to the Future," *Daily Variety,* October 14, 2003, 7; and "And Then There Were Five," *Emmy,* May/June 2006, 152–56.
58. "Sorting Through the Fallout of Cable Programming," *Broadcasting,* October 17, 1983, 29.
59. "Cable's Court Is in Session," *Broadcasting & Cable,* July 8, 1991, 49; and "CTV: Punch Lineups Unveiled for Comedy Service," *Broadcasting & Cable,* March 4, 1991, 54.
60. "Cable Nets Do Digital," *Broadcasting & Cable,* December 1, 1997, 6.
61. "Public Access, Spotty Success," *Los Angeles Times,* July 16, 1999, B-2.
62. "Time Warner to Unplug FSN," *Electronic Media,* May 5, 1997, 8.
63. "HDTV: From 1925 to 1994," *TV Technology,* August 4, 2004, 42; "FCC Starts Digital Clock," *Daily Variety,* April 4, 1997, 1; and "Broadcasters Agree to Go All Digital," *Los Angeles Times,* July 13, 2005, C-1.
64. "Local Stations Multiply," *Broadcasting & Cable,* March 10, 2008, 16–17; and "Cable Nets Sue Over Must-Carry," *Hollywood Reporter,* February 5, 2008, 2.
65. "AT&T, Verizon in Airwaves Grab; A Win for Google?" *Wall Street Journal,* March 21, 2008, B-1.
66. "Low-Power Broadcasters Worried About Digital," *Broadcasting & Cable,* May 26, 2008, 8.
67. "EchoStar Raises Stakes with Slingbox Deal," *TV Technology,* October 27, 2007, 24; and "'Beam' Me Up—in Analog," *Broadcasting & Cable,* February 26, 2006, 16.
68. "Top MSOs Own 90% of Subs," *Broadcasting & Cable,* May 24, 1999, 34.
69. "Remember the V-chip? TV Guide Cuts Icons," *Daily Variety,* September 16, 2003, 5; and "First Amendment End Run," *Broadcasting & Cable,* June 21, 2004, 8.

Chapter 5

RADIO

Radio knows how to reinvent itself. At present it consists largely of disc jockeys announcing music, talk-show hosts engaging in controversial discussions, and newscasters giving the latest information. This, however, has not always been the case.

> It is inconceivable that we should allow so great a possibility for public service as broadcasting to be drowned in advertising chatter.
>
> **Herbert Hoover, while serving as secretary of commerce**

During the 1930s and 1940s, radio was the main source of national entertainment programming. Most of the models of entertainment and information that are common to the media today were formed by radio during those years. When television took away radio's audience in the 1950s, some believed radio would die, but today it is a viable medium that enters homes, automobiles, and many other places people inhabit.

5.1 Early Inventions

The beginnings of radio are veiled in dispute. People living in various countries devised essentially the same inventions. Ironically, this was partly because no communication system was available for people to learn what others were inventing. This led to numerous claims, counterclaims, and patent suits.

Many people believe that radio originated in 1873 when James Clerk Maxwell, a British physics professor, published his theory of electromagnetism, which predicted the existence of **radio waves** and how they should behave. During the 1880s, a German physics professor, Heinrich Hertz, undertook experiments to prove Maxwell's theory. Hertz actually generated the radio energy that Maxwell had theorized at one end of his laboratory and transmitted it to the other end.[1]

Maxwell and Hertz

Guglielmo Marconi (see Exhibit 5.1) expanded upon radio principles. Marconi, the son of a wealthy Italian father and an Irish mother, was scientifically inclined from an early age. Fortunately, he had the leisure and wealth to pursue his interests. Soon after he heard of Hertz's ideas, he began working fanatically, finally reaching a point where he could transmit the dots and dashes of Morse code using radio waves. Marconi wrote to the Italian government in an attempt to interest it in his project but to no avail. His determined mother decided he should take his invention to England. There, in 1897, he received a patent and the financial backing to set up the Marconi Wireless Telegraph Company, Ltd. Under the auspices of this company, Marconi continued to improve on **wireless** technology and began to supply equipment to ships. In 1899, he formed a subsidiary company in the United States.[2]

Marconi

People became intrigued with the idea of transmitting voices. In 1904, John Fleming of Britain took a significant step in this direction when he developed the **vacuum tube,** which led the way to voice transmission. Three years later, American inventor Lee De Forest (see Exhibit 5.2) patented the **audion tube,** which was an improvement on Fleming's tube in that it was capable of amplifying sound to a much greater degree than was previously possible. De Forest, like Marconi, was fascinated with electronics at an early age and later secured financial backing to form his own company. However, he experienced management and financial problems that frequently rendered him penniless and led him to sell

Fleming

De Forest

Exhibit 5.1
Guglielmo Marconi shown here with wireless apparatus (about 1902).

(Smithsonian Institution, Photo No. 52202)

Exhibit 5.2

Lee De Forest, shown here with wireless apparatus (about 1920).

(Smithsonian Institution, Photo No. 52216)

his patent rights. De Forest strongly advocated voice transmission for entertainment purposes. In 1910, he broadcast the singing of Enrico Caruso from the New York Metropolitan Opera House. Several years later, he started a radio station of sorts in the Columbia Gramophone Building, playing Columbia records (he later referred to himself as the first disc jockey) in hopes of increasing their sales.[3]

5.2 The Sinking of the *Titanic*

ship radios

During its early stages, radio grew with very few government controls. The first congressional law to mention radio was passed in 1910; it required all ships holding more than 50 passengers to carry radios for safety purposes.

***Titanic* calls**

The rules concerning this safety requirement were not very effective, as was proven with the 1912 sinking of the *Titanic*. As the "unsinkable" *Titanic* sped through the night on its maiden voyage, radio operators on other ships warned it of icebergs in the area. The *Titanic*'s radio operator, concerned with transmitting messages for the many famous passengers, passed the warnings on to the captain, who disregarded them. The wireless operator transmitted SOS signals when the *Titanic* struck the fatal iceberg about midnight on April 14, but none of the nearby ships that could have helped save some of the 1,500 passengers and crew who died heard the distress calls because their wireless operators had signed off for the night.

The operators on land who had been receiving the passengers' messages also heard the distress calls, so for the first time in history people knew of a distant tragedy as it was happening. One wireless operator decoding the messages in New York was reputed to be David Sarnoff (see Exhibit 5.3), who later became president of RCA. He and others relayed information about the rescue efforts to anxious friends and relatives and to newspapers. This brought wireless communication to the attention of the general public for the first time.

Exhibit 5.3
David Sarnoff working at his radio station position atop the Wanamaker store in New York where he later said he heard the *Titanic*'s distress call.

(Courtesy of RCA)

Soon after the sinking of the *Titanic,* Congress passed the Radio Act of 1912; it emphasized safety and required everyone who transmitted on radio waves to obtain a license from the secretary of commerce. The secretary could not refuse a license but could assign particular wavelengths to particular transmitters. Thus, ship transmissions were kept separate from amateur transmissions, which, in turn, were separate from government transmissions. All this was done without any thought of broadcasting as we know it today.[4]

licenses

5.3 World War I

In 1917, during World War I, the U.S. government took over all radio operation. Prior to this, Marconi's company, still the leader in wireless, had aroused the concern of American Telephone and Telegraph (AT&T) by suggesting the possibility of starting a wireless phone business. AT&T, in an effort to maintain its supremacy in the telephone business, had acquired some wireless patents, primarily those of Lee De Forest. The stalemate that grew out of the refusal of Marconi's company, AT&T, and several smaller companies to allow one another to interchange patents had stifled the technical growth of radio communications. Because of the war, these disputes were set aside so the government could develop the transmitters and receivers needed.

patent problems

World War I also ushered two other large companies into the radio field: General Electric (GE) and Westinghouse, both established manufacturers of lightbulbs. GE and Westinghouse assumed responsibility for manufacturing vacuum tubes because both lightbulbs and radio tubes require a vacuum. General Electric had also participated in the development of the **alternator,** used to improve long-distance wireless.

GE and Westinghouse

The patent problem returned after the war, and GE began negotiating with Marconi's company to sell the rights to its alternator. The navy, which had controlled radio during the war, feared this sale would enable the Marconi company to achieve a monopoly on radio communication. The navy did not want radio controlled by a foreign company, so it convinced GE to renege on the Marconi deal. This cancellation left GE sitting with an expensive patent from which it

navy intervention

could not profit because GE did not control other patents necessary for its utilization, but the patent placed GE in an excellent negotiating position.[5]

5.4 The Founding of RCA

company involvement

What ensued was a series of discussions among Marconi, AT&T, GE, and Westinghouse that culminated in the formation of Radio Corporation of America (RCA) in 1919. The Marconi American subsidiary, realizing that it would not receive navy contracts as long as it was controlled by the British, transferred its assets to RCA. AT&T, GE, and Westinghouse bought blocks of RCA stock and agreed to make patents available to one another, thus averting the patent problem and allowing radio to grow. This was undertaken with ship-to-shore transmission in mind, not entertainment broadcasting.[6]

Sarnoff's memo

One person who saw entertainment possibilities was David Sarnoff, a Russian immigrant who at age 15 had become an employee of Marconi and at age 21 is said to have received distress messages from the *Titanic*. Apparently, at the age of 24, he wrote a memo to Marconi management suggesting entertainment radio that read in part as follows:[7]

> I have in mind a plan of development which would make radio a "household utility" in the same sense as the piano or phonograph. The idea is to bring music into the home by wireless. . . . The receiver can be designed in the form of a simple "Radio Music Box" and arranged for several different wavelengths, which should be changeable with the throwing of a single switch or pressing of a single button. . . . The same principle can be extended to numerous other fields as, for example, receiving lectures at home which can be made perfectly audible; also, events of national importance can be simultaneously announced and received. Baseball scores can be transmitted in the air by the use of one set installed at the Polo Grounds. . . . This proposition would be especially interesting to farmers and others in outlying districts removed from cities. By purchase of a "Radio Music Box," they could enjoy concerts, lectures, music recitals, etc. . . .

This idea was not acted upon, and Sarnoff, who joined RCA when RCA bought out Marconi, had to wait for a more propitious time.

5.5 Early Radio Stations

Conrad

Meanwhile, many amateur radio enthusiasts began undertaking experiments. One was Frank Conrad, a physicist and an employee of Westinghouse in Pittsburgh (see Exhibit 5.4). From his garage he programmed music and talk during his spare time. A local department store began selling wireless reception sets and placed an ad for these in a local newspaper, mentioning that the sets could receive Conrad's broadcasts. One of Conrad's superiors at Westinghouse saw the ad and envisioned a market. Until this time, both radio transmission and reception had been for the technical-minded who could assemble their own sets. It was obvious that sets could be preassembled for everyone who wished to listen to what was being transmitted.

Exhibit 5.4
Frank Conrad, who worked with experimental equipment and supervised the construction of KDKA.

(Courtesy of KDKA, Pittsburgh)

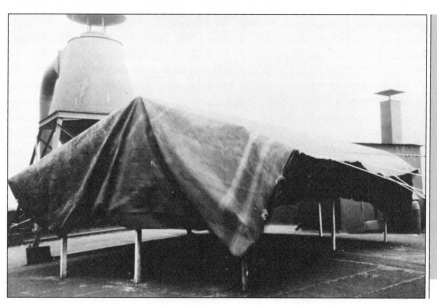

Exhibit 5.5
A tent atop a Westinghouse building in East Pittsburgh served as KDKA's first studio. It caused some of early radio's unusual moments, such as the whistle of a passing freight train heard nightly at 8:30 and a tenor's aria abruptly concluded when an insect flew into his mouth.

(Courtesy of KDKA, Pittsburgh)

Conrad was asked to build a stronger transmitter at the Westinghouse plant, one capable of broadcasting on a regular schedule so that people who purchased receivers would be assured listening fare. In 1920, Westinghouse applied to the Department of Commerce for a special type of license to begin a broadcasting service. The station was given the call letters KDKA (see Exhibit 5.5) and was authorized to use a frequency away from amateur interference. Because

KDKA

Westinghouse was the first to acquire this type of license, KDKA is generally considered the first radio station, but other stations that were experimenting at the same time also lay claim to the title of "first." KDKA launched its programming schedule with the Harding-Cox election returns, interspersed with music, and then continued with regular broadcasting hours. Public reaction could be measured by the long lines at department stores where radio receivers were sold.[8]

Dempsey-Carpentier fight

KDKA's success spurred others to enter broadcasting, including David Sarnoff, who now received more acceptance for the ideas in his memo. He convinced RCA management to invest $2,000 to cover the Jack Dempsey–Georges Carpentier fight on July 2, 1921, through a temporary transmitter set up in New Jersey. Fortunately, Dempsey knocked Carpentier out in the fourth round, for soon after, the overheated transmitter became a molten mass. This fight, however, helped to popularize radio, and both radio stations and sets multiplied rapidly.

1923 status

By 1923, radio licenses had been issued to more than 600 stations, and receiving sets were in nearly 1 million homes.[9] The stations were owned and operated primarily by those who wanted to sell sets (Westinghouse, GE, RCA) and by retail department stores, as well as radio repair shops, newspapers, and universities that wanted to offer college credit courses that people could listen to in their homes. Unfortunately, all stations were on the same frequency—360 meters (approximately 830 on the AM dial). Stations in the same reception area worked out voluntary arrangements to share the frequency by broadcasting at different times of day. However, as more stations went on the air, interference became common. This was particularly hard on students trying to hear their lessons for college credit courses, and many universities had to cease this form of instruction.

talent

Programming was no problem for these early radio stations. People were mainly interested in the novelty of picking up any signal on their battery-operated crystal headphone receivers. Programs consisted primarily of phonograph record music, call letter announcements, and performances by endless free talent who wandered in the door eager to display their virtuosity on this new medium (see "Zoom In" box).[10]

music

One programming form of the era was dubbed **potted palm music**—the kind played at teatime by hotel orchestras usually flanked by potted palms (see Exhibit 5.6). Sometimes a vocalist was featured, and sometimes a pianist or small instrumental group played. Sopranos outnumbered all other "potted palm" performers. Often the performers who "appeared" on radio wore tuxedos or evening gowns.

Drama was also attempted, even though engineers at first insisted that men and women needed to use separate microphones placed some distance from each other. Performers found it difficult to play love scenes this way. Finally engineers "discovered" that men and women could share a microphone.

other programming

From time to time radio excelled in the public-affairs area, broadcasting political conventions and presidential speeches. Religious broadcasts were also part of early radio. Evangelist Aimee Semple McPherson operated a station in Los Angeles that frequently wandered off frequency. When the secretary of commerce threatened to shut down her station, she wired back, "Please order your minions of Satan to leave my station alone. You cannot expect the Almighty to abide by your wavelength nonsense."[11]

ZOOM IN: The Joys of Early Radio Programming

Sometimes the use of amateurs on early radio created awkward situations. For example, a woman who was a strong advocate of birth control asked to speak on radio. The people at the station were nervous about what she might say, but when she assured them that she only wanted to recite some nursery rhymes, they allowed her into the studio. She then broadcast, "There was an old woman who lived in a shoe/She had so many children because she didn't know what to do." She was not invited back.

A Chicago man wanted to discuss Americanism over a Chicago station and even submitted a script ahead of time. He appeared at the station with a group of bodyguards, who assured no buttons were pushed to take him off the air. It turned out that he was a potentate of the Ku Klux Klan, and, digressing from the script, he extolled the virtues of white supremacy.

A young man in New Jersey wanted to let his mother know how he sounded over the air, so he dropped in at WOR, which had just opened a studio near the music department of a store. The singer the studio was expecting had not arrived, so this young man was put on the air before he even had time to notify his mother. He sang to piano accompaniment for more than an hour as a messenger rushed sheet music from the music counter to the studio.

A carbon microphone of the type used by many early performers.

(Courtesy of KFI, Los Angeles)

Exhibit 5.6

Los Angeles's first station, KFI, began broadcasting in 1922. Although the audience heard only the audio, special attention was given to the studio's decor, including the potted palms.

(Courtesy of KFI, Los Angeles)

5.6 The Rise of Advertising

As the novelty of radio wore off, people were less eager to perform, and some means had to be found for financing programming. Many ideas were proposed, including donations from citizens, tax levies on radio sets, and payment from radio set manufacturers. Commercials evolved largely by accident.

WEAF "toll" experiment

AT&T was involved mainly in the telephone business and, although it was a partner with RCA, was reluctant to see radio grow because such growth might diminish the demand for wired services. One of its broadcasting entries was closely akin to phone philosophy. It established station WEAF in New York as what it termed a **toll station:** AT&T would provide no programming, but anyone who wished to broadcast a message could pay a "toll" to AT&T and then air the message publicly, in much the same way as private messages were communicated by dropping money into pay telephones. In fact, the original studio was about the size of a phone booth. The idea did not take hold. People willing to pay to broadcast messages did not materialize. AT&T realized that before people would pay to be heard, they wanted to be sure that someone was listening. As a result, WEAF began broadcasting entertainment material, drawing mainly on amateur talent found among its employees. Still there were no long lines of people willing to pay to have messages broadcast.

Queensboro ad

Finally, on August 22, 1922, WEAF aired its first income-producing program: a 10-minute message from the Queensboro Corporation, a Long Island real estate company, which paid $50 for the time. The commercial was just a simple courtesy announcement because AT&T ruled out direct advertising as in poor taste. Many people of the era said that advertising on radio would never sell products. Every dollar of income that WEAF obtained was a painful struggle.

WEAF frequency change

Eventually, AT&T convinced the Department of Commerce that WEAF should have a different frequency. The argument was that other broadcasters were using stations for their own purposes, while WEAF was for everyone and therefore should have special standing and not be made to broadcast on 360 meters. As a result, WEAF and a few other stations were assigned to the 400-meter wavelength. This meant less interference and more broadcast time. The phone booth was abandoned, a new studio was erected, and showmanship took hold.[12]

5.7 The Formation of Networks

AT&T's network

AT&T began using phone lines for remote broadcasts because it was still predominantly in the phone business. It aired descriptions of football games over its long-distance lines and established toll stations in other cities that it interconnected by phone lines—in effect, establishing a network. During this time, AT&T did not allow other radio stations to use phone lines and also claimed sole rights to sell radio toll time. At first, other stations were not bothered because they were not considering selling ads. In fact, there was an anti-advertising sentiment in the early 1920s. For example, people thought that toothpaste should never be advertised because it was an intimate product. As the AT&T toll network emerged and began to prosper, however, other stations became discontented with second-class status. The fires were further fanned by a Federal Trade Commission inquiry that accused AT&T, RCA, GE, and Westinghouse of creating a monopoly in the radio business.

A series of closed hearings held by the major radio companies resulted in the 1926 formation of the National Broadcasting Company (NBC)—owned by RCA, GE, and Westinghouse. AT&T agreed to withdraw from radio programming in exchange for a long-term contract assuring that NBC would lease AT&T wires. This agreement earned the phone company millions of dollars per year. NBC also purchased WEAF from AT&T, thus embracing the concepts of both toll broadcasting and networking.

formation of NBC

In November 1926, the NBC Red Network, which consisted of WEAF and a 23-station national hookup, was launched in a spectacular debut that aired a symphony orchestra from New York, a singer from Chicago, a comedian from Kansas City, and dance bands from various cities throughout the nation. A year later, NBC's Blue Network was officially launched, consisting of different stations and different programming.

In 1932, GE and Westinghouse withdrew from RCA, largely because of the U.S. attorney general's order that the group should be dispersed and partly because David Sarnoff, now president of RCA, believed his company should be its own entity. Another series of closed-door meetings resulted in a divorce settlement. RCA became the sole owner of NBC, and GE and Westinghouse received RCA bonds and some real estate. In retrospect, it appears that RCA walked off with the lion's share of value. But all this happened during the Depression, and GE and Westinghouse were not eager to keep what they thought might be an expensive broadcasting liability.[13]

withdrawal of GE and Westinghouse

Paley and CBS

ABC

Both RCA and NBC have an interesting parentage. NBC was originally owned by RCA, GE, and Westinghouse, which ousted AT&T in forming NBC. RCA was formed by GE, Westinghouse, and AT&T, which ousted Marconi during RCA's formation. The exact details of all these corporate maneuvers will probably never be known.

What eventually became the Columbia Broadcasting System (CBS) was founded in 1927 by a man who wanted to supply radio talent to stations. His plans did not work out, and his failing company was bought by the family of William S. Paley (see Exhibit 5.7). Paley became president and built a radio network that was similar in organization to NBC in that it consisted of a chain of stations. The network became successful, and during the 1930s and 1940s, Paley lured much of the top radio talent, including Jack Benny, from NBC to CBS.[14]

The American Broadcasting Company (ABC) came into existence because of Federal Communications Commission (FCC) actions in the early 1940s. By 1940, the networks had established a power base that the FCC thought could be detrimental, so it attempted to limit their power by issuing rules that prohibited any company from owning and operating more than one national radio network. Thus in 1943, NBC had to sell its Blue Network to a group of investors who, in 1945, changed the network's name to the American Broadcasting Company.[15]

A fourth radio network, Mutual Broadcasting Company, was formed in 1934 when four stations decided to work jointly to obtain advertising. Unlike the other networks, Mutual owned no stations. Instead, it sold ads and bought programs and then paid the stations in its network

Exhibit 5.7
William S. Paley originally worked with his father, who owned a cigar company. The younger Paley took a six-month leave to get CBS going, but he never returned to the cigar business.

(UPI/Bettman News Photos)

to carry the programs and the network ads. It also allowed stations to sell their own ads.[16]

Mutual

Mutual, too, was involved with FCC regulations. NBC and CBS affiliate contracts stipulated that the local stations could not carry programs from a different network. In 1938, Mutual gained exclusive rights to broadcast the World Series, but the NBC and CBS contracts would not allow their affiliated stations to carry these games, even in cities where there were no Mutual stations. The people wanted the games, the stations wanted to carry them, and advertisers wanted to pay for the coverage. Nevertheless, many Americans did not hear the 1938 World Series. The FCC determined that this type of program thwarting was not in the public interest and issued a ruling that no station could have an arrangement with a network that hindered that station from broadcasting programs of another network.

5.8 Chaos and Government Action

The problem of broadcast frequency overcrowding continued to grow during the 1920s. Secretary of Commerce Herbert Hoover was besieged with requests that the broadcast frequencies be expanded and that stations be allowed to leave the

Hoover and radio

360-meter frequency band on which most of them were broadcasting. Hoover made attempts to improve the situation by altering station power and broadcast times, and he called four national radio conferences to discuss problems and solutions to the radio situation, but he was unable to deal with the problem in any systematic manner because he could not persuade Congress to give him the power to do so.

educational stations

One ramification of the frequency situation was that commercial stations overpowered educational radio stations. In the early 1920s, educational and commercial stations often alternated hours on a shared frequency. If the commercial station decided it wanted a larger share of the time, it would petition the government and usually win because it could afford an expensive, time-consuming hearing in Washington while the educators could not. In addition, Hoover urged people who wished to enter broadcasting to buy an existing station rather than add one to the already overcrowded airwaves. As a result, many educational facilities were propositioned by commercial ventures and sold out.[17]

Radio Act of 1927

By 1925, the interference problem was so widespread that Hoover threw up his hands and told radio station operators to regulate themselves as best they could. During 1926–27, 200 stations were created, most of them using any frequency or power they wished and changing at whim. With so much chaos, Congress finally acted and passed the **Radio Act of 1927.** The act proclaimed that radio waves belonged to the people and could be used by individuals only if they had a license and were broadcasting in the "public convenience, interest, or necessity." All previous licenses were revoked, and applicants were allowed 60 days to apply for new licenses from the newly created Federal Radio Commission (FRC). The commission then granted 620 licenses in what is now the **amplitude modulation (AM)** band.[18]

Communications Act of 1934

Several years later Congress passed the **Communications Act of 1934,** which created the Federal Communications Commission. This act was passed primarily because Congress decided that all regulation of communications should rest with one body. The FCC was given power over not just radio but also telephone and other forms of wired and wireless communications. As the act was being formulated, educators lobbied for 15 percent of the frequencies to be reserved for educational radio; however, they were unsuccessful in their bid, and today there are few if any educational stations left in the AM band.[19]

5.9 The Golden Era of Radio

With the chaotic frequency situation under control, radio was now ready to enter the era of truly significant programming development—a heyday that lasted some 20 years. Improvements in radio equipment helped (see Exhibit 5.8). **Earphones** that only one person could use had already been replaced by **loudspeakers** so that

equipment improvements

a.

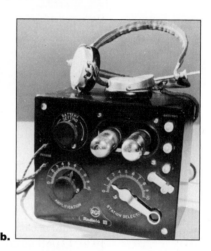
b.

c.

d.

Exhibit 5.8

a. An early station setup that included a multi-tubed audio board, a carbon microphone, and Westinghouse receivers. **b.** A battery-operated radio receiver from about 1923. **c.** A home radio receiver with speaker from about 1924. **d.** Early backpack equipment for remote radio broadcasting.

(a, d: courtesy of KFI, Los Angeles; b, c: courtesy of RCA)

Exhibit 5.9

Amos 'n' Andy as it appeared when broadcasting from Studio B in NBC's Hollywood Radio City. Freeman Fisher Gosden is on the left side of the table with Madaline Lee; Charles J. Correll is on the right. Sitting in the left foreground is the announcer, Bill Hay.

(Courtesy of KFI, Los Angeles)

the whole family could listen simultaneously. The early **carbon microphones** were replaced by **ribbon microphones,** which had greater fidelity. Battery sets were introduced for portability and use in automobiles. (The first portables, however, were cumbersome because of the size of early batteries.)

Radio became the primary entertainment medium during the Depression. In 1930, 12 million homes were equipped with radio receivers; by 1940, this number had jumped to 30 million. During the same period, advertising revenue rose from $40 million to $155 million. In 1930, NBC Red, NBC Blue, and CBS offered approximately 60 combined hours of sponsored programs a week. By 1940, the four networks (Mutual had been added) carried 156 hours.[20]

Amos 'n' Andy

The first program to generate nationwide enthusiasm was *Amos 'n' Andy* (see Exhibit 5.9). It was created by Freeman Fisher Gosden and Charles J. Correll, who met while working in vaudeville. Gosden and Correll, who were white, worked up a blackface act for the company and later tried it on WGN radio in Chicago as *Sam 'n' Henry*. When WGN did not renew their contract, they took the show to WMAQ in Chicago and changed the name to *Amos 'n' Andy* because WGN owned the title *Sam 'n' Henry*.

Correll and Gosden wrote all the material themselves and played most of the characters by changing the pitch and tone of their voices. Gosden always played Amos, a simple, hardworking fellow, and Correll played Andy, a clever, conniving, and somewhat lazy individual who usually took credit for Amos's ideas. According to the scripts, Amos and Andy had come from Atlanta to Chicago to seek their fortune, but all they had amassed was a broken-down automobile, known as the Fresh-Air Taxicab Company of America. Much of the show's humor revolved around a fraternity-type organization called the Mystic Knights of the Sea headed by a character called Kingfish, who was played by Gosden. The show's success caught the attention of the NBC Blue Network, which hired the two in 1929. Their program, which aired from 7:00 to 7:15 P.M. Eastern time, became such a nationwide hit that it affected dinner hours, plant closing times, and even the speaking schedule of the president of the United States.[21]

Many other comedians followed in the wake of the success of Correll and Gosden—Jack Benny, Lum and Abner, George Burns and Gracie Allen, Edgar Bergen and Charlie McCarthy, Fibber McGee and Molly (see Exhibit 5.10). Music, especially classical music, was also frequently aired. Broadcasts featured

comedy

music

a.

b.

Exhibit 5.10

a. Jack Benny with his wife and costar, Mary Livingston, in 1933. The *Jack Benny Show,* sponsored for many years by Jell-O and Lucky Strike cigarettes, featured such surefire laugh provokers as an ancient Maxwell automobile that coughed and sputtered, Benny's perennial age of 39 years, a constant feud with Fred Allen, and Benny's horrible violin playing.

(Courtesy of NBC)

b. Lum and Abner, played by Chester Lauck *(left)* and Norris Goff. This comedy took place in the Jot 'Em Down grocery store in the fictional town of Pine Ridge, Arkansas. In 1936 the town of Waters, Arkansas, changed its name to Pine Ridge in honor of Lum and Abner.

(Courtesy of KFI, Los Angeles)

c.

c. George Burns and Gracie Allen. Many jokes in this program were plays on words based on Gracie's supposed empty-headedness. At one point Gracie started searching for her "lost brother" by suddenly appearing on other shows to inquire about him.

(Courtesy of NBC)

Exhibit 5.10
(*continued*)

d. Charlie McCarthy
and ventriloquist
Edgar Bergen *(right)*
with W. C. Fields *(left)*
and Dorothy Lamour.
Charlie had a running
feud with W. C. and a
love affair with just
about all the 1930s and
1940s beauties and
even the 1950s movie
idol Marilyn Monroe.

(Courtesy of NBC)   **d.**

e.

e. Marian and Jim Jordon as Fibber McGee and Molly. The
commercials were integrated directly into the program when
the announcer visited the McGee home and talked about
Johnson's Wax. The two also had a famous overstuffed
closet. Whenever one of them opened it, a raft of sound
effects would indicate that all sorts of things had fallen out.

(Courtesy of NBC)

**amateur
participation**

children's shows

New York Philharmonic concerts and performances from the Metropolitan Opera House. NBC established its own orchestra led by Arturo Toscanini. *Your Hit Parade,* which featured the top-selling songs of the week, was introduced in 1935, and people who later became well-known singers, such as Kate Smith and Bing Crosby, took to the air. The big bands of the 1940s could also be heard over the airwaves (see Exhibit 5.11).

Some programs featured the general public. Among many talent shows, the most famous was *The Original Amateur Hour* hosted by Major Edward Bowes, which, like today's *American Idol,* spawned stars—including Beverly Sills, Teresa Brewer, and Frank Sinatra. Quiz shows, such as *Professor Quiz,* rewarded people for responding with little-known facts. Stunt shows, such as *Truth or Consequences,* which prompted people to undertake silly assignments if they answered questions incorrectly, attracted large and faithful audiences.

Many programs were developed for children, including *Let's Pretend,* a multi-segment program that emphasized creative fantasy; *The Lone Ranger,* a western; *Quiz Kids,* a panel of precocious children who answered questions; and *Little Orphan Annie,* a drama about a child's trials and tribulations. During the day, many stations broadcast continuing dramas. These programs, called soap operas because soap manufacturers were frequent sponsors, always ended with an unresolved situation to entice the listener to "tune in tomorrow." Most did. The

a.

b.

Exhibit 5.11

a. Arturo Toscanini and the NBC orchestra. Toscanini was coaxed out of retirement in Italy by David Sarnoff, head of NBC and classical music lover. A special studio, 8H, was built for the orchestra; because of its unique constru- ction, it was referred to as the world's only floating studio.

(Courtesy of NBC)

b. Tommy Dorsey's Band. At a time when cigarette companies backed many of the swing bands, Raleigh- Kool sponsored the Dorsey musicians. Because the radio programs were performed before live audiences, the huge cigarette packs did make an impact.

(Courtesy of KFI, Los Angeles)

scripts for a major portion of the soap operas were developed by a husband-wife team, Frank and Ann Hummert. They defined the basic idea for each series, wrote synopses of programs, and then farmed the actual script writing to a bevy of writ- ers around the country, some of whom they never met.

soap operas

In the area of drama, the networks first tried to rebroadcast the sound of Broadway plays but discovered that this was akin to sitting in a theater blind- folded. So the networks hired writers such as Norman Corwin, Maxwell Ander- son, and Stephen Vincent Benet to script original dramas for radio. These dramas usually used many sound effects and were sponsored by one company that often incorporated its name into the program, such as *Lux Radio Theater* or *Collier's Hour.* In 1938, Orson Welles produced *The War of the Worlds,* a fantasy about a Martian invasion in New Jersey. Upon hearing the broadcast, an estimated

drama

THE INCOMPARABLE AMOS 'N' ANDY, returning to air via NBC Friday Oct. 8.

NBC
PROUDLY
Presents:

RAMP, TRAMP, TRAMP.
It's NBC's parade of stars marching along to open the fall and winter season of happy listening.

For dialers, the biggest news of all is the return of Amos 'n' Andy. The two old favorites introduce a brand new show on Friday, October 8, complete with guest stars, music and the kind of laughter which made Freeman Gosden and Charles Correll famous.

TRAMP, TRAMP, TRAMP.

"The Great Gildersleeve" started the parade by huffing and puffing his way back to his fans late in August. This is Hal Peary's third season on the air with his own program, and from the way the polls were going when he went off in June, it looks like his biggest.

TRAMP, TRAMP, TRAMP.

Fanny Brice, with more antics of

THE INCORRIGIBLE BABY SNOOKS and Frank Morgan on Maxwell House show Thursday, 8:30 p. m.

THE INVENTIVE ARKANSAS TRAVELER, Bob Burns, back on air Thursday night.

THE INGENIOUS JACK BENNY, airing with all the gang at 4 p. m., Sunday.

THE INFALLIBLE H. V. KALTENBORN, commentator, heard four afternoons weekly.

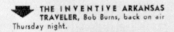

Page Four

Exhibit 5.12

A wartime plug for NBC's programs.

(Courtesy of KFI, Los Angeles)

A Star-Bedazzled Parade Of Fast-Stepping Radio Entertainers, on March To Storm Your Listening

THE IMPERTINENT CHARLIE McCARTHY and Bergen for Chase & Sanborn, Sunday, 5 p. m.

her inimitable Baby Snooks, and Frank Morgan, with a new batch of tall stories, followed "Gildersleeve" to NBC microphones on the first Thursday in September.

TRAMP, TRAMP, TRAMP.

Edgar Bergen and Charlie McCarthy marched back from Newfoundland, where they entertained the troops stationed there. This season they are presenting Victor Moore and William Gaxton, in addition to Ray Noble's orchestra and the songs of Dale Evans.

TRAMP, TRAMP, TRAMP.

That bad little boy, Red Skelton, was next in line, and with him were the popular members of his cast— Harriet Hilliard and Ozzie Nelson and his band.

TRAMP, TRAMP, TRAMP.

An account of Bob Hope's travels while away from his radio show for the summer sounds like a review of the war headlines—England, Bizerte, Tunis, Algiers, Sicily. Back on the air with him for the new season comes another grand trouper, Frances Langford, who also went into the battle areas with Bob. And, of course, Jerry Colonna and Vera Vague will be on hand.

TRAMP, TRAMP, TRAMP.

And so they come. Those two top

comedians, Jim and Marian Jordan, who have more delightful sessions with "Fibber McGee and Molly" ready for their listeners.

Eddie Cantor with another season of Wednesday night laughfests.

Jack Benny, another of radio's globe-trotters, only recently returned from the European and North African battlefronts.

And, of course, there are all the favorites who have been on NBC this summer and who will continue to make radio listening America's Number One pastime—"One Man's Family;" Bing Crosby; the Standard Symphony Hour; H. V. Kaltenborn and the other commentators who bring the world into our homes; Kay Kyser's "College of Musical Knowledge;" Ginny Simms; the Joan Davis-Jack Haley show; and the Sunday morning Westinghouse program.

Happy listening? Yes, indeed!

THE INGRATIATING FATHER AND MOTHER BARBOUR of "One Man's Family," Sunday, 5:30 p. m.

THE INIMITABLE FIBBER McGEE AND MOLLY, back with their friends Tuesday, 6 p. m.

THE INEXHAUSTIBLE LAUGH CREATOR, Bob Hope, back in his regular Tuesday spot, 7 p. m.

THE IRREPRESSIBLE LITTLE KID as played by Red Skelton, Tuesday, 7:30 p. m.

1.2 million people succumbed to hysteria. They panicked in the streets, fled to the country, and seized arms to prepare to fight—despite the fact that the *Mercury Theater* program included interruptions to inform the listeners that the presentation was only a drama.

commercials

The Depression spurred the growth of commercials. During the 1920s, advertisements were brief and tasteful, and price was not mentioned. As radio stations and all facets of the American economy began digging for money in any way they could, the commercial standards dissolved. Some advertisers believed commercials should irritate, and broadcasters, anxious for the buck, acquiesced. The commercials became long, loud, hard-driving, and cutthroat.

Most radio programs were produced not by the networks but by **advertising agencies.** These agencies found that personal help programs could effectively promote products. Listeners would send letters to radio human relations "experts" detailing traumas, crimes, and transgressions and asking for help. Product box tops accompanying a letter qualified it for an answer; or the suggested solution might involve the sponsor's drug product or the contentment derived from puffing on the sponsor's brand of cigarette. By 1932, more airtime was spent on commercials than on news, education, and religion combined. The commercials brought in profits for networks and stations and for the advertising agencies that were intimately involved in most details of programming, including selecting program ideas, overseeing scripts, and selling and producing advertisements for the shows (see Exhibit 5.12).

stunt broadcasts

There were also many events that could be termed **stunt broadcasts** (see Exhibit 5.13), such as those from widely separated points, gliders, and underwater locations. A four-way conversation involved participants in Chicago, New York, Washington, and a balloon. One music program featured a singer in New York accompanied by an orchestra in Buenos Aires. These stunt broadcasts paved the way for the broadcasting of legitimate public events from distant points. In 1931, 19 locations around the world participated in a program dedicated to Marconi. People heard the farewell address of King Edward VIII when he abdicated the British throne and the trial of the man who kidnapped aviator Charles Lindbergh's baby.

politics

Radio also figured in politics of the day. President Franklin Delano Roosevelt effectively used radio for his **fireside chats** (see Exhibit 5.14) to reassure the nation during the Depression. Louisiana's firebrand Governor Huey Long was often heard on the airwaves, and Father Charles E. Coughlin, a Detroit priest, tried to build a political movement through radio.[22]

Exhibit 5.13
The NBC radio mobile unit making contact with an airplane. This 1929 experimentation led to future possibilities for news coverage.

(Courtesy of NBC)

5.10 The Press-Radio War

News was destined to become one of radio's strongest services, but not without a struggle. At first, announcers merely read newspaper headlines over the air, but gradually networks began purchasing news from **wire services.** In 1932, the Associated Press sold presidential election bulletins to the networks, and programs were interrupted with news flashes. Newspapers objected to this on the grounds that news on radio would diminish the sale of papers. From 1933 to 1935, a **press-radio war** ensued.

A meeting of newspaper publishers, network executives, and wire service representatives, held at the Biltmore Hotel in New York in 1933, established the **Biltmore Agreement.** It stipulated that networks could air two five-minute newscasts a day, one in the morning after 9:30 A.M. and one in the evening after 9:00 P.M. so they would not compete with the primary hours of newspaper sales. No "hot-off-the-wire" news was to be broadcast, and newscasts were not to have advertising support because this might detract from newspaper advertising. Newspaper pub-

Exhibit 5.14

President Franklin Delano Roosevelt delivering a "fireside chat."

(Courtesy of NBC)

lishers ensured that these provisions appeared in the Biltmore Agreement because they were the most numerous, most powerful, and wealthiest of the meeting participants.

Biltmore Agreement

But the ink on this agreement was barely dry when its intent began to be subverted. The newspaper publishers had agreed to allow radio stations and networks to have **commentators.** Radio took advantage of this provision, and often these commentators became thinly disguised news reporters. NBC and CBS began their own news-gathering activities. At NBC, one person gathered news simply by making telephone calls. Sometimes he scooped newspaper reporters because almost anyone would answer a call from NBC. In addition, he could reward news sources with highly prized tickets to NBC's top shows. Most of the material he collected was broadcast by NBC's prime commentator, Lowell Thomas (see Exhibit 5.15). CBS, whose prime news commentator was H. V. Kaltenborn, set up a larger news force that included **stringers**—reporters paid only for material actually used.

commentators and stringers

The public became increasingly aware of news as world tensions grew prior to World War II. Advertisers became interested in sponsoring news radio programs because of the growing potential listener market. At one point, two services agreed to make their news available to advertisers, which would then broadcast it over radio, but they would not make it available to radio stations directly. This arrangement led to a breakdown of broadcast news blackouts, and radio began to develop as an important news disseminator. Americans heard actual sounds of Germany's march into Austria and the voices of Adolf Hitler and Benito Mussolini.[23]

broadcast news

Exhibit 5.15

Lowell Thomas began broadcasting in 1929 with one of the earliest programs, called *Headline Hunters*, and remained on the radio regularly until after his 80th birthday. For many years he preceded *Amos 'n' Andy* with the news, prompting him to say of himself, "Here is the bird that everyone heard while waiting to hear *Amos 'n' Andy*."

(Courtesy of Lowell Thomas)

5.11 World War II

war efforts

news function

The government did not take over broadcasting during the 1940–45 World War II period as it had during World War I. It did, however, solicit radio's cooperation for bond purchase appeals, conservation campaigns, and civil defense instructions. Among the most famous of these solicitations were singer Kate Smith's marathon broadcasts for war bonds, which sold more than $100 million worth. Many of the plays and soap operas produced during this period dealt with the war effort, and some even tried to address segregation, which was an issue because of racial separation in the armed forces.

The news function greatly increased as up-to-date material was broadcast at least every hour. At one point H. V. Kaltenborn didn't leave the CBS studios for 18 days and went on air 85 times to analyze news from Europe. One of the best-known voices heard from overseas was that of Edward R. Murrow (see Exhibit 5.16), whose broadcasts from London detailed what was happening to the English during the war.[24]

One result of the war was the perfection of **audiotape recorders.** Before the war, NBC and CBS policies forbade the use of recorded material for anything other than sound effects, and even most of those were performed live. This policy was abetted by the musicians' union, which insisted that all broadcast music utilize musicians rather than phonograph records. The

Exhibit 5.16

Correspondent Edward R. Murrow broadcasting on CBS.

(Globe Photos, Inc.)

recording technique used before the audiotape recorder usually employed phonograph discs because the only magnetic recording known in America before World
War II was **wire recording.** To edit or splice, a knot had to be tied in the wire and
then fused with heat, making it a cumbersome technique. During the war, American troops entering German radio stations found them operating without any people. The broadcasting was handled by a machine that used plastic tape of higher
fidelity than Americans had ever heard from wire. This plastic tape could be cut
with scissors and spliced with adhesive. The recorders were confiscated, sent to
America, and improved, and they eventually revolutionized programming procedures because material could be recorded for later playback.[25]

**audiotape
recorders**

Radio stations enjoyed great economic prosperity during the war. About 950
stations were on the air when the war began. No more were licensed during the
war, so these 950 received all the advertisements. A newsprint shortage reduced
ad space in newspapers, and some of that advertising money was channeled into
broadcasting. As a result, radio station revenue increased from $155 million in
1940 to $310 million in 1945.[26]

**economic
prosperity**

5.12 Postwar Radio

Postwar radio prospered. Advertisers were standing in line, and the main programming problem was finding a way to squeeze in the commercials. To the networks, especially NBC, this boon provided the necessary capital to support the
then-unprofitable television development. To invest even more in the new technology, nonsponsored public-affairs radio programs dropped by the wayside, as
did some expensive entertainment. Radio fed the mouth that bit it. On the local
level, this prosperity created a demand for new radio station licenses as both entrepreneurs and large companies scrambled to cash in on the boom. The 950
wartime stations expanded in rabbitlike fashion to more than 2,000 by 1950. Advertising revenues increased from $310 million in 1945 to $454 million in 1950.[27]

**station
expansion**

The bubble burst, however, as advertisers deserted radio to try TV. This left
radio networks as hollow shells. The 2,000 local stations found that the advertising dollars remaining in radio did not stretch to keep them all in the black. In
1961, almost 40 percent of radio stations lost money.[28]

TV takeover

After the war, radio networks returned to prewar programming—comedy,
drama, soap operas, children's programs, news. However, a new phenomenon appeared on the scene—the disc jockey (DJ). Several conditions precipitated this
emergence. A court decision in 1940 ruled that if broadcasters purchased a
record, they could then play it. Previously, records had been stamped "not
licensed for radio broadcast." Removing this restriction added legal stature to
disc jockey programs. During the mid-1940s, the musicians' union, which had
previously voted to halt recording, was appeased with a musicians' welfare fund
to which record companies contributed. This opened the door to mass record production. This mass production of records led to a symbiotic relationship between
radio and the record business.

**changes in the
record business**

The beginnings of this relationship can be traced to several people, most notably Alan Freed, Gordon McLendon, and Todd Storz. During the early 1950s,

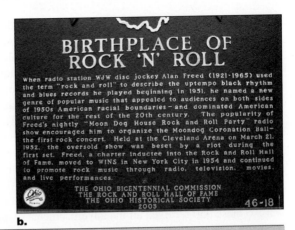

BIRTHPLACE OF ROCK 'N' ROLL

When radio station WJW disc jockey Alan Freed (1921-1965) used the term "rock and roll" to describe the uptempo black rhythm and blues records he played beginning in 1951, he named a new genre of popular music that appealed to audiences on both sides of 1950s American racial boundaries—and dominated American culture for the rest of the 20th century. The popularity of Freed's nightly "Moon Dog House Rock and Roll Party" radio show encouraged him to organize the Moondog Coronation Ball—the first rock concert. Held at the Cleveland Arena on March 21, 1952, the oversold show was beset by a riot during the first set. Freed, a charter inductee into the Rock and Roll Hall of Fame, moved to WINS in New York City in 1954 and continued to promote rock music through radio, television, movies, and live performances.

THE OHIO BICENTENNIAL COMMISSION
THE ROCK AND ROLL HALL OF FAME
THE OHIO HISTORICAL SOCIETY
2003   46-18

a.   b.

Exhibit 5.17

Because rock and roll was first broadcast by Alan Freed in Cleveland, the Rock 'n' Roll Museum (a) was built in that city. Its plaque (b) tells of the early beginnings of the format.

Top 40

Freed, a Cleveland DJ, began playing a new form of music he called rock and roll (see "Zoom In" box and Exhibit 5.17). The music caught the fancy of teenagers and gave radio a new primary audience and a new role in society—a mouthpiece for youth. McLendon and Storz were both station owners who began programming Top 40 music. According to radio lore, Storz was in a bar one night trying to drown his sorrows over the sinking income of his radio stations when he noticed that the same tunes seemed to be played over and over on the jukebox. After almost everyone had left, a waitress went over to the jukebox. Rather than playing something that had not been heard all evening, she inserted her nickel and played the same song that had been selected most often. Storz decided to try playing the same songs over and over on his radio stations. Top 40 radio was born. McLendon programmed the same Top 40 format and promoted it heavily.[29]

mobility

At the same time that recorded music was being introduced on radio, radios were becoming more portable and Americans were becoming more mobile. The public (especially the young) appreciated the DJ shows, which could be enjoyed while listeners were engaged in other activities, such as studying, going to the beach, or talking with friends. Another important reason for the rise of the DJ was

lower overhead

that station management appreciated the lower overhead, fewer headaches, and higher profits associated with DJ programming. A DJ did not need a writer, a bevy of actors, a sound effects person, an audience, or even a studio. All that was needed were records, and these were readily available from companies that would eagerly court DJs in the hope that they would plug certain tunes, thus assuring sales of the records.[30]

payola

This courtship slightly tarnished the DJs' image during the late 1950s when it was discovered that a number of DJs were engaged in **payola,** accepting money or gifts in exchange for favoring certain records. To remedy the situation, Congress amended the Communications Act so that if station employees received money from individuals other than their employers for airing records, they had to disclose

ZOOM IN: **The Father of Rock and Roll**

Alan Freed, who referred to himself as the Father of Rock and Roll, wasn't the first to air the type of rhythm and blues music that became rock, and he wasn't the first to use the term rock and roll, but he put the two together. In the 1950s, rhythm and blues was considered African American music that whites shouldn't listen to and "rock and roll" was a term used in relation to the sex act. By popularizing rhythm and blues with white audiences, Freed has been credited with breaking down racial barriers among young people of the 1950s decade.

Freed, who was born in 1921, became interested in radio while in college. After World War II, he landed jobs in small radio stations and then, in 1950, went to Cleveland, where he was an afternoon movie show host for a TV station and also played rock and roll late at night on WJW, which was otherwise a classical music station. He called himself Moondog, after an instrumental that he liked. He organized a rock and roll concert at a large Cleveland stadium that had to be shut down because of near-riot overcrowding. That brought him fame, and WJW gave him more air time.

His reputation traveled to New York, and in 1954 WINS hired him to turn that station into a rock and roll station. He continued to arrange concerts to showcase rock and roll musicians, some of which became rowdy. In 1958 he was indicted for inciting a riot at a Boston concert, and although charges were eventually dropped, WINS did not renew his contract, and he moved to WABC, New York. In 1959 he got caught up in a payola scandal, and when he wouldn't sign

a form saying he had never accepted payola, WABC fired him. In 1960 he pleaded guilty to several counts of commercial bribery. Although the penalty he received was mild, the bad publicity ruined his career. He had a few more short radio jobs but was basically unemployable and died in 1965 at the age of 43 of cirrhosis of the liver brought on by alcohol.

When the Rock and Roll Hall of Fame opened in 1986, he was one of the first inductees, and in 1991 he received a star on the Hollywood Walk of Fame.

Alan Freed.

(AP/Wide World Photos)

that before broadcast time under penalty of fine or imprisonment. This helped control the practice of payola, but every now and then it still rears its head.[31]

The stations' need for the networks declined as the stations courted DJs and top talent left radio for TV. The increasing number of stations also meant that more stations existed in each city, so more of them were programming independently of networks. Therefore, the percentage of network-affiliated stations decreased dramatically. The overall result was a slow but steady erosion of network entertainment programming.

decline of networks

5.13 FM Radio Development

Armstrong

During the early 1930s, David Sarnoff mentioned to Edwin H. Armstrong that someone should invent a black box to eliminate static. Armstrong did not invent just a black box, but a whole new system—**frequency modulation (FM).** He wanted RCA to back its development and promotion, but Sarnoff had committed RCA funds to television and was not interested in underwriting a new radio structure despite its obviously superior fidelity.

Armstrong continued his interest in FM, built an experimental 50,000-watt FM station in New Jersey, and solicited the support of GE for his project. During the late 1930s and early 1940s, an FM bandwagon was rolling, and some 150 applications for FM stations were submitted to the FCC. The FCC altered channel 1 on the TV band and awarded spectrum space to FM. It also ruled that TV sound should be frequency modulated. Armstrong's triumphant boom seemed just around the corner, but World War II intervened, and commercial FM had to wait.

moving FM

After the war, the FCC reviewed spectrum space and decided to move FM to another part of the broadcast spectrum, ostensibly because it thought sunspots might interfere with FM. Armstrong protested this move because it rendered all prewar FM sets worthless and saddled the FM business with heavy conversion costs. Armstrong was further infuriated because, although FM sound was to be used for TV, RCA had never paid him royalties for the sets it manufactured. In 1948, he sued RCA. The suit proceeded for more than a year, and the harassment and illness it caused Armstrong led him to leap from the window of his 13th-floor apartment to his death.[32]

FM's slow start

FM continued to develop slowly. With television on the horizon, there was little interest in a new radio system. Many of the major AM stations acquired FM licenses as insurance in case FM replaced AM, as its proponents were predicting. AM stations simply duplicated their AM programming on FM, which did not increase the public's incentive to purchase FM sets. In fact, for a while an industry joke ran, "What do the letters 'FM' in FM radio stand for?" The answer was "Find me."

noncommercial radio

One brighter spot was on the education front. In 1945, educators convinced the FCC to reserve the 20 FM channels between 88.1 and 91.9 exclusively for noncommercial radio. Most of these stations were used by universities, although some were owned by nonprofit community groups, such as the Pacifica Foundation, and others were operated by religious organizations. In 1949, the National Association of Educational Broadcasters (NAEB) formed a **bicycle network** to provide programming for educational stations. The programs were duplicated at a central location and then sent by mail from one station to another on a scheduled round-robin basis.[33]

FM success

As general interest in high-fidelity music grew, FM's interference-free signal became an asset to commercial stations. In 1961, the FCC authorized stereophonic sound transmission for FM, which led to increased awareness of the medium. At first, classical music dominated the FM airwaves. When hi-fi equipment became inexpensive enough to be purchased by teenagers, rock music became prominent FM fare. This led to an increased number of listeners, followed by an increased number of advertisers.

A further aid to FM's success was a 1965 FCC ruling that in cities of more than 100,000 population, AM and FM stations with the same ownership had to have separate programming at least 50 percent of the time. This helped FM gain a foothold because it now developed its own distinctive programming.

During the 1970s, FM developed so successfully that it began taking audience away from AM. In 1972, AM had 75 percent of the audience, and FM had a paltry 25 percent. By the mid-1980s, those percentages had reversed, and AM stations were the ones losing money.[34]

The switch to FM was mostly caused by the superior sound quality of the medium, including the capability for **stereo** sound. AM proponents tried to combat this by developing an AM stereo system. During the late 1970s and early 1980s, several companies proposed stereo transmission systems that, unfortunately, were not compatible with each other. AM stations hoped the FCC would choose a standard, as it had for FM, but in 1982, the FCC refused to rule on one common standard, stating instead that the marketplace should decide. The marketplace was not quick to decide. By the 1990s, a system developed by Motorola seemed to have become the de facto standard, so the FCC rubber-stamped it, but very few stations switched to stereo. AM stereo sound is not as good as FM and did not give AM a competitive boost.[35]

AM stereo

5.14 The Restructuring of Public Radio

The educational stations at the lower end of the FM dial received some sprucing up with the passage of the Public Broadcasting Act in 1967. This act was adopted mainly to benefit educational television, but radio was also included. The term "educational" was dropped and "public radio" was used instead. The act resulted in funding for a network, National Public Radio (NPR), which began operating from Washington, D.C., in 1970. It replaced the bicycle network of the NAEB and today delivers programming by satellite. NPR's mission was to upgrade the quality of public radio programming with news, Senate hearings, music, talk shows, documentaries, and programming from other countries. One of the most popular early programs was the in-depth evening news program *All Things Considered,* which is still on the air (see Exhibit 5.18).

NPR

Over the years, some structural elements of NPR were not popular with many of the stations, most notably the fact that most of the programming was produced in Washington, D.C. Local stations thought their programs deserved wider dissemination. A group of public stations in 1982 formed American Public Radio, which in 1994 changed its name to Public Radio International (PRI). PRI is an independent, nonprofit network that does not receive direct federal funding. Station members pay fees that support the national office in Minnesota and the satellite distribution system. In 2004, Minnesota-based American Public Media (APM) decided to distribute programs it owned and produced itself rather than through PRI, so there are now three distributors of public radio programming—NPR, PRI, and APM.

PRI

APM

The public broadcasting networks do not fill all the airtime on the 750-plus public radio stations that affiliate with them. Local programming includes

music—classical, jazz, swing, and other forms not generally heard on commercial radio. Some stations also produce their own public-affairs programs, dramas, and children's programs.[36]

5.15 College Radio

**network
affiliated**

A number of colleges and universities own NPR-, APM- and/or PRI-affiliated stations that program primarily network material and operate much like public radio stations owned by community groups in that they have a paid staff and no involvement from the student body. Other colleges affiliate with NPR and PRI but allow for some student involvement, such as airing programs produced by students in radio production classes or offering student internships.

student run

However, what the term "college radio" usually conjures up is an independently operated radio station owned by the college but operated by students as both a learning experience and a form of expression (see Exhibit 5.19). In addition to the college stations, more than 300 high school stations operate in this manner. Many of these high school stations offer computer-programmed music during school hours and live DJ programming after school. These student stations, like other noncommercial stations, are located at the lower end of the FM band. They originally arose because of a 1948 FCC ruling authorizing low-powered, 10-watt educational stations that generally reached only a 2- to 5-mile radius and were inexpensive to operate. These **10-watter** college and high school radio stations grew rapidly, and most have since increased their power. The student-run stations are important to musicians because they are the main outlet for alternative music and are often the first to air new groups and new sounds.

In addition to FM stations, some colleges and high schools have closed-circuit stations that operate only on campus, programming primarily to the cafeteria or the dorms. Others operate low-power AM or FM stations that are designated

Exhibit 5.19
Student stations, such as this one at John Carroll University in Ohio, usually have many posters and stickers to enhance the radio station atmosphere.

to provide traffic and weather information. The student groups supplement this programming material with announcements about campus events. Many colleges have placed their FM stations on the internet, and a number have established internet-only student stations (see Chapter 2).[37]

5.16 The Changing Structure of Commercial Radio

On the commercial front, the business structure of radio underwent a number of significant changes in the latter part of the 20th century. Radio networks, which hit a low in the 1960s, began to reemerge, but in a different form. They no longer brought common programming to the nation but instead developed numerous satellite services with different features and formats, each intended for a particular niche of stations. Some stations used only bits and pieces from the network, some used the network music along with local personalities, and others used a preponderance of network material—often obtaining it from several different networks. Obviously the old rules against one company owning more than one network were dropped.[38]

reemergence of networks

Network ownership also changed. A new radio network, Westwood One, purchased Mutual in 1985 and NBC radio networks in 1987. Westwood One then entered into an arrangement whereby it was managed by Infinity Broadcasting, which later became part of CBS. CBS and Westwood One still have ties, although they are looser than they used to be. In 1995, Disney bought ABC, including its radio properties, and then in 2006 Disney sold the radio interests to Citadel. CBS radio underwent several ownership changes as its parent organization merged with various companies, but today CBS radio remains a major part of the overall CBS structure. More than 500 companies now distribute radio programming, primarily by satellite. Some call themselves *networks,* and others call themselves *syndicators* or *program producers.*[39]

deregulation

Another big change in the structure of the radio business involved **deregulation,** particularly that encompassed in the **Telecommunications Act of 1996** that greatly relaxed station licensing procedures and ownership restrictions. License renewal, which used to be a very complicated procedure occurring every three years, is now a simplified process that occurs only every eight years. In previous decades, a particular company could own no more than seven AM and seven FM stations and could have no more than one of each in any listening area.

ZOOM IN: Minority Leadership

Catherine Hughes is the founder and chairperson of Radio One, the largest African American–owned station group in the United States. Hughes entered radio in 1973 as general sales manager at WHUR-FM Howard University Radio in Washington, D.C. She then became the first female vice president and general manager of a station in the nation's capital and made WHUR one of the most listened to stations in the D.C. area.

In 1979, she and her husband bought a small Washington radio station and founded Radio One. When her marriage ended, she bought her husband's share, but for a while she had to give up her apartment and she and her son, Alfred Liggins, slept at the station to make ends meet. Eventually the station became profitable, and she purchased other stations; Radio One now has 54 stations in 17 markets. She was the first African American woman to head a publicly traded firm. Alfred is now CEO of Radio One, and he and his mother have taken the business into TV by forming TV One, a cable TV channel aimed at African Americans.

In April 2001, the National Association of Broadcasters bestowed its Distinguished Service Award on Hughes. In her acceptance speech, she told of a time when she needed money to close a deal. "There was only one problem. I needed $1.5 million to close the deal, and I was only $1.5 million short." She credits a Puerto Rican woman banker with believing in her enough to find her the funding. Her speech also urged those in broadcasting to make room for more women and minorities.

The FCC had rules that gave priority to women and minorities in terms of station ownership, but these rules no longer exist. Do you think they should they be reinstituted?

Catherine Hughes.

(Courtesy of Catherine Hughes)

Now there are no national ownership limits, and local ownership limits have been changed so that, in some markets, one company can own eight stations. This deregulation has resulted in companies with large financial resources buying many radio stations. For example, Clear Channel Communications, which in 1995 owned only 45 stations, now owns in excess of 1,200.[40]

Ethnic radio has shown great growth recently (see "Zoom In" box). Black-oriented stations have been popular for decades, and foreign-language stations existed near the periphery of radio for many years. But by the mid-1990s, more than 400 stations were operating with an ethnic format. In Los Angeles, for example, Spanish-language stations often rate number one during rating periods.[41]

ethnic radio

Talk radio has become important to the social fabric of America. Talk-show hosts help set and challenge public policy, and listeners freely express their opinions and viewpoints. Shock jocks stir controversy with their sexual and scatological language and their often outlandish antics on and off the radio. But the personality who was heard on the most radio stations until his death in 2009 at age 90 was Paul Harvey, who had his own brand of news and commentary for ABC and who started his radio career in 1933.[42]

talk radio

5.17 Satellite and HD Radio

Satellite and HD radio both use digital technology. **Satellite radio,** also known as **digital audio radio service (DARS),** started in 2001 when XM Satellite Radio began broadcasting, and eight months later Sirius Satellite Radio initiated a similar service. Both companies had received permission to develop their services in 1997 when they paid the FCC more than $80 million for the frequencies.

satellite radio

One of the first things each did was gain partners among automobile manufacturers (for example, GM and Honda for XM and Ford and Chrysler for Sirius). They needed car companies to make satellite radio receivers an option for automobiles because satellite radio cannot be received on conventional radios. They wooed different car manufacturers because XM and Sirius broadcast on different frequencies and with different **encryption,** making it difficult to design a radio to receive both. However, because the programming comes from satellite, someone driving across the country can hear any channel continuously without it fading in and out. In addition to paying about $100 for the radio, subscribers pay about $13 a month for the programming.

Each service developed more than 100 channels with programming from established networks and programming specially developed by their companies, much of it free of advertising. Sirius captured "shock jock" Howard Stern, who switched from conventional radio in 2006, in part because the FCC was cracking down on indecency, and government agencies have no authority over satellite radio programming content. Shortly after Stern started on Sirius, XM announced it had struck a deal with Oprah Winfrey for a series called *Oprah and Friends.* The big-name stars and the quality and convenience of satellite radio garnered close to 15 million subscribers between the two companies, but that was not enough to turn a profit, so Sirius and XM came up with a plan to merge. After considerable controversy, the government approved the merger in 2008.[43]

HD radio

Conventional broadcasters are not standing idly by and allowing satellite radio to steal the digital thunder. They were the main ones opposing the merger, but they have their own digital system referred to as **HD radio**—sometimes called high-definition radio or **in-band on-channel (IBOC)** radio. What this means is that the digital signals can be heard on the same radios and at the same dial position as conventional analog radio signals. Technology developed by iBiquity squeezes the digital signal into the same frequencies that the station uses for AM or FM broadcasting. IBOC, which was officially launched in 2004, allows for better reception, higher fidelity, the possibility to add services such as a second channel or traffic information on demand, and the ability to pause and rewind music. People must buy a special radio for about $100 in order to hear HD radio, but then the programming is free. About 1,300 radio stations have added an HD service, and about 1 million sets have been sold. Unlike TV, a switch to digital radio has not been mandated by the government; it is purely voluntary. AM and FM broadcasting will continue in their present form regardless of the fate of HD radio.[44]

5.18 Issues and the Future

broadcaster opposition

The conventional terrestrial radio business has been consistently opposed to satellite radio because it could drain listeners and advertisers from them. Their opposition has not been particularly effective, however, and satellite radio has continued to grow. The merger of Sirius and XM may increase broadcasters' fears because it will make satellite radio even stronger. However, there are many issues to be resolved with the merger, not the least of which is the type of transmission and encoding that will be used, given that the two systems use different technologies.

other competition

Satellite radio is not the only competition conventional radio faces. Much of radio content is music, and there are many ways for people to obtain this music—without commercials. They can download it from the internet, buy CDs, or obtain it for their portable devices. In addition, all types of radio programming are available on the internet. Regular radio stations put their programming on the internet, so the amount they collect in advertising revenue for their internet service goes to the same company bottom line as what they collect for terrestrial broadcasting. But there are also internet-only radio stations that can take away audience (and dollars) from conventional radio (see Chapter 2).

consolidation

Many bemoan the consolidation of radio in the hands of a few companies. They point to a sameness of radio programming and the increasing difficulty that new music groups have in getting airplay because the big companies want to play it safe with what has worked in the past. They sense stagnation in radio programming. Clear Channel has been accused of "stomping on the competition, destroying artistic integrity, and making mush out of the little guy."[45] Yet its defenders point out that it owns less than 10 percent of radio stations.

A totally different issue that radio must deal with involves ethics and **indecency.** The violent lyrics of music and the sexual chatter of DJs and talk-show hosts frequently raise the ire of the FCC, Congress, and some members of the public. Other members of the public tune in to listen to this, however, allowing the stations to charge high advertising rates because the ratings are high. It is not only sex and violence that have gotten radio personalities into hot water. Don Imus's 2007 racial slur related to the Rutgers University women's basketball team cost him his job. Ethics and indecency are thorny subjects that are covered more thoroughly in Chapters 11 and 12.

indecency

Radio's history is full of cataclysmic changes. Its resilience in reinventing itself during the 1960s could serve as a model as it marches forward in the 21st century.

5.19 Summary

Radio has survived periods of experimentation, glory, and trauma. Early inventors, such as Maxwell, Hertz, Marconi, Fleming, and De Forest, would not recognize radio in its present form. Many people who knew radio during the 1930s and 1940s do not truly recognize it today. Radio has endured and along the way has chalked up an impressive list of great moments: picking up the *Titanic*'s distress calls, broadcasting the Harding-Cox presidential election returns, broadcasting the World War II newscasts of Edward R. Murrow, and surviving the television takeover.

Government interaction with radio illustrates the medium's growth as a broadcasting entity. Early laws dealt with radio primarily as a safety medium. The fact that government took control of radio during World War I but not during World War II indicates that radio had grown from a private communication medium to a very public one that most Americans relied on for information. The need for the government to step in to solve the problem of overcrowded airwaves during the late 1920s proved the popularity and prestige of radio. The ensuing Communications Act of 1934 and the various FCC regulations helped solidify the government's role in broadcasting. When Congress passed the Public Broadcasting Act of 1967, it set the groundwork for the reorganization of public radio. Government was involved in the development of FM and the initiation of satellite radio and HD radio. It also deals with issues in radio such as payola and indecency. The lessening of ownership regulations stipulated in the Telecommunications Act of 1996 is further evidence that radio has survived.

Companies from the private-enterprise sector have also played significant roles in the history of radio, starting with the Marconi company and progressing through the founding of RCA by AT&T, GE, and Westinghouse. These early companies contributed a great deal in terms of technology, programming, and finance. The networks and stations, each in their own peculiar way, set the scene for both healthy competition and elements of unhealthy intrigue. Intrigue also characterized the rivalry between newspapers and radio in the pre–World War II days, and free enterprise in its purest sense altered the format of radio when television stole its listeners. The formation of new networks such as Westwood One and the revitalization of networks and syndicators are further proof that radio has survived.

Radio programming is indebted to early pioneers who filled the airwaves with boxing matches, "potted palm" music, and call letters, and to *Amos 'n' Andy,* Jack Benny, and others who are remembered for creating the golden era of radio. Today countless DJs, talk-show hosts, and newscasters let us know that radio has survived.

Suggested Websites

www.clearchannel.com (site of Clear Channel, the company that owns the most radio stations)

www.mtr.org (the Paley Center for Media—formerly the Museum of Television and Radio—that has facilities in New York and Los Angeles with exhibits related to the history of radio)

www.otr.com/index.shtml (Old Time Radio, where you can listen to and download old radio shows)

www.sirius.com and www.xmradio.com (satellite radio services Sirius and XM, which will be combined into one service)

www.westwoodone.com (site for Westwood One radio networks)

Notes

1. Orrin E. Dunlap, Jr., *Radio's 100 Men of Science* (New York: Harper and Brothers, 1944), 65–68, 113–17.
2. Degna Marconi, *My Father Marconi* (New York: McGraw-Hill, 1962).
3. Dunlap, *Radio's 100 Men of Science,* 90–94; and Lee De Forest, *Father of Radio: The Autobiography of Lee De Forest* (Chicago: Wilcox and Follett, 1950).
4. Erik Barnouw, *A Tower in Babel: A History of Broadcasting in the United States to 1933* (New York: Oxford University Press, 1966), 77.
5. Ibid., 37–38, 47–56.
6. Ibid., 57–61.
7. There is some question as to whether this memo was actually written in 1915. Some sources say 1916; others say the ideas may not have come until 1920. Sarnoff himself said many times that he wrote the memo in 1915. The dispute over the memo is chronicled in Louise M. Benjamin, "In Search of the 'Radio Music Box' Memo," *Journal of Broadcasting & Electronic Media,* Summer 1993, 325–35; and Louise Benjamin, "In Search of the Sarnoff 'Radio Music Box' Memo: Nally's Reply," *Journal of Radio Studies,* June 2002, 95–106.
8. Yochi J. Dreazen, "Pittsburgh's KDKA Tells Story of How Radio Has Survived," *Wall Street Journal,* May 15, 2001, A-1.
9. Robert E. Summers and Harrison B. Summers, *Broadcasting and the Public* (Belmont, CA: Wadsworth, 1966), 34.
10. Barnouw, *A Tower in Babel,* 85, 86, 100.
11. Ibid., 180.

12. William P. Banning, *Commercial Broadcasting Pioneer: WEAF Experiment, 1922–1926* (Cambridge, MA: Harvard University Press, 1946), 150–55.

13. "The First 60 Years of NBC," *Broadcasting,* June 9, 1986, 49–64.

14. Robert Metz, *CBS: Reflections in a Bloodshot Eye* (Chicago: Playboy Press, 1975); and "Farewell to the Man in the CBS Eye," *Broadcasting,* November 5, 1990, 35–39.

15. "The Silver Has Turned to Gold," *Broadcasting,* February 13, 1978, 34–46.

16. "How Sweet It Was," *Broadcasting & Cable,* April 19, 1999, 74–76.

17. J. Wayne Rinks, "Higher Education in Radio 1922–1934," *Journal of Radio Studies,* December 2002, 303–16.

18. Louise Benjamin, "Working It Out Together: Radio Policy from Hoover to the Radio Act of 1927," *Journal of Broadcasting & Electronic Media,* Spring 1998, 221–36; and *Public Law No. 416,* June 19, 1934, 73rd Congress (Washington, DC: Government Printing Office, 1934), sect. 303.

19. John Witherspoon and Roselle Kovitz, *The History of Public Broadcasting* (Washington, DC: Current, 1987), 8–9.

20. Summers and Summers, *Broadcasting and the Public,* 45–50.

21. Charles J. Correll and Freeman F. Gosden, *All About Amos 'n' Andy* (New York: Rand McNally, 1929); and Bert Andrews and Ahrgus Julliard, *Holy Mackerel: The Amos 'n' Andy Story* (New York: Dutton, 1986).

22. For more information on radio programming, see Michele Hilmes, *Radio Voices: American Broadcasting 1922–1952* (Minneapolis: University of Minnesota Press, 1997); and Vincent Terrace, *Radio's Golden Years: The Encyclopedia of Radio Programs 1930–1960* (San Diego: A.S. Barnes, 1981).

23. "Genesis of Radio News: The Press-Radio War," *Broadcasting,* January 5, 1976, 95.

24. Paul White, *News on the Air* (New York: Harcourt, Brace, 1947).

25. Eric Barnouw, *The Golden Web: A History of Broadcasting in the United States 1933–1953* (New York: Oxford University Press, 1968), 46.

26. Summers and Summers, *Broadcasting and the Public,* 46.

27. Ibid., 70; and *Broadcasting Yearbook,* 1975 (Washington, DC: Broadcasting Publications, 1975), C-289.

28. Ward L. Quaal and James A. Brown, *Broadcast Management* (New York: Hastings House, 1976), 292.

29. Michael C. Keith, *Radio Programming* (Boston: Focal Press, 1987), 3.

30. For more insight into the rise of DJs, see Arnold Passman, *The DJs* (New York: Macmillan, 1971).

31. "Gov't Gets 1st Payola Conviction," *Daily Variety,* May 23, 1989, 1; "Record Label Exec Agrees to Plead Guilty to Payola," *Los Angeles Times,* July 1, 1999, 1; "FCC Launches Payola Probes of 4 Radio Giants," *Los Angeles Times,* April 20, 2006, 1.

32. "Armstrong's Legacy," *TV Technology,* October 17, 2001, 14–15.

33. Donald N. Wood and Donald G. Wylie, *Educational Telecommunications* (Belmont, CA: Wadsworth, 1997).

34. "AM Radio Plays Hard to Be Heard Again As Audience Dwindles," *Insight,* July 4, 1988, 44.

35. Bruce Klopfenstein and David Sedman, "Technical Standards and the Marketplace: The Case of AM Stereo," *Journal of Broadcasting & Electronic Media,* Spring 1990, 171–94; and "AM Stereo Rears Its Divided Head," *Broadcasting,* April 12, 1993, 61.

36. J. W. Mitchell, *Listener Supported: The Culture and History of Public Radio* (Westport, CT: Praeger, 2005); and "New Name, Focus for American Public Radio," *Daily Variety,* December 20, 1993, 52.

37. Jim McCluskey, "An Examination of New and Existing FCC Policies and Procedures Affecting Student/Noncommercial Radio Stations," *Feedback,* Fall 1998, 32–36; and "These DJs Are Well-Schooled," *Los Angeles Times,* May 22, 2004, E-19.

38. "New Radio Networks Spring Up," *New York Times,* January 11, 1980, III-6.

39. "Westwood One to Buy NBC Radio Networks for $50 Mil," *Daily Variety,* July 21, 1987, 1; and "Citadel Jumps to Top of Radio Dial," *Hollywood Reporter,* February 7–13, 2006, 12.

40. Wenmouth Williams, "The Impact of Ownership Rules and the Telecommunications Act of 1996 on a Small Radio Market," *Journal of Radio Studies,* Summer 1998, 8–18; and "'96 Act Killed Radio's Star," *Hollywood Reporter,* December 14, 2006, 6.

41. "Radio Rides Hispanic Population Boom," *Broadcasting & Cable,* October 6, 1997, 46; and "Clear Channel: Spanish Revolution," *Hollywood Reporter,* September 17–19, 2004, 8.
42. Jack Kay, George W. Ziegelmueller, and Kevin M. Minch, "From Coughlin to Contemporary Talk Radio: Fallacies and Propaganda in American Populist Radio," *Journal of Radio Studies,* Winter 1998, 9–21; and Stephen Earl Bennett, "Americans Exposure to Political Talk Radio and Their Knowledge of Public Affairs," *Journal of Broadcasting & Electronic Media,* March 2002, 72–86.
43. "Voices from the Heavens," *Los Angeles Times,* November 1, 2001, T-1; "Stern Has Sirius Intentions," *Hollywood Reporter,* October 7, 2004, 1; and "FCC Staff Back XM-Sirius Merger," *Wall Street Journal,* June 16, 2008, B1.
44. "iBiquity Signals Digital Shift for Terrestrial Radio," *Hollywood Reporter,* April 9–15, 2002, 1; and "New Wave of Radio Choices," *Los Angeles Times,* May 27, 2007, C-3.
45. Christine Y. Chen, "The Bad Boys of Radio," *Fortune,* March 3, 2003, 119–22.

Chapter 6

MOVIES

H istorically, movies have had the most captive and attentive audience of all the media. When people go to a movie theater, they are away from distractions such as email and family chores. Because people pay to see a movie in a theater, they usually watch the entire show. The dark theater and large screen can be totally absorbing, allowing the viewers to leave the "real world" and enter a fantasy world where they can experience events vicariously.

Today, many people watch movies on a television or computer screen, either by tuning into a broadcast or cable network, by downloading from an internet site, or by buying or renting DVDs. This chapter deals with movies primarily in the context of a movie theater environment, but also considers the impact of the newer distribution forms on the traditional movie theater.[1]

Things you would never know if it weren't for the movies:

Once applied, lipstick will never come off, even when swimming under water.

All grocery bags contain at least one stick of French bread.

When staying in a haunted house, women should investigate any strange noises in their most revealing underwear.

It is always possible to park directly outside the building you are visiting.

A man will show no pain while taking the most ferocious beating but will wince when a woman tries to clean his wounds.

When a car is being driven down a dark road at night, the light under the dashboard will always be lit.

Anonymous Internet Humor

6.1 Early Developments

stroboscopic toys

The motion picture business grew out of photography and stroboscopic toys. One of these toys, the 1825 Thaumatrope, was a flat disk with the picture of a parrot on one side and the picture of a cage on the other. When someone spun the toy, the parrot appeared to be in the cage. This demonstrated a phenomenon called **persistence of vision** in which the human eye retains images for short periods. That is why film, which is actually a series of projected still images, appears as moving pictures.[2]

Stanford bet

In 1877, Leland Stanford, the governor of California, had a $25,000 bet with a friend that when a horse gallops, all four of its feet are off the ground at the same time. He hired Eadweard Muybridge, who was experimenting with continuous motion photography, to prove his theory. Muybridge rigged 24 cameras that were triggered in synchronized fashion by string stretched across a racetrack. The photos showed that Stanford was right (see Exhibit 6.1), and he won the bet—although the experiment cost him $40,000.[3]

Edison, Dickson, and Eastman

The major developments in early film in the United States were undertaken in the 1880s by Thomas Edison, with his assistant W. K. L. Dickson, and George Eastman. Eastman had undertaken experiments with rolls of celluloid film that he intended for still cameras, but the flexible film lent itself well to what eventually became movies. Edison was looking for a way to have pictures accompany his phonograph and was persuaded by Dickson to use celluloid film that they purchased from Eastman. So, in a way, the first movie concept was a music video. Dickson went on to invent an electrically powered camera and a battery-powered

Exhibit 6.1

Some of the photos taken by Eadweard Muybridge to enable Leland Stanford to prove that a horse has all four feet off the ground while galloping.

(© Corbis)

viewer. Edison thought that images should be projected to one person at a time rather than to a large audience because the image would be clearer and because he believed he could earn more money this way. So the viewer that Dickson invented was a peephole machine called a **Kinetoscope** (see Exhibit 6.2). In the 1890s, Kinetoscope parlors appeared around the country; for a few pennies, people could watch a movie that was about 20 seconds long.[4]

Dickson and Edison had abandoned sound for their movies by this time, but Edison still thought his phonograph was a much more important invention than anything related to movies. They built a small studio to shoot movies for the parlors. Its exterior was tar paper to keep out unwanted light, and it soon became known as the Black Maria (see Exhibit 6.3). The roof opened to let in sunlight, and the whole building rotated so it could catch the light as the day progressed. The camera in the room could move, but Dickson never changed its position when shooting any one film.

Projection and portability were developed by others, primarily the Lumière brothers (Auguste and Louis) of France. They created one of the first functional projectors, and in 1895, the first movie theater open to the paying public showed movies in the basement of a Paris café. The Lumières also utilized a smaller hand-cranked camera that they took outside to film everyday events.[5]

Exhibit 6.2

A Kinetoscope. People viewed through the peephole on the top.

(Photo by Steve Gainee, ASC, Courtesy of American Society of Cinematographers)

6.2 The First Movies

The first known Edison-Dickson movie is *Fred Ott's Sneeze* (1894), a short scene of one of Edison's mechanics sneezing. Other movies consisted of jugglers, animal acts, and dancers who paraded through the Black Maria. The Lumière movies showed factory workers leaving work, a baby eating, and a train pulling into a station. The train had people in the theater ducking as it came toward them on the screen. One Lumière film was a precursor to slapstick comedy. In it a boy is standing on a hose. A man picks up the hose, the boy steps off the hose, and the man's face is pelted with water. Projection caught on in the United States, too, and an 1895 Edison-Dickson film called *The Kiss* caused quite a stir. As the name implies, it showed two people kissing, something that wasn't publicly accepted in that era.[6]

The very early movies were shot with no camera movement and no editing. Gradually more production values crept in. Frenchman George Méliès, who was a magician, experimented with special effects such as split screens, double exposures, and stop motion. In 1902, he employed special effects in the name of "science fiction" for his best-known

Black Maria

Lumière brothers

Exhibit 6.3

The Black Maria was covered with tar paper and rotated to follow the sun.

Exhibit 6.4

A scene from Edwin S. Porter's *The Great Train Robbery.*

Méliès and Porter

Biograph and Vitagraph

MPPC

movie, *A Trip to the Moon,* based on a Jules Verne novel. Edwin S. Porter, who worked as a director for Edison, extended the art in films such as *Life of an American Fireman* (1903) and *The Great Train Robbery* (1903) (see Exhibit 6.4). Among his techniques were allowing for a break in time, cutting back and forth from one scene to another, and panning the camera. He also used some medium close-ups, such as a shot of an outlaw pointing a gun at the audience that made viewers jump in their seats.[7]

6.3 Studio Beginnings

Two other companies joined the moviemaking ranks—Biograph and Vitagraph. Biograph was formed by W. K. L. Dickson, who left Edison and became his rival. Because Edison (as Dickson's former employer) held the patents on Dickson's inventions, Dickson really had to invent the motion picture camera twice.[8]

Biograph, Vitagraph, and Edison competed fiercely. For example, in 1899 Biograph set up a large number of lights to film the Jeffries-Sharkey boxing match at Coney Island. When the Biograph people discovered a Vitagraph camera filming the fight several rows back, they sent detectives to confiscate Vitagraph's equipment. A fight ensued that many people thought was better than the fight in the ring. Eventually Vitagraph succeeded in recording the fight and took the film back to its lab. The next morning the Vitagraph people found that the film had been stolen during the night by folks from Edison.[9]

In 1909, because of all the bad blood and lawsuits among the film companies, nine New York companies, led by Edison, decided to form the Motion Picture Patents Company (often referred to as the Trust or the MPPC). The nine companies were to share patent rights and also keep all other companies from entering the film production business. To accomplish this, they agreed not to sell films to distributors who bought from anyone else, and they also arranged with George Eastman's company, Kodak, to sell film only to these nine companies.[10]

Distributors had become important because movie theaters were springing up around the country. Most of these were storefronts with a few chairs and a screen, but some were plush and accompanied the silent films, now often a half-hour long, with piano music performed live as the movie was rolling. The theaters originally charged customers a nickel to view a film, so they became known as **nickelodeons.** Most of the early film viewers were working-class people, but the wealthy joined in when the theaters became fancier.[11]

The MPPC didn't keep its lock on film forever. Filmmakers who didn't knuckle under to the Trust became independent producers and distributors, and some of them moved to faraway California to escape the eye of the New York–based companies. They could also purchase film cheaply, some of it manufactured in Mexico.

William Fox started what became 20th Century Fox, and Carl Laemmle formed the company that later became Universal.[12]

6.4 Griffith and His Contemporaries

David Wark (D. W.) Griffith, a would-be actor, became a talented director who worked briefly for Edison and then switched to Biograph. He experimented with many aspects of filmmaking, such as lighting subjects from below, moving the camera in long tracking shots, and creating flashbacks. He also used close-ups and medium shots, which most previous directors had avoided, thinking that people wouldn't pay to see only part of an actor. Many of his movies contained last-minute rescue scenes, which allowed him to employ **parallel editing** in which he cut back and forth between the victim and the rescuer. Griffith regularly cast the same talented actors—Lionel Barrymore, Lillian Gish, Mary Pickford—and he worked with a creative cameraman, "Billy" Bitzero. (See Exhibit 6.5.)

Eventually Griffith left Biograph because his movies became too long and expensive for Biograph's tastes. In the silent film era, the director could talk to the actors as they were acting, and the owners of Biograph became upset when Griffith wanted to spend time and money on rehearsals. Griffith went to an independent company, Mutual, where he made a deal that allowed him to make one movie a year of his choice in addition to the movies Mutual assigned him.

In 1915, Griffith, a Southerner by birth, released *The Birth of a Nation,* which cost $115,000. It was a historical film about the Civil War in which the heroes were Ku Klux Klan members. The movie advanced filmmaking in that it had huge sets, battle scenes, many rehearsals, and dynamic editing. The racist subject matter, however, caused riots throughout the country. Although the movie was banned in a number of cities, it made a great deal of money and could be considered the first blockbuster. Griffith's next film, *Intolerance* (1916), cost $2 million but was not successful. It intercut four stories and four last-minute rescues that occurred in different centuries, something that was too difficult for audiences to follow. After that, Griffith's Victorian sensibilities seemed to be out of touch with the Roaring Twenties, and his movies did not generate large audiences.[13]

During this period, Mack Sennett and Charlie Chaplin experimented with comedy. Although they worked together for a while, they did not get along because they had different views of comedy. Sennett preferred sight gags such as people bumping into things or falling into swimming pools. Frequently he spoofed Griffith's last-minute rescues. Chaplin preferred more character-driven comedy, albeit with a few fine sight gags. He developed and acted the character of the Little Tramp, an immigrant worker who was at the fringe of society but who aspired to be with the rich and powerful. The character allowed Chaplin, who mainly improvised rather than working from a completed script, to show both the good and bad aspects of wealth as well as his gifted abilities with comedic pantomime. Chaplin knew both wealth and poverty, having been at one point in a home for destitute children and at another point the first actor to sign a million-dollar contract (see "Zoom In" box).[14]

Exhibit 6.5
D. W. Griffith directing a silent movie.

techniques

The Birth of a Nation

Chaplin and Sennett

ZOOM IN: **Accusations and Applause**

Charlie Chaplin was one of the actors who became caught up in the blacklisting of the 1950s (see Chapter 3). The accusations that he was a Communist stemmed in part from the fact that he was a foreigner, having been born in England in 1889, where he became a popular child actor before coming to the United States to pursue his career in 1910. The FBI claimed that Chaplin was sympathetic to leftist beliefs and put together a document of more than 1,900 pages against him. Accusations included that he had given a reception for a prominent labor leader and that he was often the guest of a millionaire whose son became a Communist. When Chaplin was questioned about being a Communist, he had no concrete answer except that he didn't want to create any revolution; he just wanted to create a few more films. He was quoted as saying "I am a citizen of the world," and he participated in War Bond drives for the allies. The FBI finally admitted in 1948 that, despite its 1,900-page document, it had no evidence that Chaplin was a Communist.

All the accusations, however, affected Chaplin and his career. He stopped impromptu performances and often stayed away from the studio, lonely and depressed. In 1952, with McCarthyism in bloom, Chaplin took a trip back to his native England. The FBI, under J. Edgar Hoover, planned to bar his reentry through U.S. customs, despite its lack of evidence. Rather than face the situation, Chaplin made his home in Switzerland with his wife Oona (daughter of playwright Eugene O'Neill) and their eventual eight children.

In 1972, the Academy of Motion Picture Arts and Sciences gave Chaplin an honorary award for "the incalculable effect he has had in making motion pictures the art form of this century." At the Academy Awards ceremony, he received an unprecedented five-minute standing ovation. To view Chaplin receiving his award, go to www.youtube.com/watch?v=J3Pl-qvA1X8.

Charlie Chaplin.

(VPPA/IPOL/Globe Photos)

6.5 World War I Developments

The American film industry became predominant during World War I because the war in Europe made film production difficult overseas. The industry also became more prestigious in America because film was used in the war effort. Under the government Committee on Public Information, George Creel set up a Division of Films and used motion pictures for public information, legitimizing the medium.[15]

The New York film production power base deteriorated during this period. In 1917, the MPPC was outlawed by the courts, but it was already losing its dominant

Creel

film production role to the independents, which preferred to produce long films to be shown in comfortable theaters rather than short films for nickelodeons. These longer films became quite popular with the public, but the Trust thought they were just a passing fad.

The indies also allowed the stars more personal recognition, whereas the Trust rarely gave screen credits. The stars appreciated this new approach, so the "Biograph Girl" left Biograph and went to an independent where she could be known by her real name, Florence Lawrence. "Little Mary" became Mary Pickford.

indies

6.6 Hollywood during the Roaring Twenties

As the stars assumed their own personas, the public became interested in their personal lives. The era of the Roaring Twenties was not the height of morality, and Hollywood exaggerated the characteristics of the time. Although many movies had moralistic, sentimental plots, others hinted at lust and promoted materialism. The press highlighted stories of drugs, alcohol abuse, and divorce among Hollywood luminaries. In 1920, Mary Pickford, who was then "America's Sweetheart," went to Las Vegas and divorced her husband, Owen Moore. Three weeks later she married Douglas Fairbanks (also divorced), who played gallant, pure, exuberant roles. This marriage shocked (and captivated) the nation (see Exhibit 6.6). The public interest proved that movies had left the novelty phase and become an important social force.[16]

Pickford and Fairbanks

Hays Office

In 1922, partly in reaction to stories of sex and sin in Hollywood, the industry formed the Motion Picture Producers and Distributors of America (MPPDA). It was headed by Will Hays, a Presbyterian elder and ex–postmaster general, who took on the charge of cleaning up Hollywood movies and Hollywood's image. Hays headed the agency for 23 years, so it was often referred to as the Hays Office. His efforts were only partially successful during the 1920s, but in 1930, pressured by religious leaders and the threat of government censorship, the movie business adopted the Motion Picture Production Code to which producers voluntarily submitted movies. It contained many provisions, but mainly it stated there were to be no sexual acts or innuendos in movies and bad characters were to be punished.[17]

Exhibit 6.6

Mary Pickford and Douglas Fairbanks on the deck of a ship during their honeymoon.

(© Bettmann/Corbis)

Zukor

Movie companies proliferated during the 1920s. Adolph Zukor's Paramount Pictures became very powerful. Zukor acquired a number of motion picture production companies and set up a distribution arm for the films his combined company produced. He established the practice of **block booking,** in which a movie theater had to buy lesser-quality Paramount films to be able to show the more popular features. This angered theater owners, but Zukor had the might to enforce it. Eventually he purchased theater chains so that it was easy for Paramount to produce, distribute, and exhibit all its movies. In a similar fashion, Marcus Loew established Metro-Goldwyn-Mayer by buying several film production companies, and he also owned the Loews theaters. This **vertical integration** of companies that produced films, distributed them, and then exhibited them in theaters grew during the 1920s. One variation on the theme was United Artists, formed in 1919 by three of the most important actors (Douglas Fairbanks, Mary Pickford, and Charlie Chaplin) and the most important director (D. W. Griffith). Each principal in the company produced his or her own films, and United Artists distributed them.[18]

vertical
integration

Because most movie producers were in California, they took advantage of the landscape and the sun and started making westerns; John Ford became a particularly well-known director of this genre. Comedy continued, too, with Harold Lloyd, Buster Keaton, and Laurel and Hardy as headliners. A few documentaries were produced in the 1920s, most notably Robert Flaherty's *Nanook of the North* (1922), a feature-length story of an Eskimo family.[19]

genres

theaters

The theaters themselves became larger and plusher, with ushers and ornate decorations that catered to rich and poor alike. Two of the most elaborate were Radio City Music Hall in New York and Grauman's Theater in Los Angeles, both of which hired orchestras to play music to accompany the silent films. Other theaters employed piano or organ players, and some even had people backstage who spoke words of dialogue. But that would soon change.

6.7 Sound

As mentioned previously, Edison's early idea was that movies would have sound—or, more accurately, that sound would have pictures. Others, including Lee De Forest, experimented with sound. In fact, it was De Forest's **audion tube** used for radio (see Chapter 5) that allowed sound to be amplified for movies. Before synchronized talk occurred, some films had synchronized music to fit the action. Some newsreels shown in theaters of the time had synchronized sound; for example, Fox Movietone showed George Bernard Shaw talking.

The Jazz Singer

But Warner Bros.' *The Jazz Singer* (see Exhibit 6.7), which opened on October 6, 1927, was the first full-length movie to use sound in such a way that it was part of the narrative story. Part of the movie was silent, and part was sound. The sound parts were very static visually—for example, a static shot of Al Jolson in blackface singing "Mammy" in a theater. But Jolson exuded warmth and style when he talked and sang, and audiences loved it.

conversion
to sound

Much of the film community pooh-poohed sound, considering it to be a passing fad, but the public demanded it. By 1929, most movies were sound. The conversion to sound was difficult technically. Cameras had to be placed in an isolation booth so that their noise would not be recorded. Microphones were hidden on sets, and actors couldn't move far away from them or the sound would not be picked up. Eventually, someone thought of putting a microphone on a pole above the actors and having a technician move it as the actors moved. Also, cases built for cameras silenced them. As technology improved, production values returned. Theaters had to convert from silent to sound, but with audience members flocking to sound movies, it behooved theaters to convert quickly.

It was more difficult for many of the actors and actresses who were silent stars. Some could not make the conversion to sound because of their voice quality. Talking movie scripts were less visual and less slapstick. New stars such as Clark Gable, Katharine Hepburn, Humphrey Bogart, Claudette Colbert, John Wayne, and Bette Davis captured the nuances needed for sound and replaced the older stars.[20]

6.8 The "Golden Years" of Moviemaking

The 1930s and 1940s (roughly from the beginning of sound to the beginning of television) were powerful years in the moviemaking business and are sometimes referred to as the "Golden Years." Part of this time is also often called the **studio years.** Major Hollywood studios (MGM, Paramount, 20th Century Fox, Warner, RKO, Universal, Columbia) mass-produced movies in almost assembly-line fashion. Directors were assigned scripts, sets, actors, and crew. Actors were under contract to individual studios, and the studios prepared materials that would utilize their talents and make them more valuable to the bottom line. Crew members had very specific jobs that they undertook from one movie to another. Lighting was created to make the stars look good, not to aid the drama of the movie. But despite all these restrictions, craftsmanship emerged, and many of the movies of this period are still considered classics.[21]

Another element that had an effect on moviemaking was the MPPDA (Hays Office) Code. As mentioned previously, a voluntary code had been established in 1930. It might not have been enforced except that in 1933 the Catholic Legion of Decency threatened to have Catholics boycott movies. Although banned by the voluntary code, sexual innuendos still appeared, but now they were verbal as well as visual. Mae West (see Exhibit 6.8), in particular, used her throaty voice and undulating hips to sing songs with titles such as "I Like a Guy Who Takes His Time" and to make comments such as "Are you packing a rod or are you just glad to see me?"

the Code

Exhibit 6.7
The story of *The Jazz Singer* was such that Jolson was singing "Mammy" as a stage actor, but within the movie, he was singing for his mother and his girlfriend.

(Hulton Archive/Getty Images)

Exhibit 6.8

Mae West.

(Supplied by Globe Photos, Inc.)

box office health

genres

The Catholic leadership was strong enough to influence many of the MPPDA provisions, such as one that stipulated that a producer who released a movie without getting approval through the MPPDA would be fined $25,000. To obtain approval, the movie had to pass the code's moral restrictions. Sexual language, promiscuity, and "unnatural" sex were not permitted, but even hints of sex between married couples was frowned upon. Long-wedded Nick and Nora Charles of *The Thin Man* (1934) each had a separate twin bed. Many words were prohibited in scripts—not only *damn* and *hell,* but also words such as *guts* and *louse,* which were considered lacking in gentility. Anything illegal could not be shown as pleasant or profitable, so gangsters no longer lived fancy lives before meeting their comeuppance. Mae West's career plummeted.[22]

This led to movies that were basically escapist fare and showed life in a good light, even though it was the Great Depression. Audiences enjoyed movies as an inexpensive form of entertainment that carried them away from their day-to-day troubles. Theaters started showing double features during the 1930s because this sounded like a better deal to moviegoers. Usually at least one of the features was a **B picture** (a movie that was relatively inexpensive to produce because it didn't have big-name stars or expensive sets), but people still flocked to see them. The box office was healthy during most of the Depression and was particularly healthy right after World War II. In 1946, an average of more than 90 million tickets a week were sold, the highest of any year in film history. Today's figure is about 26 million.[23]

Films of various genres were popular during the 1930s and 1940s. Westerns and comedies continued to be popular. The former frequently accentuated the values of the common folk and criticized people in power positions. Charlie Chaplin continued to produce, direct, and act in comedies after sound entered the scene, but he never really developed a talking character. The Marx Brothers performed crazy, unpredictable antics that were action for action's sake. Frank Capra directed movies about life in America such as *You Can't Take It With You* (1938), *Mr. Smith Goes to Washington* (1939), and *It's a Wonderful Life* (1946), which again glorified simple people struggling against those with power and money.

Now that song and dance could be synchronized, musicals became a well-known genre. MGM specialized in them, and its director/choreographer, Busby Berkeley, cranked them out (see Exhibit 6.9). The early musicals were mostly filmed versions of Broadway plays; later ones tended to be "backstage" stories of life in the theater business. They were thin on plot but contained dazzling musical numbers with complicated choreography and varied camera angles. There were musicals for children (often starring Shirley Temple), musicals on ice, and musicals under water, among others.

Exhibit 6.9

An elaborate musical number from Busby Berkeley's *Golddiggers of 1933.*

(Photofest)

Gangster films, albeit with gangsters who fit the provisions of the code, became popular during these years. Some, such as *The Front Page* (1931), featured newspaper reporters. Disney produced animated family movies, first short films such as *The Three Little Pigs* (1933) and then full-length features, the first of which was *Snow White and the Seven Dwarfs* (1937). Patriotic war movies were produced during World War II. Some of these were documentaries directed by major feature-film directors, such as John Ford's *The Battle of Midway* (1942) and Frank Capra's *Why We Fight* series (1942). Feature war movies often had a documentary feel to them. Also during the war, "film noir" became a popular genre, with its dark, conflicted heroes and heroines. One of the classic films of this genre was *Casablanca*.[24]

6.9 Hitchcock and Welles

Two of the most influential directors of the Golden Years were Alfred Hitchcock and Orson Welles. Hitchcock was one of a number of directors who started in England and moved to the United States. His movies—with such well-known actors as Jimmy Stewart, Cary Grant, Ingrid Bergman, and Grace Kelly—include alarming crimes that occur in public places and humor in the face of impending doom. Tiny details reveal clues, and normal and abnormal behaviors are separated by a very thin line. He was a master at using editing to increase the sense of horror without actually showing gory events, such as in the shower scene from *Psycho* (1960) (see Exhibit 6.10). Some of his other well-known films include *Spellbound* (1945), *Strangers on a Train* (1951), *Rear Window* (1954), and *Vertigo* (1957).[25]

Hitchcock

Welles

Orson Welles did not produce a large body of work as Hitchcock did. His most famous film is *Citizen Kane,* released in 1941, which is based on the life of William Randolph Hearst. The Hearst family was less than pleased with his portrayal of the lead character, and the movie did not have a Hollywood "happily-ever-after" ending. It didn't do very well at the box office, but it did extend the art of filmmaking. The movie used **deep focus** shots to show alienation and **low angle** shots to accentuate abuse of power. It included a "fake" newsreel, and it combined animation and live footage. Welles used newly developed incandescent lights to accent depth and shadow. With his background in radio drama (see Chapter 5), he also advanced sound with echoes, well-placed sound effects, and spatial cues. The film has garnered critical acclaim and was named the greatest film in movie history by the American Film Institute (see Exhibit 6.11).[26]

Exhibit 6.10

In the shower scene in *Psycho,* Hitchcock cut together 72 short shots without actually showing the brutal slaying, but the blood trickling down the drain at the end lets the audience know exactly what has happened.

(Supplied by Globe Photos, Inc.)

6.10 Color

Some of the early films had been delicately hand colored frame by frame, and some were tinted to convey a particular mood, such as red for passion and blue for night. In 1917, Technicolor was founded and established a monopoly over experiments regarding color. By the 1920s, it

AFI's List of the Greatest Movies of All Time

1. *Citizen Kane* (1941)
2. *Casablanca* (1942)
3. *The Godfather* (1972)
4. *Gone with the Wind* (1939)
5. *Lawrence of Arabia* (1962)
6. *The Wizard of Oz* (1939)
7. *The Graduate* (1967)
8. *On the Waterfront* (1954)
9. *Schindler's List* (1993)
10. *Singin' in the Rain* (1952)

Exhibit 6.11

The American Film Institute nominated 400 movies to be on a list of the all-time best movies and then asked 1,500 members of the film community to select the 100 best. The criteria were historical significance, critical recognition and awards, and popularity. This list shows the ones that made the top 10.

stable color

star costs

Exhibit 6.12

The 1939 film *Gone with the Wind*, starring Vivien Leigh and Clark Gable, was fourth on the AFI all-time best movie list.

(ADH/Globe Photos, Inc.)

had developed a two-color process that involved blue-green and red-orange. This was adequate for clothing and sets, but it made people look a strange orange color.

In 1933, Technicolor introduced a three-color process that used three lenses and three separate rolls of film (one sensitive to cyan, another to magenta, and the third to yellow) that were joined when the film was processed. Shooting with this color process was tedious and expensive, so it was reserved for very special films such as the 1939 classics *The Wizard of Oz* and *Gone with the Wind* (see Exhibit 6.12). Later the three rolls of film were joined during shooting, but the color faded quickly. It wasn't until the late 1960s, when a stable color system was developed, that shooting in color became the norm rather than the exception.[27]

6.11 Hard Times

When World War II ended in 1945, the movie business was booming. As previously mentioned, 1946 was the all-time high year for ticket sales. By 1953, the weekly attendance rate had dropped 75 percent from what it had been immediately after the war. The movie industry was in a slump that lasted until about 1963. Several factors caused this downturn.[28]

The creative people within the movie industry didn't help. Stars and directors, in particular, knew their value to the studio and began demanding higher pay. This increased the costs of moviemaking beyond what the studios could bear, and eventually they began letting the stars' contracts expire. This led to a major change in the studio system because talented people were now free to work for anyone, not just the studio that employed them.

In addition, the movie industry suffered a legal setback in 1948. For years the major studios engaged in vertical integration, producing movies, distributing them, and exhibiting them in theaters that they owned. This helped the studios to become very profitable because they could guarantee that their movies would play, even if they were less than wonderful. In 1938, the Justice Department began looking into this practice, using a test case of *United States v. Paramount*. In 1948, after 10 years of litigation, the Supreme Court agreed with the

Justice Department and outlawed vertical integration. The studios had to give up either production, distribution, or exhibition. Most chose exhibition and sold their theaters. This changed the economic model of the studio system and contributed to its collapse. Now the movie theater owners decided what films would be shown. This increased the risk in producing movies because they were no longer guaranteed a box office. B pictures disappeared, and eventually so did the double feature.[29]

vertical integration outlawed

In the midst of coping with the changes brought about by *United States v. Paramount,* the movie industry lost some of its best talent because of the **blacklisting** of the McCarthy era (see Chapter 3). In 1947, the House Un-American Activities Committee called 47 screenwriters, actors, and directors to Washington to answer accusations about leftist leanings. Ten of the witnesses, referred to as the **Hollywood 10,** refused to answer what, in retrospect, were inane, insulting questions, and were jailed for contempt. For the most part, their careers were ruined because the studios, fearful that the investigations would extend to them, did not hire those who had been tainted by the accusations. It has been estimated that more than 200 people, the majority of them writers, suddenly could not find work despite previously successful careers.[30]

blacklisting

Another culprit was television. People could stay in their homes and be entertained—for free. As the postwar population moved from the cities to the suburbs, they preferred this stay-at-home mode to driving into the central city where the movie theaters were located. At first the movie industry reacted very defensively to television. It wouldn't allow TV sets to be shown in films, and studios tried to prevent the actors and actresses they had under contract from appearing on TV. Eventually they learned "if you can't fight them, join them," and the studios found ways to make money from video enterprises. But that was a long time coming.[31]

television

The movie industry's initial response to the curse of television was technological. It incorporated large screens and special effects that were beyond the capability of the home and the television set (see "Zoom In" box). Scary **three-dimensional (3-D)** movies, such as *Creature from the Black Lagoon* (1954), graced a few screens, but the novelty soon wore off. Filming in 3-D used two cameras placed slightly apart. To show the film, two projectors threw the images on the screen and audience members wore special glasses with red and blue lenses that combined the two images to make them look like one three-dimensional shot.

technological attempts

Cinerama was popular for a bit longer, but it, too, was not successful at getting people to leave their couches to enjoy a theater experience. Cinerama consisted of three interlocking projectors and a wraparound screen that made the movie extend to viewers' peripheral vision (see Exhibit 6.13). This was accompanied by six-track stereo sound. The first Cinerama movie, *This Is Cinerama,* was shown in 1952; Stanley Kubrick's *2001* was the last in 1968, but Cinerama was essentially dead before that.

Another 1950s idea was **CinemaScope.** A special **anamorphic** lens squeezed a wide picture onto a frame of film; then an anamorphic lens on the projector unsqueezed it and showed it on a curved screen. This was cheaper than 3-D or Cinerama because it did not involve multiple cameras and projectors, but theaters wishing to show CinemaScope movies had to invest in anamorphic

ZOOM IN: **Grandmaster of Gimmicks**

Many gimmicks were used to bring people into theaters during the 1950s and early 1960s, but some of the best came from producer-director William Castle. Born in New York in 1914, as a teen Castle worked (often by lying his way in) in just about every aspect of Broadway theater—lighting, makeup, set building, directing, writing, acting—before coming to Hollywood. He made use of his theatrical experience to add fun to the moviegoing experience. He was the ultimate showman and a great pitchman. Most of his movies were B picture thrillers with spooky premises and hyped advertising, but moviegoers were happy to suspend belief and fall for his spiels. Here are some of his movies and their gimmicks.

- *House on Haunted Hill* (1959): During points in the movie, inflatable plastic skeletons attached to a guy wire flew over the audience.
- *The Tingler* (1959): Some of the seats in the movie theater were wired with a motor that gave the seat occupant a

tingling sensation. The premise of the movie was that the tingler was a creature that attached to the spines of people who were really scared. One of the movie characters was trying to catch the tingler, and at an appropriate point in the movie, the seat motors activated. The people in the selected seats reacted, of course, often jumping out of their chairs screaming. The other moviegoers were left to wonder what was going on or to laugh at their friends who were being "attacked."

- *13 Ghosts* (1960): As publicity, Castle had floats drive along Hollywood Boulevard with "ghosts" promoting the movie. Moviegoers were also given two pieces of colored cellophane. They could look through one if they wanted to see the ghosts or look through the other if they were "chicken" and didn't want to see the ghosts.
- *Mr. Sardonicus* (1961): Before entering the theater, people were given glow-in-the-dark pieces of paper that indicated thumbs up or thumbs down. During the movie, there was a Punishment Poll in which people held up their papers to indicate whether they thought the villain should be punished. It really didn't matter how the audience voted; it was always thumbs down because there was only one ending for the movie.
- *Zotz* (1962): Castle distributed magic Zotz coins—which did absolutely nothing.
- *Macabre* (1958): The advertising read, "If it frightens you to death, you'll be buried free of charge." There was also a "Fright Break" during the movie so if you were really scared you could go sit in the "Cowards Corner." Yeah, sure!

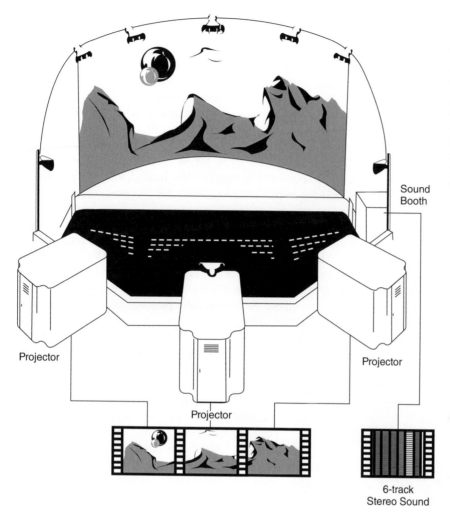

Exhibit 6.13
How Cinerama
worked. Each
projector threw a
different image onto
part of the screen,
and the overall
image melded
together. The
speakers allowed
for six-track stereo
sound.

Sound
Booth

Projector

Projector

Projector

6-track
Stereo Sound

lenses, curved screens, and stereo sound. Eventually the wide-screen concept caught on, but mainly by shooting on 70 mm film instead of 35 mm.[32]

6.12 The Road Back

Although the number of movie tickets sold per week has never surpassed the 1946 high, the movie business started to improve gradually during the 1960s. A significant factor was a change in the basic structure of the industry that finally led to profitability. The combination of low movie attendance, abolishment of vertical integration, and high costs of production led to a downgrading of power for the major studios. They could no longer afford to keep high-maintenance actors and directors on contract, and they could not ensure that the movies they produced would actually be shown in theaters. As the studio

**independent
producers**

system crumbled, **independent** production companies rose, slowly but steadily, to fill the vacuum.

These independent companies did not keep talent under contract. They pulled together actors and crew to make a movie and then disbanded. They had very little overhead and a great deal of flexibility. They did have problems, however. For one, they had to raise money to make the movie. In the past, banks had lent money to the big studios, knowing they would stay in business. Banks were more leery of small companies that were likely to evaporate. With less money, independents steered clear of the extravaganzas that the major studios produced, such as *Quo Vadis,* a 1951 film from MGM that used 5,500 extras, and *Cleopatra,* a Fox film that cost $44 million in 1963. Instead, they concentrated on movies that were stronger on plot and sensibilities but that used only a few characters and unpretentious sets and costumes.

The independents also gave the artists more power. No longer were directors and actors given unalterable scripts. They were encouraged to stamp the movies with their artistic style. Some of the independents were directors who created their own material. They were well aware of European filmmaking (see Chapter 14) and its **auteur** (French for "author") theory of the director as the prime creative force, which encouraged films with more personal reflections and statements by directors (see Exhibit 6.14).[33]

youth

The films made by independents helped bring about a movie renaissance because they caught on with audiences, albeit different audiences than attended in

code

the 1930s and 1940s. Movies became more geared toward the young, who were eager to leave the house and have a place to go that was not under the watchful eye of elders. The subject matter that appealed to these younger audiences revolved around sex and violence. In previous decades, the code would have prevented this subject matter, but in 1951, the Supreme Court had ruled that movies were considered part of the press and as such were protected by the First Amendment. Before this (based on a 1915 decision), movies had been considered art and were not protected by the First Amendment.

Exhibit 6.14

Woody Allen is representative of the independent directors. He openly disdained the studio system and worked out of New York rather than Hollywood. Most of his movies involve an outsider, neurotic character that he uses to explore social structures such as sexuality, conformity, and religious stereotypes. Here, Allen is shown in his 1973 film *Sleeper.*

(Supplied by Globe Photos, Inc.)

The code didn't disappear, but it was revised several times to allow for changing mores. The Motion Picture Producers and Distributors of America had been renamed the Motion Picture Association of America (MPAA), and Jack Valenti headed it for many years. In 1968, the MPAA established the backbone of the present rating system— G for general audience; PG for parental guidance; R for restricted (those under 17 need to attend with an adult); and X, no one under 17 allowed. Later, in 1984, PG-13 was added for slightly racier PG films, and in 1990, the MPAA created NC-17 that meant no one under 17 was allowed because X had been appropriated by producers of pornography.[34]

Films of the 1950s included quite a few westerns and musicals, but they were stronger on story than their predecessors. Both these genres have had to struggle, and, with a few exceptions, have not reached the status of their glory days in the 1950s. *The Sound of Music,* released in 1965, did become a big hit. Many producers at the time thought the musical had made a comeback and came out with expensive copycat musicals, but they all failed at the box office. The 1950s also included the rise of Marilyn Monroe's career (see "Zoom In" box) and the comedies of Dean Martin and Jerry Lewis and Bob Hope and Bing Crosby. Another popular genre was the rebel youth film. Some examples are *On the Waterfront* (1954) with Marlon Brando and *Rebel Without a Cause* (1955), starring James Dean.

changing content

By the 1960s, the films had much more social commentary and psychology, including antiheroes and attacks on American social mores. For example, Stanley Kramer's 1967 *Guess Who's Coming to Dinner?* dealt with racial tension, and

 ZOOM IN: **Why Marilyn?**

Wander through the tourist shops in Hollywood today and you will see a large number of souvenirs related to Marilyn Monroe. There are Marilyn paperweights, Marilyn T-shirts, Marilyn refrigerator magnets, Marilyn license plates, and so on. Books, websites, and documentaries about Marilyn abound. The skirt-blowing photo from *The Seven Year Itch* has become legendary. Why does the public still have such a fascination with Marilyn? It has been more than half a century since she started in movies. She only starred in a handful of films, and often she was late coming to the set and couldn't remember her lines. True, aspects of her personal life were movielike:

- Illegitimate birth as Norma Jeane Mortenson on June 1, 1926.
- Childhood with foster care, an orphanage, various friends and relatives, and occasionally her mentally unstable mother.
- "Discovered" while working in a parachute factory when a photographer took an army picture of her to show women working to support the war effort.

- Marriages to Joe DiMaggio and Arthur Miller and "close associations" with Jack and Bobby Kennedy.
- Drugs.
- Untimely, mystery-shrouded death (suicide? overdose? foul play?) on August 5, 1962.

Why has this 1950s icon survived in such a high-profile manner through several generations? Why not Katharine Hepburn, who was a far superior actress? Why not Betty Grable, who had better legs? Why not Zsa Zsa Gabor, who had more husbands? Why Marilyn?

Souvenirs related to Marilyn Monroe.

Stanley Kubrick's 1964 *Dr. Strangelove or How I Learned to Stop Worrying and Love the Bomb* was a dark comedy that attacked America's war interests.

other positive factors

Reasons other than the rise of the independents also enabled the movie industry to get back on its financial feet. The nemesis, television, actually became a benefactor. Television stations and networks began programming old movies (see Chapter 3). The fees they paid were an unexpected, welcome addition to the studios' coffers. The theaters also moved closer to where the people were. First were the drive-ins that fit with Americans' love for their cars. Then came movie complexes in suburban shopping areas. These were not the large, lush palaces of the older city theaters but multiplexes of small theaters that could show a variety of movies appealing to different audiences. Of course, the price of movie tickets rose and rose, so although attendance has never surpassed the 1946 level, revenues far exceed the levels of previous years.

6.13 Mythmakers Lucas and Spielberg

Lucas

Spielberg

The movie business was doing well enough by the 1970s that the inexpensive production elements of the 1960s could be set aside and movies could again employ expensive production techniques. Special effects became popular, especially with the 1977 release of George Lucas's *Star Wars* (see Exhibit 6.15). Lucas's incorporation of flying spaceships and **Dolby** noise reduction ushered in a new form of movie experience.

He and Steven Spielberg led moviemaking away from the dark, negative films of the 1960s into films that were more like the older Hollywood escapist fare. They embraced elements of myths, such as strong heroes, questlike journeys, happy endings, and innocence, and they made movies with entertainment value as a primary purpose. Spielberg's *Indiana Jones* films, starring Harrison Ford, personify the hero who overcomes unbelievable obstacles (see Exhibit 6.16).

Lucas stopped directing after his 1983 *Return of the Jedi* and did not return to that aspect of his career until 1999, when he released the first of three **prequels** to *Star Wars*. During his hiatus from directing, he concentrated on the development of special effects through his Industrial Light + Magic company, located in northern California. This company and several others have taken digital special effects to new levels that make the original *Star Wars* pale by comparison.

Exhibit 6.15

Because of the clear delineation of "good guys" and "bad guys," some people referred to the *Star Wars* movies as westerns set in outer space. The characters Storm Trooper (*left*) and Darth Vader are examples of this delineation.

(Supplied by Globe Photos, Inc.)

Spielberg remains prolific, usually directing several films a year. He has made some of the highest grossing films, including *Jaws* (1975), *E.T.: the Extra-Terrestrial* (1982), and *Jurassic Park* (1993). Some of his films lean toward social messages, such as his 1993 Oscar-winning *Schindler's List,* which deals with the Holocaust, and *Munich* (2005), which deals with the aftermath of the murder of 11 Israeli athletes at the 1972 Olympics. In 1994, Spielberg, along with Jeffrey Katzenberg and David Geffen, established a movie studio, DreamWorks SKG, the first new studio in decades. At first it stood on its own, but eventually it merged functions with other studios.[35]

6.14 Moviemaking Today

Today most movie studios have adopted an organizational model that is a combination of the independent era and the studio era. The major studios (Disney, Fox, NBC Universal, Paramount, Sony/MGM, and Time Warner) do not have stars or others under contract. They hire people film by film, using the independent model. However, they give independent film companies space in their facilities, help underwrite their films, and then act as distributors for the films in which they have an interest. Sometimes several film companies join to finance a film, as in the collaboration of Paramount and Fox on James Cameron's 1997 blockbuster *Titanic* (see Exhibit 6.17). Banks sometimes underwrite consortiums of companies to produce a variety of films, knowing that some will be hits and others duds.[36]

Another important factor in today's moviemaking is the **aftermarket.** Movie companies make more money from DVD sales and rentals, international distribution, airplane showings, television rights, video game rights, pay-per-view, the internet,

Exhibit 6.16

Harrison Ford as Indiana Jones in *Raiders of the Lost Ark* constantly overcame obstacles.

(Supplied by Globe Photos, Inc.)

studio structure

aftermarket

The All-Time Top Grossing Films as of 2008

Rank Gross (in millions)

1 — $600.8 *Titanic* (1997)
2 — $522.1 *The Dark Knight* (2008)
3 — $461.0 *Star Wars: Episode IV–A New Hope* (1977)
4 — $437.2 *Shrek 2* (2004)
5 — $435.0 *E.T. the Extra-Terrestrial* (1982)
6 — $431.1 *Star Wars: Episode I–The Phantom Menace* (1999)
7 — $423.4 *Pirates of the Caribbean: Dead Man's Chest* (2006)
8 — $407.7 *Spider-Man* (2002)
9 — $380.3 *Star Wars: Episode III–Revenge of the Sith* (2005)
10 — $377.0 *The Lord of the Rings: The Return of the King* (2003)

Exhibit 6.17

Titanic tops the list of all-time top-grossing films. As the costs for the film rose, Fox, which was the original studio for the movie, asked for financial help from Paramount. When the movie became extremely popular and brought in huge earnings, Paramount made a very good return on its investment. The income from the movie far exceeded its $200 million budget.

and the like than they do from theatrical showings. As a result, movies are often released to home video shortly after or at the same time as they are released to theaters. In this way, studios can gear their promotional efforts simultaneously to people who might go to the theater and those who prefer to watch in their homes.[37]

independents

Independent filmmaking has also profited by direct-to-DVD distribution. By-passing a theatrical release, these movies (which are often low budget or films aimed at adults rather than teens) go straight to the video store shelves. In fact, the video chain Blockbuster has financed some independent movies. Independents can also get the word out about their films cheaply through the internet; sometimes they even allow people to download movies for free for a short period of time in order to create a buzz before the movie is released through other avenues.[38]

international

Because of the importance of the worldwide box office, movies are much more international than in the past when there were different national styles of filmmaking (see Chapter 14). Now these styles are merging as filmmakers in different parts of the world adopt techniques from each other. Movies are often co-produced with contributions from various countries, something that improves their chances of being successfully distributed in all the participating countries.[39]

content

From a content point of view, movies continue to evolve. Today's technology enables many special-effects–laden movies. In fact, some would say that movies are so special effects driven that story is given little attention—but then lack of story is not a new criticism for the movie industry. Animation has particularly profited from improved technologies that have been showcased in such movies as *Cars, Ratatouille,* and *Wall-E.* Movie producers seem to be paying more attention to females than in the past, when movies were aimed mostly at teenage males. For example, *What a Girl Wants, Legally Blonde 2,* and *Mona Lisa Smile* all did well in 2003 as "chick flicks," and pregnancy became an "in" subject for comedy in 2007 with *Juno* and *Knocked Up.* Some **sequels,** such as those for *Shrek* and *Pirates of the Caribbean,* have done better than the originals, a phenomenon that was not true in the past, and series movies such as *The Lord of the Rings* and the *Harry Potter* films have done very well. Even documentaries became popular theater fare when Michael Moore's *Fahrenheit 9/11* won the top prize at the Cannes Film Festival in 2004.[40]

6.15 Issues and the Future

The future for movies is digital. The death of film has been predicted for decades. No one challenges the idea that well-produced, entertaining, visual stories will survive. But the production, distribution, and exhibition systems that have formed the underpinning of the movie industry for years are changing.

production

Celluloid film, which has been the staple for shooting movies since their beginnings, is gradually being replaced by digital media, such as digital videotape and computer hard drives, especially now that high-definition video cameras yield quality far superior to that of previous video cameras. Movies are edited digitally using **nonlinear** computer software. Doing so allows for easier insertion and manipulation of special effects than did the older, film-based methods. Some movies are made totally of computer-generated images. Others use **motion-capture** technology to record actors' movements, which are then used to create characters and movement in a computer.[41]

Theatrical movie distribution and exhibition are also edging toward digital. The distribution system for theatrical movies that has been in place for years is clumsy and expensive. Thousands of copies of a film must be made so that it can open in theaters around the country, and sometimes around the world, at the same time. It costs the studios about $1,200 for each print that goes to a theater; if the movie is placed on a digital medium such as a hard drive, that cost goes down to about $300. The expense is in converting theaters so that they can project a digital movie rather than one that is on film—a cost of about $80,000 per theater.[42]

distribution

Movie exhibition is currently not a particularly profitable business. Theaters make more money on popcorn sales than they do on movie tickets, and a number of theater chains have gone bankrupt. Theater companies cannot afford to buy the new equipment needed for digital projection. Besides, it is the studios that profit most from the switch to digital because it lowers their costs. The major studios have agreed to help pay for theaters to convert to digital, and the process is proceeding—slowly.

exhibition

First, technical standards needed to be adopted so that equipment manufacturers could create products that could be used in movie theaters throughout the world. Such standards were adopted in 2005 by the Digital Cinema Initiatives, a coalition of the major movie studios. Then the equipment, such as projectors and servers, had to be developed and tested. Currently about 5,000 of the 37,000 movie theaters in the United States have converted to digital, with many others planning to do so in the near future. Of course, digital technology keeps improving, so new applications have appeared within the parameters of the technical standards. For example, one of these improvements easily facilitates the showing of 3-D movies, which may lead to a resurgence of 3-D, something that has been tried numerous times in the past with only short-lived success. However, the movie theaters that converted to digital early would need to invest even more to equip themselves to project 3-D. Movies, whether as film prints or on hard drives, now travel through an elaborate airplane-train-truck-hand delivery system. It would be much easier and cheaper to deliver a movie in its digital form by satellite or the internet to theaters everywhere. This may happen at some future stage.[43]

processes

If movies are produced, distributed, and exhibited digitally, however, **piracy** problems may increase. Illegal distribution of films has been a continuing issue in the movie business, but at least pirated copies in the past were often of low quality because they were distributed in analog form, such as on VHS tape. Digital material can be reproduced without loss of quality, so if a movie is intercepted somewhere in the digital production-distribution-exhibition chain, high-quality illegal copies can be made and sold. However, digital

piracy

technologies are here in forms other than digital motion picture theater presentation. Digital copies of movies are illegally distributed on DVDs and over the internet, even by members of the film industry. In 2003, when members of the Academy of Motion Picture Arts and Sciences were sent DVD **screeners** of movies so they could judge them for Oscars in the comfort of their homes, several of the movies made their way onto the internet. They were identified and traced because they carried coded markings identifying them as videos sent to an Oscar voter.[44]

One obvious conclusion to be drawn from all of this is that the issues for movies intertwine with the issues for other media, such as DVDs, satellites, and the internet. Movies are likely to be incorporated within any media structure because of their proven ability to entertain.

6.16 Summary

Movies have undergone many changes through the years, in terms of both technology and content.

Technologically, the first moving pictures were stroboscopic toys that utilized persistence of vision. Thomas Edison and his assistant, W. K. L. Dickson, invented the film camera and the Kinetoscope for viewing films; they made movies in a studio called the Black Maria. The Lumière brothers of France made the first projector, and eventually nickelodeons and elaborate theaters appeared to show the movies. There were so many companies with competing interests that in 1909 the Motion Picture Patents Company was created, with New York companies Edison, Biograph, and Vitagraph as the main participants.

Equipment continued to improve, and film companies moved to California where they could take advantage of the sunny weather and get away from the New York stranglehold on production and film stock. Sound was developed and became very popular after the 1927 showing of *The Jazz Singer*. Several different forms of color were developed in the 1920s and 1930s, but color did not become stable until the late 1960s.

During the 1950s, the movie industry tried technological wizardry to stem the erosion of the audience to television. Cinerama, CinemaScope, and 3-D were tried, but to little avail. People have been more receptive to the technology of later decades that has centered on special effects. Electronic editing and digital video cameras have become part of the production process, and digital electronics are now playing a big role in the future of distribution and exhibition.

In terms of content, the stroboscopic "movies" were simply moving images, such as a parrot in a cage. Edison and Dickson's movies were mostly vaudeville acts, while the Lumières took the camera outside and filmed everyday events. Early filmmaker George Méliès experimented with special effects, and Edwin S. Porter cut back and forth between scenes. D. W. Griffith developed the narrative film and produced the controversial *The Birth of a Nation* in 1915. Charlie Chaplin developed comedy with his Little Tramp character. During the silent film era of the 1920s, movie stars came to the fore, and fans became interested in their lives. Movie content was somewhat risqué, and Will Hays was brought in to head the MPPDA and clean up movies.

When film studios were set up in California, they began filming westerns because the land and climate lent itself to this genre. Once sound was introduced, the studios were able to produce musicals and fare that was more dramatic than in the past. The MPPDA Code's moral restrictions influenced movie content, but people flocked to the movies in record numbers. Two important directors of the studio era were Alfred Hitchcock, who used editing to increase tension, and Orson Welles, who used camera angles and focus to accentuate emotion.

When the movie industry hit bad times because of television and the outlawing of vertical integration, movie content changed. Independent producers came to the fore; they made cheaper films by hiring people for only one film at a time, thereby reducing overhead. Their films, released mainly in the 1960s, were more artistic and personal and contained the stamp of the auteur. A court ruling put movies under the First Amendment, so they were able to contain more sex and violence.

Movies of the 1970s reverted back to more positive themes and happy endings. Lucas and Spielberg led the way with *Star Wars* and the *Indiana Jones* films. In their turn, girl films, sequels, series movies, and documentaries have become popular. The content of films today is affected by the needs of the aftermarket and the influence of other countries.

Suggested Websites

www.oscars.org (the Academy of Motion Picture Arts and Sciences, the organization that gives out Oscars)

www.afi.com (the American Film Institute, the organization that devised the list of the greatest movies of all time)

www.imdb.com (the Internet Movie Database, which gives facts about almost all movies)

www.dreamworks.com (DreamWorks, the company founded by Spielberg, Geffen, and Katzenberg)

www.mpaa.org (the Motion Picture Association of America, which lobbies for the movie industry and oversees the ratings system)

Notes

1. Many excellent books cover film history, including David Bordwell and Kristin Thompson, *Film History: An Introduction* (New York: McGraw-Hill, 2002); Gerald Mast and Bruce F. Kawin, *A Short History of the Movies* (Boston: Allyn and Bacon, 2001); and Jack C. Ellis, *A History of Film* (Englewood Cliffs, NJ: Prentice Hall, 1990).
2. Joseph McBride, ed., *Persistence of Vision* (Madison: University of Wisconsin Press, 1968).
3. Kevin MacDonnell, *Eadweard Muybridge* (Boston: Little, Brown, 1972).

4. Gordon Hendricks, *The Edison Motion Picture Myth* (Berkeley: University of California Press, 1961); and Ray Zone, "Vintage Instruments," *American Cinematographer,* January 2003, 70–77.
5. Georges Sadoul, *Louis Lumière* (Paris: Seghers, 1964).
6. Benjamin B. Hampton, *A History of the American Film Industry from Its Beginnings to 1931* (New York: Dover, 1970); and Joseph H. North, *The Early Developments of the Motion Picture, 1887–1909* (New York: Arno Press, 1973).
7. John Frazer, *Artificially Arranged Scenes: The Films of Georges Méliès* (Boston: Twayne, 1980); and Charles Musser, *Before the Nickelodeon: Edwin S. Porter and the Edison Manufacturing Company* (Berkeley: University of California Press, 1991).
8. Anthony Slide, *The Big V: A History of the Vitagraph Company* (Metuchen, NJ: Scarecrow, 1996).
9. Mast and Kawin, *A Short History of the Movies,* 45–46.
10. Robert Anderson, "The Motion Picture Patents Company: A Reevaluation," in Tino Balio, *The American Film Industry* (Madison: University of Wisconsin Press, 1985).
11. The "odeon" part of "nickelodeon" stood for admission to a theater, from the Greek word *odeon.*
12. "The First Moguls," *Hollywood Reporter,* May 26, 2005, S1–S8.
13. Tom Gunning, *D. W. Griffith and the Origins of American Narrative Film: The Early Years at Biograph* (Chicago: University of Illinois Press, 1991); and Martin Williams, *Griffith: First Artist of the Movies* (New York: Oxford University Press, 1980).
14. Charles Chaplin, *My Autobiography* (New York: Simon and Schuster, 1964); Mack Sennett, *King of Comedy* (New York: Doubleday, 1954); and David Robinson, *Chaplin: His Life and Art* (New York: McGraw-Hill, 1985).
15. Larry W. Ward, *The Motion Picture Goes to War* (Ann Arbor, MI: UMI Research Press, 1985).
16. Robert Windeler, *Sweetheart* (New York: Praeger, 1974); and Booton Herndon, *Mary Pickford and Douglas Fairbanks* (New York: Norton, 1977).
17. Leonard J. Leff and Jerold L. Simmons, *The Dame in the Kimono: Hollywood, Censorship, and the Production Code from the 1920s to the 1960s* (New York: Grove Weidenfeld, 1990); and Richard S. Randall, *Censorship of the Movies* (Madison: University of Wisconsin Press, 1968).
18. Tino Balio, *United Artists: The Company Built by the Stars* (Madison: University of Wisconsin Press, 1976); William K. Everson, *American Silent Film* (New York: Oxford University Press, 1978); and "Mind of the Mogul," *Hollywood Reporter,* July 2000, 44.
19. Tag Gallagher, *John Ford* (Berkeley: University of California Press, 1985); Daniel Moews, *Keaton* (Berkeley: University of California Press, 1977); Adam Reilly, *Harold Lloyd* (New York: Collier, 1977); Charles Barr, *Laurel and Hardy* (Berkeley: University of California Press, 1968); and Erik Barnouw, *Documentary: A History of the Non-Fiction Film* (New York: Oxford University Press, 1992).
20. Evan W. Cameron, ed., *Sound and the Cinema: The Coming of Sound to the American Film* (Pleasantville, NY: Redgrave, 1980); and Harry M. Geduld, *The Birth of Talkies* (Bloomington: Indiana University Press, 1975).
21. Roger Dooley, *From Scarface to Scarlett: American Film in the 1930s* (New York: Harcourt Brace Jovanovich, 1981).
22. Paul W. Facey, *The Legion of Decency: A Sociological Analysis of the Emergence and Development of a Pressure Group* (New York: Arno, 1974).
23. "The Numbers," http://www.the-numbers.com/market/2007.php (accessed July 5, 2008).
24. Jane Feuer, *The Hollywood Musical* (Bloomington: Indiana University Press, 1993); and "Of Mice, Moguls, and Cartoon Stars," *Los Angeles Times,* March 21, 2004, E-26.
25. Donald Spoto, *The Art of Alfred Hitchcock* (New York: Anchor, 1992); and Robin Wood, *Hitchcock's Films Revisited* (New York: Columbia University Press, 1989).
26. Robert Carringer, *The Making of* Citizen Kane (Berkeley: University of California Press, 1985); Charles Higham, *The Films of Orson Welles* (Berkeley: University of California Press, 1970); and www.afi.com/tv/lists.asp (accessed June 26, 2004).
27. Rod Ryan, *A History of Motion Picture Color Technology* (New York: Focal Press, 1978); and "The Color-Space Conundrum," *American Cinematographer,* January 2005, 88–110.
28. Ellis, *A History of Film,* 252.
29. Michael Coant, *Antitrust in the Motion Picture Industry* (Berkeley: University of California Press, 1960).

30. Larry Ceplair and Steven Englund, *The Inquisition of Hollywood: Politics in the Film Community* (New York: Doubleday, 1980).

31. Ken Auletta, *Three Blind Mice: How the TV Networks Lost Their Way* (New York: Random House, 1991), 76.

32. Dan Symmes, *Amazing 3-D* (Boston: Little, Brown, 1982); David W. Samuelson, "Golden Years," *American Cinematographer,* September 2003, 70–77; and Dennis P. Kelley, "When Movies Played Downtown: The Exhibition System of the San Francisco Theater Row, 1945–1970," *Journal of Film and Video,* Summer/Fall 2002, 71–89.

33. John Baxter, *Hollywood in Sixties* (Cranbury, NJ: A. S. Barnes, 1972).

34. Valenti served as head of the MPAA from 1966 to 2004, when he turned over the reins to Dan Glickman. "Valenti Passes the Torch," *Daily Variety,* July 2, 2004, 1.

35. Thomas G. Smith, *Industrial Light and Magic* (New York: Del Rey, 1986); Donald R. Mott and Cheryl McAllister Saunders, *Steven Spielberg* (Boston: Twayne, 1986); "Darth Vader's New Offices," *Newsweek,* June 27, 2005, E32; "Paramount Confirms Studio Deal," *Los Angeles Times,* December 12, 2005, C-1; and "A Spielberg Blockbuster," *Hollywood Reporter,* July 3, 2008, 1.

36. "Money Crunch," *Hollywood Reporter,* August 2001, 121; and "All Time 1000 Top Grossing Films," http://www.movieweb.com/movies/boxoffice/alltime.php (accessed October 11, 2008).

37. "For Blockbuster Games, EA Goes Hollywood," *Los Angeles Times,* January 30, 2003, C-1; and "Cart before the 'Jackass,'" *Hollywood Reporter,* December 13, 2007, 1.

38. "Blockbuster Breaks Away," *Wall Street Journal,* April 22, 2002, B-1; and "Love at First Site," *Hollywood Reporter,* August 2006 Marketing Special, 23.

39. "Turns Out World Really Is Flat," *Hollywood Reporter,* May 2, 2008, 5.

40. "String of Box-Office Hits Shows Filmmakers 'What a Girl Wants,'" *Los Angeles Times,* April 4, 2003, C-1; "Spidey Swings to New Heights," *Daily Variety,* July 6, 2004, 1; "Moore Power to Ya!" *Daily Variety,* June 11, 2004, 4; and "The List," *Los Angeles Times,* July 6, 2008, C-12.

41. "All Aboard," *American Cinematographer,* November 2004, 64-75; and "Red One Faces Proof of Post," *TV Technology,* June 25, 2008, 29.

42. "Hollywood's Digital Delay," *Wall Street Journal,* November 27, 2005, B-1; and "Carmike Theaters to Convert to Digital," *Los Angeles Times,* December 20, 2005, C-1.

43. Matthew Doman, "AMC Will Close 249 Screens," *Hollywood Reporter,* June 1–3, 2001, 10; "Studios Settle on Technical Standards for Digital Shift," *Los Angeles Times,* July 27, 2005, C-1; "D-Cinema Closer to Close-up," *Hollywood Reporter,* August 15, 2008, 6; and "Coming at You: Dueling 3-D Films," *Los Angeles Times,* April 20, 2007, C-1.

44. "Latest Plot Twist for 'Star Wars': Attack of the Cloners," *Los Angeles Times,* May 10, 2002, 1; "Second Oscar 'Screener' Finds Its Way Onto Internet," *Los Angeles Times,* January 14, 2004, C-1; and "High-Tech Sleuths Fight Pirates," *DVD Exclusive,* September 2004, 3.

ELECTRONIC MEDIA FUNCTIONS

The various forms of electronic media all must operate within the confines of certain guidelines. They should handle their finances responsibly; they must supply a product that someone wants to hear or see; they need to find someone to pay for their product; it behooves them to know their audience size and composition; they must abide by certain rules and regulations and by dictates of conscience; they must make sure the product reaches its destination. The nuances of the various functions vary from one form of electronic media to another. The first chapter in this section deals with the careers that support the functions; the following chapters deal with functions such as programming, sales, advertising, promotion, audience research, legal actions, ethical behavior, technical processes, and international influence.

Chapter 7

CAREERS IN ELECTRONIC MEDIA

I've been fired twice, canceled three times, won some prizes, owned my own company, and made more money in a single year than the president of the United States does. I have also stood behind the white line waiting for my unemployment check. Through television I met my wife, traveled from the Pacific to the Soviet Union, and worked with everyone from President John F. Kennedy and Bertrand Russell to Miss Nude America and a guy who played "Melancholy Baby" by beating his head. With it all, I never lost my fascination for television nor exhausted my frustration.

Bob Shanks, then vice president of ABC

The information age is here, making jobs in electronic media very important in terms of social significance and impact. The field is expanding as new processes and distribution means come to the fore. Whereas electronic media used to involve only radio and television, they now cover a vast array of areas including video game production, DVD sales, website creation, satellite radio promotion, internet advertising, movie research, cell phone technical development, and satellite TV legislation. It is an excellent time to be interested in a career in electronic media. Books, articles, and websites that deal with the best jobs for the future usually list entertainment and information among the top.[1]

However, there are many people seeking electronic media jobs. Gaining employment in this field has never been easy because more people would like to be employed in it than there are jobs. It is difficult to gather reliable statistics about the number of people employed in electronic media. Directors cross over from corporate production to broadcast fare to cable programs to movies; people who work for advertising agencies spend some time on radio, TV, and the internet and some on print; some American media experts work mainly in other countries. Likewise, it is difficult to determine just how many people aspire to jobs in the field. More than 550 colleges have programs related to the field, but some of their graduates will enter newspapers, magazines, the theater, or fields totally unrelated to their major. Conversely, students from other majors will enter electronic media.

If you aspire to enter the field, the amount of competition you will face is relevant but not overly important. What really matters is how well you prepare yourself to enter the field and how capably and flexibly you perform your duties once you have a job.

7.1 Desirable Traits for Electronic Media Practitioners

Electronic media–related jobs are not for everyone. Those who crave routine, want security, or like working by themselves had best look elsewhere. Because the field is so competitive, people who do not *really* want to work in the business usually fall by the wayside. A number of traits are particularly important for media people. You don't need to have all of them for every job, but if you don't have most of them, this business probably isn't for you (see Exhibit 7.1).[2]

Interpersonal skills. Most media projects involve teamwork, so you must be able to get along well with others. There may be 200 people on a movie or TV set, all needing to interact with each other to accomplish a special effect. In an office environment, the programming executives must

Exhibit 7.1

News reporting is a good example of a job that requires the traits discussed here. Can you think of reasons why these reporters covering a bank robbery would need interpersonal skills, flexibility, versatility, dependability, positive attitude, energy, resilience, ability to "have a life," and skill to do the job?

interpersonal skills

work closely with the sales and promotion departments to ensure a successful program launch. Reporters interact with newsmakers and members of the public. The engineers who designed the internet had to work with each other both in person and over long distances. Sometimes you need to be a leader and sometimes a follower; you must be able to handle both roles with equal ease. One director selects his crew by interviewing the most talented people he can find for each position and then selecting the ones he would most like to go camping with—not a bad idea, because some of the lifestyle related to electronic media is not too far removed from camping.

flexibility

Flexibility. Because the media world changes so often and so rapidly, you, too, must be willing to change. If you are in any job that involves equipment, you must be able to learn new operating methods and new computer programs several times a year. You can't count on your employer to train you; usually you need to learn on your own. Networks often need to make changes when the ratings are low; if you fight the changes, you can find yourself in the unemployment line. When a promising new program distribution form, such as the internet, comes along, you should be nimble and position yourself to take advantage of its capabilities. Another element of flexibility is geographic flexibility. You may need to spend six months on a movie location in Malaysia. You may need to "commute" between New York and Los Angeles. Travel is broadening and enjoyable for many people, but it does require flexibility.

versatility

Versatility. Closely related to flexibility is versatility. It behooves you to be able to do more than one thing. If you are an excellent studio camera operator, but your facility changes to **robotic cameras** that do not need operators, you need to find something else to do. If you know how to operate all the equipment, if you know how to maintain equipment, if you understand budgeting, you are more likely to remain employed. As media splinter and costs rise, people are asked to take on more roles, so the ability to multitask helps.

dependability

Dependability. Because the media industry is so people driven, those in it depend on each other. After you work with someone for a while, you develop a mutual trust, and that trust is a large part of what makes the media business function. Punctuality is a very important part of dependability. Many functions of the media are run by the clock—the news starts at 5:00, not 5:07.

positive attitude

Positive attitude. People don't like to be around someone with a sour disposition, someone who is constantly complaining. Many things are wrong with the media business, but if you dwell on them, there is little chance that you will help to correct them. A good sense of humor helps you and others have a positive attitude.

energy

Energy. Hours are long, and activity is high. There aren't many jobs in the media world where you sit at a desk all day. Often you need to take the initiative to undertake some task that needs attention but that other people are too busy to do. In most jobs, you will gain energy from your fellow workers. The camaraderie that develops when working on a movie, an internet site, or a radio sales project is part of what makes the industry so much fun.

resilience

Resilience. Just about everyone has down periods during a career. You may make a mistake that gets you fired, but often a lack of employment is not your fault. When the economy is in trouble, executives may renege on an innovative idea you are proposing. When the actors go on strike, many people are out of work. When sitcoms are out of favor, writers lose their jobs. You can't let unfortunate circumstances keep you down. You must realize that security is not part of the media business, dust yourself off, and try again.

Ability to "have a life." The media business can devour you. It is difficult to keep relationships when you are traveling constantly and working long hours. You need to take care in selecting your significant others, and you need to communicate well with them so that they understand the needs of your job and are accepting of those needs. The nature of the business can make it difficult to give proper attention to spouses and children; on the plus side, it's a fascinating business, and close relatives can share it with you if you are open to devising ways that they can become involved. Try to engage in some activities other than work, such as a hobby or charity work, so you don't put all your emotional energy into your job. In addition, save some money so you are not in dire financial straights if you lose your job.

ability to have a life

skills

Skills to do your job. You do need to be able to accomplish what you are asked to do. A cinematographer must have an eye for picture composition and know the capabilities of the camera. A programmer must understand ratings. A reporter must know the legalities related to journalism and have excellent writing skills. But most people can master the skills needed for a career in electronic media; it's not nuclear physics or brain surgery. What is important is the ability to use your skills in a constructive manner that enhances your interpersonal skills, flexibility, versatility, dependability, positive attitude, energy, resilience, and ability to have a life.

7.2 College Preparation

If you feel you have the traits needed for a career in electronic media, an excellent way to begin your preparation is to enroll in one of the colleges or universities that offer courses in the field. If you are in high school, you might want to consider first attending one of the many community colleges that have excellent hands-on media programs. If you are looking for a four-year college, you want to consider many personal factors such as location, cost, and extracurricular opportunities, but you can find colleges with media programs by surfing the internet or by consulting books that describe the offerings (see Exhibit 7.2).[3]

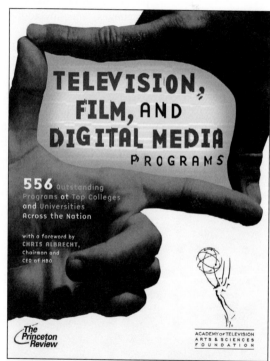

Exhibit 7.2

This book, compiled by the Academy of Television Arts & Sciences Foundation and the Princeton Review, contains facts about various college media programs.

(Photo courtesy of the Academy of Television Arts & Sciences Foundation)

courses

Often the coursework you are looking for is in a radio-TV-film department, but courses and majors are also found in many other departments—communications, theater arts, English, journalism. In reality, a college education is not essential for most entry-level jobs, but people who do the hiring for entities such as radio stations, TV networks, movie productions, cable systems, satellite operations, and internet start-ups are interested in hiring people with promotion potential and generally believe that a college education makes people well rounded and builds good work habits. Some executives do not even bother to look at résumés of people who do not have college degrees.[4]

other coursework

Although courses that specifically cover electronic media are advisable, you should also have a broad knowledge in other fields. If you want to enter the news field, you will be almost useless if you are a good camera operator but know nothing about national or international affairs. Political science, history, and writing courses are a must for reporters. Likewise, accountants and salespeople should take business courses; engineers should know electronics and computer science; people wanting to work in international aspects of media should study foreign languages; and directors should be knowledgeable about drama, music, and psychology.

You should learn as much as you can about budgeting and finance because you will be working in an environment where costs are tight, and people who can draw up accurate budgets are valued. You definitely need computer skills because just about everyone employed in electronic media works with computers. They are used to design sets, edit videos, bill clients, schedule programs, write news stories, and accomplish a host of other tasks. And, although it may sound redundant, people aspiring to communications jobs should know how to communicate. Take courses to improve your writing and speaking skills because one of the main complaints of employers is that employees cannot express written or spoken thoughts clearly.[5]

experience

Exhibit 7.3
Working at a student radio station is a good activity to undertake while in college.

7.3 Outside Activities

Engaging in media activities beyond the classroom enables you to come closer to "real world" work experiences that will help you obtain a job. If your university has a radio station or TV facility, become involved (see Exhibit 7.3). Try to exhibit the same interpersonal skills, flexibility, versatility, dependability, and other personal qualities that you will need for a job. If you can find a part-time job in the media business, do so. It will build your work ethic and add a valuable line to your résumé. If you have a job that is not related to media, volunteer to take on an extra task that is communications oriented. Perhaps you can design a flyer or produce a video that the organization can use to promote itself.

Another way to make yourself salable on the job market is to win awards. Students who win

Exhibit 7.4
Winning a contest provides a good entry for a résumé.

Contests

Academy of Motion Picture Arts and Sciences Student Academy Awards (www.oscars.org/awards/saa/index.html)

Academy of Television Arts and Sciences College Television Awards (http://cdn.emmys.tv/foundation/collegetvawards.php)

Broadcast Education Association Festival of Media Arts (www.beafestival.org)

The Christophers Video Contest for College Students (www.christophers.org)

David L. Wolper documentary Student Achievement Award (www.documentary.org/content/enter-ida-awards-competition-online-now)

National Religious Broadcasters Student Achievement Awards (www.nrb.org)

UFVA Student Film and Video Festival (www.temple.edu/nextframe)

prizes for productions, scripts, or articles are often highly sought after because they have proven their ability to rise above the competition. Quite a few organizations have contests or festivals especially for students, and others accept student entries to compete with entries from established professionals (see Exhibit 7.4).[6] **awards**

You should also read industry trade journals. Some, such as *Broadcasting & Cable, Variety,* and the *Hollywood Reporter,* cover many aspects of the entertainment business. Others, such as *Radio and Records, TV Technology,* and *Game Informer* are more specialized by industry or occupation. Most **trades** have websites as well as hard copies (see Exhibit 7.5). **trades**

One note of caution regarding outside activities: be careful what you post on the internet. Remember that employers surf the net, too, and that "clever" video of you acting foolish at a party or spoofing an employer might just lose you the chance for a desirable job. **internet**

Trades

Billboard (www.billboard.biz)

Broadcasting & Cable (www.broadcastingcable.com)

Emmy (www.emmys.org/emmymag)

Game Informer (www.gameinformer.com)

Hollywood Reporter (www.hollywoodreporter.com)

Journal of Broadcasting & Electronic Media (www.tandf.co.uk/journals/HBEM)

Journal of Film and Video (www.press.uillinois.edu/journals/jfv.html)

Mix (www.mixonline.com)

Radio & Records (www.radioandrecords.com/RRWebSite)

Radio Ink (www.radioink.com)

TV Technology (www.tvtechnology.com)

Variety (www.variety.com)

7.4 Internships

types

Many college electronic media programs include **internships** either as a requirement or an option. You can also obtain an internship on your own, but doing so is difficult. Many companies do not have insurance that covers interns who are not enrolled for college credit and, therefore, are reluctant to offer private internships. Also, if you undertake your internship through your college, you usually have an advisor who can help you make the most of your experience.

application

Your school will probably have a list of places that regularly accept interns, and you can either choose from that list or find some other place. Often you will need to go through an interview process very similar to that for a job. Companies do not want interns who might be disruptive or not fit into the company culture. Although paying internships are available in some industries, you probably will not receive money for your electronic media internship.

work

Most internships involve a combination of work and observation. You may be allowed to sit in on a sales meeting, watch a movie being produced, do research for a news story, proofread budget figures, and the like. You should not be confined to the copying machines, although a limited amount of photocopying, phone answering, and coffee procuring might be part of the internship. If you find that your internship experience is not valuable, talk to your college advisor.

jobs

However, most internships are very worthwhile. They give you a taste of the business and a chance to prove yourself. Even if the main thing you learn is what you *don't* want to do, that will be a valuable experience that you cannot duplicate

special
internships

in the classroom. Sometimes internships lead to jobs, so it is best to start them shortly before you plan to graduate. You are more likely to receive a job offer in good economic times or from a large company with turnover. But to do so, you must gain the trust of the people you intern with, and you must show that you possess the traits (interpersonal skills, flexibility, dependability, etc.) mentioned previously.[7]

Generally you complete your internship near your college so that you can take courses at the same time. However, there are also prestigious full-time internships in other locations that you might want to apply for. One of them is the summer internship program of the Academy of Television Arts & Sciences, a program that has been listed in the top 10 of all internship programs in the country (see Exhibit 7.6). You receive a stipend of $4,000 and have the opportunity to work

Exhibit 7.6

These students—who hailed from Southern Methodist University, University of California at Berkeley, State University of New York at Genesco, University of Miami, Western Michigan University, Occidental College, and Skidmore College—completed 2008 summer internships hosted by the TV Academy.

(Photo courtesy of the Academy of Television Arts & Sciences Foundation)

at a top Hollywood organization full-time for six to eight weeks. For more information, go to www.emmys.org/foundation/internships.php.

7.5 Networking

There is a saying in the media business that "It's not what you know but who you know." Actually, in order to succeed, you need both "what you know" and "who you know." But the importance of meeting people and keeping in touch with them cannot be overemphasized. Start with your college classmates. At least one of them may be in a position to hire you at some point. Collect contact information for everyone you are interested in working with; occasionally send an email to update them on your activities and to make sure you have their current contact information. Do the same for several of your instructors. Throughout your career, you will need letters of recommendation; by keeping your professors updated, you will be providing them with information they can include in a recommendation.

who you know

When you participate in an internship or work in a part-time media job, meet as many people as you can and add them to your contact list. It is also wise to join organizations to meet people in the field (see Exhibit 7.7). Many of these organizations have special reduced rates for students. Even though you are just a student member, you can attend meetings and join committees and in that way extend your network.[8]

organizations

7.6 Résumés and Reels

At some point you are going to need a **résumé** (see Exhibit 7.8). You may need one when applying for an internship, and you will certainly need one when you

Organizations

Academy of Motion Picture Arts and Sciences (www.oscars.org)

Academy of Television Arts & Sciences (www.emmys.org)

American Women in Radio and Television (www.awrt.org)

Broadcast Education Association (www.beaweb.org)

College Broadcasters, Inc. (www.askcbi.org)

International Radio & Television Society Foundation, Inc. (www.irts.org)

Media Communications Association (www.mca-i.org)

National Academy of Television Arts & Sciences (www.emmyonline.org)

National Association of Broadcasters (www.nab.org)

National Association of Television Program Executives (www.natpe.org)

National Cable & Telecommunications Association (www.ncta.com)

National Religious Broadcasters (www.nrb.org)

Radio-Television News Directors Association (www.rtnda.org)

Society of Motion Picture and Television Engineers (www.smpte.org)

University Film and Video Association (www.ufva.org)

USTelecom: The Broadband Association (www.ustelecom.org)

Exhibit 7.7
Joining organizations helps you network.

Exhibit 7.8

A résumé can take many forms; this is one example.

MARION RICHARDSON
1234 18th Street, Apt. 7; Anytown, AZ 12345
Cell Phone: 1-234-555-1234   Email Address: MRichardson@site.com

Job-Related Experience

Intern, KKKK-TV, Anytown. Sorted national newswire copy for the evening newscasts and suggested and researched local subjects for coverage. Helped write copy for several newscasts. Also worked in sales, assisting the sales manager by organizing leads.

Website Coordinator, Hatch Restaurant, Anytown. Wrote up changes for the restaurant webmaster so that the website remained up-to-date.

Producer/Director, "A Helping Hand." Produced the project in my advanced TV production class about the director of a Disabled Student Center.

Producer, Public Service Announcement for Children's Ways. Produced the project in my beginning TV production class and presented it to Children's Ways, which had it broadcast on Channel 7 of the Anytown Cable TV system.

Education

Bachelor of Arts degree, Communications Department, Anytown University, 2010
Associate of Arts degree, Anytown Community College, 2008

Awards

Dean's Honor Role with 3.8 grade point average, Anytown University
Honorable Mention Award of $100 from The Christophers Video Contest for Students for "A Helping Hand."
Best Producer Award for Children's Way PSA, Anytown University Awards Night

School Activities

Secretary of Publicity for the Associated Student Body, Anytown University
Treasurer of Television Society, Anytown University
Member of the Women's Basketball Team, Anytown Community College
Feature Editor of the school newspaper, Anytown High School

Employment

Hostess, The Hatch Restaurant, Anytown, 2008–2010
Sales Clerk, Evans Department Store, Anytown, 2006–2008
Youth Counselor, Park and Recreation Department, Anytown, 2004–2006

start looking for a job. A résumé is an advertisement—a way for you to present yourself to a potential employer.

structure

You should lead off your basic résumé with the strongest material. If you have had several internships, you might want those near the top of the résumé; if you have won two TV production contests, lead with those. You also need to consider what items will be most important to the person who might hire you. With computer technology it is easy to tailor your résumé to suit the needs of each job for which you apply.

presentation

One-page résumés are best for beginning jobs. Polished-looking, desktop-published résumés are so easy to compose that they are the accepted norm. As you design your résumé, keep in mind that the recipient will probably look at it for only 20 seconds. Use bold, italics, and font changes to facilitate reading rather

ZOOM IN: **What Not to Put on Your Résumé**

Following are some items that have supposedly appeared on résumés or job applications.

- Received a plague for Salesperson of the Year.
- Reason for leaving last job: maturity leave.
- Am a perfectionist and rarely if if ever forget details.
- Personal interests: Donating blood. Fourteen gallons so far.

- As indicted, I have over five years experience in computer programming.
- Instrumental in ruining entire operation of a research department.
- Finished eighth in my class of ten.

Why is it important to proofread your résumé? What other advice could you give to the "authors" of these statements?

than hinder it. Have a uniform style throughout the résumé—don't use all caps for one heading and caps and small letters for another. Spelling and punctuation errors should be nonexistent. A résumé is a sales tool; it should sell, among other things, your ability to communicate (see "Zoom In" box).[9]

Some jobs require that you submit a compilation (often called a **demo reel** or a **portfolio**) of your work. Such a compilation (be it a DVD, CD-ROM, material on a website, or a folder) should lead with your best work. If you submit an audio or video demo reel and the first 30 seconds doesn't grab a busy executive, he or she will probably look no further. Likewise, if the first several pages of a portfolio are not exciting and relevant to the job, the rest will go unseen.[10]

reel or portfolio

If you are sending out job application material for a specific job, include a short **cover letter** (see Exhibit 7.9) that highlights elements of your résumé that you feel are particularly applicable to the job at hand. You can also include information about why you are interested in the job and the organization.

cover letter

7.7 Finding the First Job

You may be one of the lucky ones to receive a job offer at the end of an internship. More likely you will need to spend time and energy looking for a job. The best method for obtaining a job is to use "who you know." Contact the people in your network and let them know you are job hunting. Even if most of them don't have any specific job openings, they can lead you to other people who may be able to employ you.

contacts

Another method is to answer ads for jobs that interest you. The internet has many sites that list electronic media job availabilities (see Exhibit 7.10). Some of the online job application forms request your résumé; other companies have their own forms for you to fill out. The latter can be very time consuming, so first select the ones that interest you the most.[11]

internet ads

You can also send cover letters and résumés to stations, networks, cable systems, and other organizations. *Broadcasting & Cable Yearbook,* an annual publication, contains names, addresses, website URLs, and facts about all radio and

sending résumés

Exhibit 7.9

A sample cover letter.

Date

Person's Name
Person's Title
Company
Street Address
City, State and ZIP Code

Dear Person's Name:

I am about to graduate as a radio-TV major from Thunderbolt University and am interested in employment as a country-western disc jockey.

As you can see from my enclosed résumé, I was disc jockey for a country-western program five hours a week on our campus FM and Internet radio station. The show gained a large loyal audience while I was on the air. I believe this success can be attributed mainly to my sincere interest in and knowledge of country-western music. I have been collecting records since I was twelve and have followed the careers of the artists carefully.

I am also an effective salesman and received a Salesperson of the Month award for selling video games and consoles. I will be happy to combine sales work with my disc jockey activities.

If you are interested in hiring a hardworking, effective disc jockey, I will be happy to send you an audition tape.

Sincerely yours,

Your Signature

Your Name
Your Street Address
Your City, State and ZIP code
Your Phone Number
Your Email Address

Exhibit 7.10

Some of these websites may help you find a job.

Sites for Finding Jobs

4 Entertainment Jobs (www.4entertainmentjobs.com)

CrewNet Entertainment Jobs (www.CrewNet.com)

Filmstaff.com: Crew Placement Services (www.filmstaff.com)

Done Deal: The Business and Craft of Screenwriting
    (www.donedealpro.com/default.aspx)

EntertainmentJobs.com (www.entertainmentjobs.com)

Hollywood Creative Directory (www.HCDonline.com/jobboard/default.asp)

JournalismJobs.com (www.journalismjobs.com)

MediaLine (www.medialine.com)

Showbizjobs.com (www.showbizjobs.com)

TV Jobs: Broadcast Employment Services (www.tvjobs.com)

TV and Radio Jobs (www.TVandRadioJobs.com)

TV stations in the country and all cable systems, advertising agencies, networks, program suppliers, equipment distributors, and other related groups.[12]

However, when you send a general application to a company, your chances of getting a reply are low. Obviously, if they do not have any positions available, they cannot hire you. If you find a company you think you would particularly like to work for, research it thoroughly on the internet and steer your cover letter to an area where you think you could be of most value.

If your university has a campus career center, visit it, because the people there have contacts that might lead to job possibilities. In some cities, there are agencies that place people in temporary jobs within entertainment-oriented companies. This is a fertile avenue because if you prove your worth in a temporary job, you may be offered a permanent position. You can also use the services of an employment agency or place a "situations wanted" ad in a newspaper or on an internet site. Neither of these approaches is likely to yield much success, however. Electronic media companies do not have trouble finding potential employees, so they rarely need to look beyond the résumés they receive. **other methods**

Many jobs, particularly in the production area, are **freelance.** You do not have a permanent job; for example, you work as a crew member on a movie, knowing that when the movie finishes shooting you will be out of a job and need to find another one. Taking a low-level (low-pay) freelance job right after college graduation is a good way to get to know other freelancers who may then recommend you for other jobs. "Paying your dues" is an accepted practice in the media field. **freelancing**

You don't need to hold out for your "dream job"; just get your foot in the door. Once you are on the inside, your options become broader. Many organizations hire from within, so you can move about within the company. Also, once you are working in the industry, you can greatly expand your "who you know" network and open up new opportunities for yourself. **moving from within**

Finding a job can be difficult, especially in poor economic times. Some people feel that finding a job is harder work than doing the job. Talk to as many people as you can who have jobs in the field and ask how they got their first job. You will hear many different stories, but some of them may be applicable to you. Also there are many books you can read that give advice for obtaining a satisfying media job.[13]

7.8 Interviewing

Most jobs require an interview. The people who are considering hiring you want to see you in person to check on your qualifications and to determine if you will fit well with others. You should prepare carefully for this interview. If you are interviewing with a station or network, watch or listen to their programming. Most companies have a website; go through it thoroughly so that you can talk intelligently about your potential role in the company and so that you know some background about the people who might be interviewing you. If you can talk about the overall media business in a knowledgeable manner, you will also receive high points. Here is where keeping up on news by reading the trades helps. **preparation**

attire

Plan what you are going to wear. The type of clothes will depend on the type of job and type of company. A suit would not be appropriate if you are interviewing for a grip position but could be appropriate for a sales job. People in small, independently owned companies usually dress more casually than people in large conglomerates. Looking at how people dress in the photos on a company website can give you ideas on what to wear.

time

On the day of the interview, give yourself plenty of time to be sure that you arrive early. Assume you will encounter traffic and parking difficulties. If you are early, you can always walk around the block—a much better alternative than rushing in late, out of breath and flustered. If you wait in a reception area, observe your surroundings. Often companies have awards or examples of their work hanging on the walls that you can refer to during the interview.

procedure

When you meet your interviewer, shake hands firmly, but if you are nervous, sit down as soon as possible so your knees don't shake. Let the interviewer begin the discussion. Small talk will often precede discussion of the actual job, and something you read in the trades or the reception area may come in handy. Throughout the interview, be positive, enthusiastic, and confident. It is appropriate to ask questions about the job, but answer more questions than you ask. Don't be concerned about matters such as vacations and fringe benefits until you are sure someone wants to hire you. Emphasize your strong points that are most closely akin to the job you are applying for, but also maintain flexibility. Occasionally people walk into an interview for one job but come out with a different one. Let the interviewer terminate the interview, but try to orchestrate the end so the interviewer feels that not hiring you will be a loss. One way to do this is to return briefly to some subject discussed earlier that seemed to particularly pique the interviewer's interest.

follow-up

A few days after the interview, send a follow-up note (either an email or a letter). This will bring you to the mind of the interviewer again and serve as a timely refresher in case he or she has been interviewing other candidates. Make the note short, summarizing points about yourself that you think the person liked the best.[14]

7.9 Diversity

The media business has been criticized for being a bastion of white males. In recent decades the percentage of women and minorities has been climbing, but there are setbacks and room for improvement (see "Zoom In" box). The area that has seen the most progress for some minorities is acting. First blacks and then Latinos experienced significant increases in movie and television roles. Asians are less prominent, and Native Americans are practically invisible. Gays have made visible gains in their portrayals, but the disabled complain because production companies often hire able-bodied people, even when the role is that of a disabled person. Most older actors, particularly women, are offered fewer parts than when they were younger.

acting

ZOOM IN: **Women Falling Back**

A study of women in the movie industry in 2007, conducted by Martha Lauzen of San Diego State University's School of Theatre, Television, and Film, found the following:

- There was a 4 percent decrease in the role of women compared to 2001.
- 21 percent of films released in 2007 employed no women as directors, producers, writers, cinematographers, or editors; 0 percent of films failed to employ a man in at least one of those roles.
- Among the 2007 films, women accounted for 6 percent of directors, 10 percent of writers, 22 percent of producers, 17 percent of editors, and 2 percent of cinematographers.

- The number of women directors (6 percent) was down from 11 percent in 2000.
- Women were most likely to work on romantic comedies, romantic dramas, and documentaries and least likely to work on horror, action adventure, and science fiction features.

Why do you think the numbers for 2007 were lower than those of previous years? Why do you think the cinematographer number, in particular, is so low? Why do you think women are more likely to work on romantic comedies than action adventure films?

Behind the scenes, both women and minorities can now move with relative ease into positions such as producers, script supervisors, station managers, and salespeople. Writing and directing are two areas in which women and minorities still lag. Women represent 18 percent of film writers and 27 percent of television writers; the comparable numbers for minorities are 6 percent and 10 percent. Discrimination against older writers is particularly severe. There are more women in top executive jobs than there used to be, but both women and minorities are scarce among union members.

behind-the-scenes

Diversity is a subject discussed frequently by media practitioners and politicians, and there are organizations that work to further career possibilities and networking opportunities for women and minorities (see Exhibit 7.11). Most of the flagrant favoritism toward white males that existed in the early decades of radio and television is gone, but overall the field has made less progress in diversifying itself than many other occupations.[15]

7.10 Unions and Agents

Some jobs require membership in a **union** or **guild.** (Unions and guilds have different historic antecedents, but in today's society they are very similar.) These organizations negotiate with networks, stations, production companies, and others to make sure their members receive proper pay and working conditions. Getting into a union can be a catch-22; you must have a job to join a union, but you must join the union to get a job. Breaking into that vicious circle is often difficult.

job requirements

Exhibit 7.11

These media
organizations help
promote diversity.

Diversity Organizations

Asian American Journalists Association (www.aaja.org)

National Association of Black Journalists (www.nabj.org)

National Association of Hispanic Journalists (www.nahj.org)

Native American Journalists Association (www.naja.com)

National Association for Multi-Ethnicity in Communications (www.namic.com)

Native American Public Telecommunications (www.nativetelecom.org)

National Black Programming Consortium (www.nbpc.tv)

National Hispanic Media Coalition (www.nhmc.org)

National Lesbian & Gay Journalists Association (www.nlgjany.org)

Unions allow nonunion workers to be hired when all the union members are employed—something that is most likely to happen when many movies and TV pilots are being produced. The people in your network can most easily hire you at that time and shepherd you through the union membership process. Most of the crew-oriented entertainment production unions (e.g., those for cinematographers, makeup artists, set designers, script supervisors, etc.) are under the jurisdiction of the International Alliance of Theatrical Stage Employees (IATSE).

guilds

The Screen Actors Guild (SAG) and the American Federation of Television and Radio Artists (AFTRA) are the main labor organizations for actors. Established directors belong to the Directors Guild of America (DGA), producers belong to the Producers Guild of America (PGA), and writers have the Writers Guild of America (WGA).[16]

nonunion

Of course, there are many electronic media jobs, such as those in management and sales, that do not require union or guild membership. Even within production and news, where the unions are most visible, you can work without being a union member. Radio and TV stations in small markets, some cable TV networks, independent movie production companies, internet organizations, video game producers—most of these are not unionized. Many recent college graduates work freelance on independent movies, albeit for much less money than union members who are working on major Hollywood movies.

agents

Actors, musicians, writers, directors, and other creative people often need agents. Their function is to find work for their clients in return for a percentage— usually 10 to 15 percent. Many production companies will not deal directly with actors and others; they deal only through agents. Newcomers have little chance of being noticed without an agent. Getting an agent when you have little experience can be difficult. It is much easier for an agent to find work for someone with an established name than someone in the "Who's she?" category. The best way to attract an agent is to have completed a sizable body of work (community playhouse, actors' workshops, scripts) to show that you are really serious about working in the business.[17]

7.11 Compensation

On the average, electronic media jobs are not especially high-paying. True, su-   **pay**
perstars and some top executives make millions, but they are few and far between. Others earn a decent but not sensational living. Two good sources for
up-to-date figures on salaries are www.tvjobs.com and www.rtnda.org. There is
no standard pattern on fringe benefits, but many employers provide medical
insurance and vacations for full-time, but not freelance, employees. People who
belong to unions can often obtain benefits from them.[18]

What undoubtedly attracts most people to electronic media is the glamour,   **other benefits**
excitement, and power of the profession. All of these are present (albeit to a lesser
degree than most people think), and they do make for a richer, more rewarding
life than many people experience in other day-to-day occupations.

7.12 Types of Job Possibilities

You don't need to decide the particulars regarding the type of electronic media job   **variations**
you want. Once you obtain a position, you can move around fairly easily. In fact, as
already implied, sometimes the industry does the moving for you. It changes so
rapidly that you are unlikely to keep the same job doing the same thing for your entire career. Also, your job may vary depending on the organization you work for.
At a small, independently owned radio station, the person hired to oversee programming might also work an on-air shift, sell ads, produce commercials, and even
sweep the floor. At a large network, the programming department may have many
people, with specific employees in charge of comedy, drama, daytime programs,
and children's programming. The type of media is another factor. The sales department at a commercial TV station is primarily concerned with selling to advertisers.
Within a satellite TV system, the selling is aimed toward subscribers.

Some of you may know exactly what you want do—"I've wanted to direct
movies since I was five years old." Even so, you need to know how the industry
operates, and you need to explore the paths you can take to reach your goal. Others may be open to any option but want to narrow the field of possibilities.

One method for determining what direction(s) you might focus your career on is   **narrowing**
to make a matrix (see Exhibit 7.12). Down the side, you can list all the media forms   **the field**
that have been discussed so far in this book. Across the top, you can list job possibilities. To get started, you can list the categories discussed in the remainder of this chapter. However, as you read this chapter and the rest of the book, you will discover
specific jobs that are excellent entry-level positions, and you can add them to your list.
Once you finish your matrix, you can combine various media and jobs and ask yourself what interests you most. Would you consider a sales job in radio? Would you like
to be an administrator of an internet start-up? Do you want to be a news reporter for a
cable TV channel? If you find there are areas that don't interest you, eliminate them.

7.12a Production

Production jobs are the ones people think of most readily when considering
job possibilities in electronic media. Everyone has heard of **actors, directors,
producers,** and **writers,** and many people are attracted to the media because they

Exhibit 7.12 Building a matrix like this may help you decide what jobs you want—and don't want.

Job Matrix						
	Production	News	Programming	Engineering	Sales	Administration
Internet						
Video games						
Commercial TV						
Public TV						
Cable TV						
Satellite TV						
Home video						
Corporate video						
Telephones						
Commercial radio						
Public radio						
Satellite radio						
Movies						

above-the-line

below-the-line

want to fill one of those roles. These four positions are generally referred to as **above-the-line** jobs and, as you might expect, they are the most difficult jobs to obtain. The good news is that there are many other creative jobs, usually referred to as **below-the-line** positions, that are excellent careers in themselves but can also lead to other jobs. (The terms "above-the-line" and "below-the-line" come from production budgets, in which costs associated with actors, directors, producers, and writers are summarized at the top. Then a line is drawn, and below that line are the costs associated with craft and technical jobs.)[19]

All you need to do is take a look at the closing credits for a movie or TV show to see that there are often hundreds of people involved in the production process. Some jobs, such as camera operator and editor, are fairly well understood, but other jobs, such as the ones listed in Exhibit 7.13 are lesser known but fulfilling. How specific jobs are executed depends, to some degree, on the **type of production**. Game shows are produced in a **studio** with multiple cameras sending signals to a nearby **control room** where the director and other crew members sit to select the pictures and sound that will be recorded. Documentaries are shot at whatever location is appropriate, usually with only one camera. Sports events come live from sports venues and involve roving cameras and a truck that acts as a control room. It is possible that during your career you will operate a camera for all three types of productions, but the techniques needed for each will be different.

type of production

other areas

Areas other than television and movies also include creative production jobs. Radio disc jockeys create their banter and interact with the music they play. DVD authors design the menus and placement of materials on DVDs. Webmasters handle the design and creation of websites (see Exhibit 7.14). Video game creators constantly push the envelope as advancing technologies allow them to be more creative.

corporate video

You might want to consider working in **corporate video,** a fertile area that job applicants often ignore. Many organizations, such as governments, churches, hospitals, charitable organizations, and educational institutions, make use of

Selected Below-the-Line Production Jobs

boom operator: holds the microphone on a long pole above the talent during production.

color timer: makes sure the colors in the frames of the final product are consistent and in line with what the director wants.

Foley artist: creates noises such as walking on gravel or clinking of glasses to accompany a movie.

gaffer: heads the lighting crew as they position the lights.

lead person: supervises the set dressers—the people who buy and arrange pillows, magazines, and other elements that give a proper ambience to the set.

location scout: finds appropriate places to shoot a production that are in line with what the script needs.

matte artist: creates computerized backgrounds that are part of visual effects.

production designer: oversees the "look" of a production, making sure the set, props, costumes, makeup, etc. work together and are consistent with the needs of the story.

property master: finds or constructs props and makes sure they are properly placed on the set.

pyrotechnician: oversees the setting of fires that are part of special effects.

rigging grip: sets up lights that are near the ceiling of a studio.

script supervisor: works closely with the director making sure all the scenes are shot and making notes about what actors are wearing and doing so that there are no continuity problems in the finished product.

sound designer: oversees how sounds such as music, sound effects, and dialogue will be incorporated into the movie and coordinate with the picture needs.

stand-by painter: touches up paint that gets damaged during rehearsal or production.

technical director: operates the switcher to select shots during a studio shoot.

telecine operator: transfers film to a digital video medium such as tape or a hard drive.

Exhibit 7.13

A list of some less commonly known movie and television production positions.

Exhibit 7.14

Designing a website demands both technical and artistic skills. One of the most recognizable sites, Google, is known for its simplicity—which it has stayed with throughout the years. This is one of its first website designs from when it started in 1998.

(Screenshot courtesy of Google, Inc. Used by permission.)

video, the internet, DVDs, and other media to enhance their operations (see Exhibit 7.15). Finding a job in this area, either freelance or full-time, is somewhat easier than finding one in conventional media, and many people are able to move back and forth between corporate video and other fields.

event video

Somewhat related to corporate video is **event video.** Even as a college student, you can earn decent pay recording and editing weddings, birthday parties, retirement parties, and the like. For production practice, you can produce material for social networking internet sites, but at present there isn't an obvious way to earn money in that field.

From One Generation to Another
"One generation will commend your works to another; they will tell of your mighty acts." Psalm 145:4

VIDEO

Here are links to five videos about Zion and about the new stewardship campaign. Enjoy!

	Click this link to play video	Description	Length
	Pastor.avi	An introduction to the stewardship campaign from Senior Pastor Mark Rossington	4:52
	Schools.avi	A video about Zion's schools	3:22
	History.avi	A video about Zion's history	6:38
	Oswald.avi	A skit in which Oswald learns about stewardship from his sidekick Eddie	5:46
	Helicopter.avi	A birds-eye fly-over of Zion's campus	1:29

Exhibit 7.15

An opening page of a church's CD-ROM, which uses photos, text, and hyperlinks to video clips.

(Courtesy of Zion Lutheran Church, Anaheim, California)

In short, there are many opportunities related to production. You need to understand the basics of equipment and aesthetics, something that is usually taught at the college level in production courses. Reading about production helps, but undertaking projects for fun or profit is what will sharpen your skills and prepare you for when that "dream job" comes along.[20]

7.12b News

News can be considered one type of production. It involves cameras, microphones, sets, and many elements common to other programs. However, most of the people in the news field are trained in broadcast journalism because news has special laws and reporting techniques. Generally a **news producer** oversees each individual newscast, organizing it and deciding what to include. There are no actors; rather there are **news reporters** and **anchor people** (see Exhibit 7.16). There may be writers, but they mostly rewrite material that has come from other sources rather than creating their own material. In addition,

Exhibit 7.16

Radio news anchor Tammy Trujillo undertakes various tasks. Obviously, she reads the news from a script, such as the one seen on the computer screen to her right. She also does her own engineering, operating an audio board and audio playback devices. She follows the list of items to be broadcast on the screen to the left and signs off that each element has aired. She also looks for new stories. She can use the monitor to the far right to check video feeds from television stations or news agencies and a monitor directly in front of her to access databases and the internet.

there are **assignment editors** who keep track of stories than need to be covered and send reporters to cover these stories. News people can work for stations and networks, or they can work for other companies that supply news, such as **news agencies** or companies with helicopters that broadcast traffic information to a number of stations. Reporters generally have no idea what they are going to do when they report to work. Their duties evolve as the events of the day evolve. Some news operations work toward specific deadlines, such as the 5:00 P.M. news. Others, such as 24-hour news services, have constant deadlines.

structure

There are differences in news procedures for different media forms. Radio is obviously different from television in that radio doesn't have pictures and therefore reporters, writers, and anchors need to provide mind's eye pictures for the listeners. For radio, reporters and anchors operate their own equipment. Most television reporters have camera operators with them, although some stations require reporters to operate their own cameras by placing them on a tripod and standing in front of them for their introductions and interviews. Anchors in a TV studio may operate the **teleprompter** on which their script is written, but there are crew members to operate most of the studio and control room equipment. Many people supply news on the internet, some of whom are **bloggers** with no training in

variations

journalism, but generally they are not paid. The people who select and organize the news for websites of news organizations are employees of those organizations.

The news business can be gritty in that you often wind up at fires, floods, or shootings, and you are expected to ask hard questions of people who do not want to be questioned. But you are at the forefront of current events, and you meet fascinating people from all walks of life.[21]

7.12c Programming

different systems

Programming jobs encompass deciding what should be distributed to the public and when. Some programmers have very little choice; they simply pass through what the network or networks send them. A programmer for a local cable TV system, for example, has almost no say as to what is shown on ESPN, MTV, or any other cable network, but often that person has a say as to which cable networks the system will carry. In other situations, particularly at the broadcast and cable networks, the programming department has a great deal of power because it is in charge of the company's primary product.

The heads of programming—usually called program directors, program managers, or vice presidents of programming—do not make major programming decisions by themselves. The sales department has input in terms of what programs will be the easiest to sell to advertisers. Top management often has a major say regarding programming; in fact, if program acquisition is fairly uncomplicated, a general manager may handle it, and there will be no program manager. But programmers

hot seat

who are on the hot seat at major networks have a hectic, and often short-lived, existence. There is no routine way to produce hits. The public is fickle, and what works one month may not work the next. The program head who is perceived to lose touch is quickly replaced. But the programming department can be an exciting, powerful place, and programmers work with very creative people as they find, develop, and schedule programs. This process will be discussed in Chapter 8.

other programming

The word *programming* has several other meanings. Within the computer world, programming refers to writing the code that controls computer functions. Programming a video game, for example, does not refer to scheduling but rather to creating it in such a way that it will respond to the actions of the players. Consumers also program their TiVos or other recording devices to record a program at a particular time and store it for later viewing. In this case, consumers are taking away some of the power of network programmers. Programming also has a different connotation on the internet. Anyone can be a programmer and place material on a website, and anyone can access website pages whenever they want.[22]

7.12d Engineering

jobs

The people who develop, operate, maintain, and install equipment and systems needed for producing and distributing material for various electronic media are often referred to as engineers. Sometimes they work within a company department called engineering or technical services. The current trend, however, is to hire many of these people, especially those who operate and maintain equipment, on a freelance rather than a permanent basis. The people who operate cameras, audio equipment, and the like are often assigned to programming or news departments, depending on where they do most of their work.

Engineers are very important when an electronic media entity is starting. **start-up**
In fact, it is technical people who usually design and test new technologies such as
the **internet, cell phones, satellites, fiber optics,** or **high-definition TV.** Equipment companies, such as Sony and Panasonic, hire engineers to develop new products and improve old ones. Distribution systems must operate technically before
programming can be sent out to listeners, viewers, or surfers (see Chapter 13).

In electronic media companies that have engineering departments, the head
of the department is called the chief engineer. Among this person's duties are **duties**
scheduling the people and equipment, tending to the transmitter or satellite
dishes, overseeing maintenance, and making recommendations to top management regarding new equipment purchases. In the cable TV business, the chief engineer oversees the installation of the cable in the community.[23]

7.12e Sales

Sales jobs are easier to attain and usually more lucrative than production jobs.
There are many jobs attached to the sales function because sales represents the
lifeblood of most media forms. Without income a business cannot exist, and people involved with sales are the main ones who bring in the money.

In many companies, sales and marketing are in the same department. Usually **marketing**
sales involves direct selling, and marketing involves everything else—hosting
lunches for potential advertising clients, preparing materials to include in telephone
bills to get people to sign up for more services, working with the programming department to develop shows that appeal to potential customers. A combined sales and
marketing department is usually headed by a vice president of marketing. The head
of sales, usually called a sales manager or sales director, is under the vice president.
This person hires, trains, and evaluates salespeople; sets sales policies; controls sales
expenses; and communicates with other departments within the company.

The main element sold in many media companies is advertising time. As a
result, audience research is often part of the sales department because **ratings** are **research**
important to selling time. Ratings are gathered by outside companies, such as
Nielsen and Arbitron (see Chapter 10), but people at a station or network need to
know how to read the reports and interpret them so that the sales force can use the
figures that most enhance the particular situation. People who are interested in audience research are highly valued in the industry and can easily find work.

Traffic is another area connected to sales—one that provides excellent **traffic**
entry-level positions. The department's responsibility includes listing all the programs and commercials to be aired each day in a **log** (see Exhibit 7.17). Traffic is
usually part of the sales department because the most important job involved with
the log is scheduling the various commercials. After sales are made, someone,
with the aid of a computer, must make sure each advertiser's schedule of spots is
actually aired under the conditions stipulated in the sales contract.

Many media entities sell hookups or subscriptions directly to consumers. For
example, cable TV systems, satellite companies, telephone companies, and computer online services sell their services to individuals. Although this is sales, companies sometimes place this function in a department called **customer service.** **customer**
This is a softer term intended to make the customers believe they are being **service**
serviced rather than sold goods. Customer service also handles complaints and

Time Scheduled	Programs/Announcements	Length	Source	Type	Actual Time	Remarks
8:00:00	News of the hour	5 min.	Net	N		
	Cancer Fund	30 sec.	LS	PS		
	Mason Ford	30 sec.	Rec	C		
8:05:00	Steve Stevens Show	23 min.	LS	E		
	Pop-a-Hop	30 sec.	Rec	C		
	Arbus Tires	30 sec.	Rec/LS	C		
	C-C Cola	15 sec.	Rec	C		
	Articycle	30 sec.	LS	C		
8:28:00	Station Break	2 min.	LS	SI		
	Crisis Helpline	30 sec.	LS	PS		
	Richers	30 sec.	Rec	C		
8:30:30	News and Weather	2 min.	LS	N		
	Sid's Market	30 sec.	LS	C		
8:32:00	Steve Stevens Show	13 min.	LS	E		
	Fuel 'n' Lube	30 sec.	Rec	C		
	Coming Events	1 min.	LS	PS		
8:34:00	Sports Special	30 min.	LR	S		

Source Abbreviations: LS-Local Studio; LR-Local Remote; Net-Network; Rec-Recorded
Type Abbreviations: PS-Public Service Announcement; C-Commercial; PR-Station Promo;
SI-Station Identification; E-Entertainment; N-News; S-Sports;
R-Religious; PA-Public Affairs

questions, so this is an easy place for someone to get a foot in the door. Cable TV systems often have both types of sales. They sell hookups to customers, and they also sell advertising for their local channels or for insertion within network shows. Sometimes both functions are placed in a sales department, but more often cable systems have both a sales department and a customer service department.

distributors   Some electronic media products are sold through distributors and retailers, just as clothing and food are. DVDs and console-oriented video games are examples. Movies are also sold through a slightly different distribution process that involves agreements between the production studios and the movie theaters.

promotion   Another area closely related to sales is **promotion,** which involves getting out the word about a program, station, network, movie, internet site, video game, telephone enhancement, and so on. Promotion is much more important today than it used to be. When three broadcast networks dominated, they could do most of their promotion during late August right before they started their new seasons. Now programs change frequently, and people have so many listening and viewing choices that all media forms must constantly remind people to partake of their product. The sales department works closely with promotion because salespeople cannot sell ads for programs that do no have an audience. Chapter 9 deals in more detail with information pertinent to sales and advertising, and Chapter 10 deals with promotion and audience research.[24]

7.12f Administration

Electronic media companies are businesses, many of them with shareholders or owners who expect a financial return for investing in the business. Therefore, media companies have many administrative positions that are similar to those of other businesses. For example, most companies have people who handle finances. Often headed by a **chief financial officer (CFO),** finance departments are involved in many of the decisions that affect all aspects of the company. They are charged with keeping companies on the straight and narrow and warning top executives if profits or stock prices appear to be in danger of falling. **Controllers** oversee profit and loss and expenditures, and **treasurers** handle cash, making sure the company has enough money on hand to pay its bills and also investing the company's money so that it will earn interest. Accountants establish and oversee computerized ledgers and print out various reports that keep track of the financial health of the company (see Exhibit 7.18). The finance department pays the bills and collects the money owed to the company. If someone (such as an advertiser) does not pay a bill, it is the finance department's job to see that the money is collected.

finance

Financing within the movie industry has a rather unsavory reputation. The business is so risky that it is hard to obtain financing, and, once obtained, the nature of moviemaking allows room for "creative bookkeeping." Actors, writers, directors, and others often negotiate to receive a percentage of the profit from the movie, and many times they believe they are cheated because the point at which the movie makes a profit is obscured in statistics. The internet business, too, has employed accounting procedures that led to both an unjustified rise and a disastrous fall in stock prices (see Chapter 2).

Some companies, especially large ones, have human resources departments. Headed by a director or manager, these departments handle procedures related to hiring and firing. They place ads to fill openings within the company, screen résumés, and recommend a number of candidates to the person who is trying to fill the vacancy. If an employee is to be fired, the human resources department makes sure proper procedures are followed, such as giving the person several notices of poor performance before the actual firing. Human resources is also responsible for handling the company benefits package and for making sure the company abides by equal employment guidelines in hiring, promotion, and layoffs. If there are unions within the company, human resources is often in charge of overseeing union–management negotiations. In smaller companies that don't hire people very often, human resource duties may be handled by a staff member who has other duties.

human resources

All companies have a "boss"—a "buck stops here" person. The title given this person varies from one situation to another. At radio and TV stations and cable systems, the person is usually called a station manager or a general manager. At networks and internet companies, the person is referred to as a president. Often the term **chief executive officer (CEO)** designates the person at the top of the organization. There are several ways to become a CEO. One is to form your own company, sometimes in conjunction with several other people. Many internet

CEOs

Exhibit 7.18

A typical balance sheet, one of the many forms the finance department prepares on a regular basis.

Balance Sheet
As of ______________
(*Date*)

	This Year	Last Year
Assets		
Current Assets		
Cash	______	______
Current investments	______	______
Receivables, less allowance for doubtful accounts	______	______
Program rights	______	______
Prepaid expenses	______	______
Total current assets	______	______
Property, plant & equipment at cost	______	______
Less: Accrued depreciation	______	______
Net property, plant & equipment	______	______
Deferred charges and other assets	______	______
Programming rights, noncurrent	______	______
Intangibles	______	______
Total assets	$______	$______
Liabilities & Stockholders Equity		
Current Liabilities		
Accounts and notes payable	______	______
Accrued expenses	______	______
Income taxes payable	______	______
Total current liabilities	$______	$______
Deferred revenue	______	______
Long-term debt	______	______
Other liabilities	______	______
Stockholders Equity		
Capital stock	______	______
Additional paid-in capital	______	______
Retained earnings	______	______
Treasury stock	______	______
Total stockholders equity	$______	$______
Total liabilities & stockholders equity	$______	$______

start-ups have been formed that way with the founders becoming instant CEOs—and in some cases instant millionaires. Another way is to inherit a company, as is the case with the current CEOs of the cable TV company Comcast, the radio company Clear Channel, and even the conglomerate News Corp. that Rupert Murdoch inherited from his father. Another method of becoming a CEO is to rise through the ranks, as the current CEOs of Time Warner, Disney, Viacom, Sony, and NBC Universal have done. You do not need to spend your entire career with one company to become a CEO. Sometimes companies bring in top people from other companies (see "Zoom In" box).[25]

ZOOM IN: **Two Paths to Glory**

There is more than one route to the top. Jeff Bezos (see Exhibit 2.11) did it by forming his own company, and Howard Stringer did it by working his way up in two major companies.

Jeff Bezos, founder of amazon.com, was born in 1964 in Albuquerque, New Mexico, and showed mechanical aptitude at a young age. After graduating from Princeton, he worked on Wall Street helping create computer networks. He had a promising future there, but he also knew that internet usage was growing and, after some methodical research, decided that book distribution would be a good internet business. In 1994 he moved to Seattle, where he could have access to a book wholesaler, and started amazon.com, named after the Amazon River with its numberless branches. The company grew quickly and inspired other online retailers but hit a rough spot when Barnes and Noble launched a rival website. Bezos remained optimistic even when amazon.com's stock dropped from $100 to $6 a share. Armed with his ability to handle numbers and statistics, he continued to grow the company and offered many more products. He became involved in other projects, such as Kindle, a portable reading device that can be used to download books from amazon.com, and a human spaceflight project. He is ranked 35 on a list of the world's wealthiest people.

Howard Stringer, chairman and chief executive officer of Sony, was born in Cardiff, Wales, in 1942 and obtained a B.A. and M.A. in modern history at Oxford University. He is actually Sir Howard Stringer, having received the title of Knight Bachelor from Queen Elizabeth in 1999. He moved to the United States in 1965 and had a 30-year career at CBS, where he earned nine individual Emmys as a writer, director, and producer. He held numerous positions at CBS News, including executive producer of the *CBS Reports* documentary unit and executive producer of the *CBS Evening News*.

Howard Stringer.

(Photo courtesy of Sony Corporation of America)

From 1988 to 1995, he was president of CBS. Among his accomplishments were convincing David Letterman to bring his show from NBC to CBS and presiding over CBS's rise from last to first place in 1993. In 1995 he left CBS to become CEO of a new company, TELE-TV, founded by three phone companies to deliver home entertainment and information services. In 1997 Sony hired him as president of Sony Corporation of America. His appointment to the position of chairman and CEO of Sony Corporation in 2005 surprised many because he is the first foreign-born CEO of a major Japanese electronic company.

If you have aspirations to head a large company, which path do you think you would prefer? Which people do you think have the greatest chance of holding their jobs the longest? Which do you feel can have more of an impact on society?

The duties of those in top management vary, but generally they provide the leadership and direction for the unit they oversee. They set goals for the company, trying to keep in mind both long-term and short-term success. They stand at the front of the firing line if a crisis occurs. They are the ones who must decide what to do if the competition is trying to hire away a well-liked anchor, if sales are lower than predicted, if the stock prices take a dangerous dip. Top management also deals with the outside world. Executives join professional and community organizations, and they try to anticipate the competition's next move. One of the most important jobs of top managers is to hire and motivate good people. They have to create and oversee a team of department heads who work together for the overall good of the company, and they must mediate if departments collide. They find means, other than pay increases, to keep employees involved and productive when times are bad.

lawyers

The electronic media business is awash with lawyers. Most companies have them on staff or on call because there are so many laws and regulations that affect media (see Chapter 11). Many producers, programmers, marketing heads, finance executives, and CEOs have law degrees because so much of what they do is of a legal nature. Having an entertainment law degree makes a person more employable in many types of media occupations.[26]

7.13 Issues and the Future

Although this is a relatively good time to enter the electronic media business, finding jobs in this area has never been easy. Hard work will continue to be the norm. The internet should grow in importance—both as a source of employment and as a method for finding jobs.

diversity

Diversity will continue to be an issue. By demonstrating their skills, women and minorities may work their way into more areas that are still the bastion of white males, but the organizations that will probably provide the most stumbling blocks are the unions.

project orientation

Some media companies are now project oriented rather than function oriented. A group of people is hired to do one public relations campaign, one program, or one sales push. At the end of the project, they are let go and other groups are hired for other projects. This form of organization presents both opportunities and challenges for those looking for work.

programming

The functions needed to operate media businesses will remain intact, but the one that is most likely to change is programming. Now that people have so many options for viewing programs when they want to view them, the programming department needs to put more effort into leading people to particular programs on all avenues of distribution. This will have an impact on sales because commercials may not always follow along with the programs. There are many opportunities for people new to the business to come up with business models that can produce profits in a constantly changing technological world.

Regardless of what job you have, you will always need to understand the broad processes that operate in electronic media. A set designer who has two job offers needs background information about the industry in order to decide which job might have the greatest longevity. A salesperson who is considering applying for a managerial position needs to know the latest trends in order to decide whether such a move is likely to be financially rewarding. The following chapters of this book deal with the major aspects of the business by discussing programming processes, sales and advertising, promotion and ratings, laws and regulations, ethics, the technical underpinnings of the media, and the important role that the world at large plays in the electronic media scene.

industry process

7.14 Summary

Whether you are interested in electronic media jobs in production, news, programming, engineering, sales, or administration, you need to first consider whether you have most of the traits needed for the field. While in college, you should take not only electronic media courses but others such as political science (for news) and accounting (for administration). A part-time job or internship while you are in college will help build your network. As you begin looking for a job, prepare a polished résumé and, if needed, a well-produced reel. Search for jobs in a multitude of ways—contacts, the internet, lists of companies, career centers. Also be aware of diversity issues and the possible need to join a union.

Don't expect high pay, especially in the most competitive and artistic areas such as production and news. In these areas, below-the-line positions are easier to obtain than above-the-line, but there are many new and interesting positions in production and news that people don't always consider. Programming is undergoing many changes, making it an exciting spot for people with innovative ideas. You have a chance of doing better financially in engineering, sales, and administration positions. Engineering requires technical ability and is at the forefront of invention. Sales and related areas are important to a company's financial health and also include exciting, little-known jobs. Administration can be rewarding, especially if you become a successful CEO, but lawyers and finance people also have productive professional lives.

Suggested Websites

www.internshipprograms.com (a site where you can search for internships of any type in any part of the country)

www.rileyguide.com/interview.html (advice for interviewing)

www.aftra.com/aftra/aftra.htm (the site for the American Federation of Television and Radio Artists, a union for actors and on-air news people)

www.dga.org (the Directors Guild of America site)

www.rab.com (the Radio Advertising Bureau site; contains information about how and why to advertise on radio)

Notes

1. J. Michael Farr and Laurance Shatkin, *Best Jobs for the 21st Century* (Indianapolis, IN: Jist Works, 2003); Linda Buzzell, "Reel Life Meets Real Life," *Emmy*, February 2002, 38–39; Michael T. Robinson, "Top Jobs for the Future," http://www.careerplanner.com/Career-Articles/Top_Jobs.htm#TopJobs (accessed June 15, 2008); and U.S. Department of Labor, Bureau of Labor Statistics, "Occupational Information Network Coverage," http://stats.bls.gov/oco/oco2007.htm (accessed June 15, 2008).

2. For other viewpoints on traits needed by electronic media practitioners, see "Get a Life," *Emmy*, November/December 2007, 32–33; "Four Kinds of People," *TV Technology*, August 22, 2007, 36; and "Getting the Ol' Crew Together," *P3*, February 2008, 12.

3. Academy of Television Arts & Sciences, *Television, Film, and Digital Media Programs* (New York: The Princeton Review, 2006); and Garth Gardner and Christina Edwards, *Garner's Guide to Selecting a Multimedia School* (Washington, DC: Garth Garner Company, 2005).

4. Ann C. Hollifield, Gerald M. Kosick, and Lee B. Becker, "Organizational vs. Professional Culture in the Newsroom: Television News Directors' and Newspaper Editors' Hiring Decisions," *Journal of Broadcasting & Electronic Media*, Winter 2001, 92–117.

5. Daniel B. Wackman, *Media Executives' Opinions about Media Management Education* (Minneapolis, MN: Media Management and Economics Resource Center, n.d.).

6. For information on contests and festivals, go to http://www.studentfilmmakers.com/classifieds/view.php?cat=46 or http://www.withoutabox.com.

7. "Interns Learn a Lot, But It's Rarely What They Were Expecting," *Wall Street Journal*, May 11, 2005, B-1; Mark Oldman, *Vault Guide to Top Internships* (New York: Vault Career Library, 2008); Garth Gardner and Bonny Ford, *Gardner's Guide to Internships in New Media 2004* (Washington, DC: Garth Gardner Co., 2004); and "America's A-List Internships," *Wall Street Journal*, February 18, 2005, W-1.

8. For a list of additional organizations, go to http://www.1800miti.com/associations/92-broad.html.

9. "Your Most Important Product," *Emmy*, February 2001, 34.

10. "A Real Good Reel," *Emmy*, February 2001, 35.

11. See also Christina Edwards, *Gardner's Guide to Finding New Media Jobs Online* (Washington, DC: Garth Garner Co, 2004); and "Job Sites," *Daily Variety*, September 15, 2004, A-8.

12. The website for information about *Broadcasting & Cable Yearbook* is http://www.bowker.com/index.php/component/content/article/33. You can also obtain station information from www.AMFMTVOnline.com.

13. Many books offer advice about getting a job in the electronic media field. Some examples are Steve Heller and David Womack, *Becoming a Digital Designer: A Guide to Careers in Web, Video, Broadcast, Game, and Animation Design* (Hoboken, NJ: Wiley, 2008); Fran Harris, *Crashing Hollywood* (Studio City, CA: Michael Wiese Productions, 2003); Chris Schneider, *Starting Your Career in Broadcasting* (New York: Allworth, 2007); and Donald E. Freguson and Jim Patten, *Opportunities in Journalism Careers* (Lincolnwood, IL: VGM Books, 2001).

14. Joyce Lain Kennedy, *Job Interviews for Dummies* (Hoboken, NJ: Wiley, 2008); and "That Dressed for Success Look," *Emmy*, February 2001, 65.

15. The facts in the box were based on "Report: Women Jobs Down," *Hollywood Reporter*, February 1–3, 2008, 103. Other sources include "The Picture Isn't Perfect, But It Is Better," *Broadcasting & Cable*, February 7, 2005, 22; "Disappearing Act," *Emmy*, November/December 2005, 86–89; and "Some Screenwriters Are More Equal Than Others," *Wall Street Journal*, October 25, 2005, D-8.

16. For more information see http://www.iatse-intl.org/home.html; "S.F. Unit Asks SAG to End Drive against AFTRA Pact," *Hollywood Reporter*, June 24, 2008, 13; and "SAG Calls Off the Wedding," *Daily Variety*, July 2, 2003, 1.

17. Lists of agents can be obtained from some of the guilds, such as the Screen Actors Guild, 5757 Wilshire Blvd., Los Angeles, CA 90036, or the Writers Guild of America, 7000 W. Third St., Los Angeles, CA 90048.

18. The full web addresses for the salary pages of the sites mentioned are http://salaries .tvjobs.com/cgi-bin/index.cgi and http://www.rtnda.org/pages/media_items/about-salary-data 489.php.

19. The video series *Journeys Below the Line* (for which the author is the Associate Producer) gives an overview of craft and technical jobs in television. It can be obtained from First Light Media Publishing, www.firstlightvideo.com. Background instructional information can be found at www.journeysbelowtheline.com.

20. Some of the books that deal with video production are David Reese, Lynne Gross, and Brian Gross, *Audio Production* (Boston: Focal Press, 2009); Stanley R. Alten, *Audio in Media* (Belmont, CA: Wadsworth, 2007); Lynne S. Gross and James C. Foust, *Video Production* (Scottsdale, AZ: Holcomb Hathaway, 2009); Herbert Zettl, *Television Production Handbook* (Belmont, CA: Wadsworth, 2005); Lynne Gross, *Digital Moviemaking* (Boston: Cengage, 2009); and Norman Medoff and Edward J. Fink, *Portable Video* (Boston: Focal Press, 2007).

21. A few of the books that deal with news are Mitchell Stevens, *Broadcast News* (Belmont, CA: Wadsworth, 2004); C. A. Tuggle, Forrest Carr, and Suzanne Huffman, *Broadcast News Handbook* (New York: McGraw-Hill, 2006); and Ted White, *Broadcast News Writing, Reporting and Producing* (Boston: Focal Press, 2005).

22. For more information, see Philippe Perebinossoff, Brian Gross, and Lynne Gross, *Programming for TV, Radio & the Internet* (Boston: Focal Press, 2005); and Susan Tyler Eastman and Douglas A. Ferguson *Broadcast/Cable/Web Programming* (Belmont, CA: Wadsworth, 2001).

23. For more on engineering, see Graham A. Jones, *A Broadcast Engineering Tutorial for Non-engineers* (Boston: Focal Press, 2005); and Jerry Whittaker, *Standard Handbook of Broadcast Engineering* (New York: McGraw-Hill, 2005).

24. Several books that deal with the sales area are Charles Warner and Joseph Buchman, *Media Selling: Broadcast, Cable, Print, and Interactive* (Ames: Iowa State Press, 2003); Paul Weyland, *Successful Local Broadcast Sales* (New York: American Management Association, 2007); Susan Tyler Eastman, Douglas A. Ferguson, and Robert Klein, *Media Promotion & Marketing for Broadcasting, Cable & the Internet* (Boston: Focal Press, 2006); and Mario Pricken, *Creative Advertising* (New York: Thames and Hudson, 2008).

25. Material for the box on Bezos and Stringer was obtained from various sources, including http://www.sony.com/SCA/bios/stringer.shtml; http://www.woopidoo.com/biography/jeff-bezos/index.htm; and http://www.achievement.org/autodoc/page/bez0bio-1.

26. Books dealing with aspects of administration include Walter McDowell and Alan Batten, *Understanding Broadcast and Cable Finance* (Boston: Focal Press, 2008); Alan B. Albarran, *Management of Electronic Media* (Belmont, CA: Wadsworth, 2005); T. Barton Carter, Juliet Lushbough Dee, and Harvery L. Zuckman, *Mass Communication Law in a Nutshell* (St. Paul, MN: West, 2007); and Peter Pringle and Michael F. Starr, *Electronic Media Management* (Boston: Focal Press, 2006).

Chapter 8

PROGRAMMING

Programming involves deciding what should be distributed to the public and when. In a limited way, you are a programmer when you decide *what* to put on your website or Facebook page. Something on the internet can be accessed by anyone at any time, so *when* is not as important. But perhaps you want to give a tribute to a friend, so you decide to post a photo of the two of you on your friend's birthday. Actual programming, of course, is a more complicated process, but it should come as no surprise that the internet has influenced the whole programming process. In fact, every time there is a successful new distribution technology, it disrupts the concepts of programming.

> Television has learned to amuse well; to inform up to a point; to instruct up to a nearer point; to inspire rarely. The great literature, the great art, the great thoughts of past and present make only guest appearances. This can change.
>
> **Eric Sevareid, longtime CBS commentator**

In the early days, when there were just three television networks, most of their strategizing revolved around competing with each other. As cable and satellite brought many more channels, programmers needed to think about how to brand their services to attract specific audiences. With the coming of the internet, programmers have had to design new ways to combine older programming practices with various facets of interactivity.

Many other factors affect programming. One of the most challenging is societal changes. The neatly packaged nuclear family sitcoms of the 1950s don't work in today's society. Audience fatigue is another factor. When MTV was founded in 1981, its programming revolved totally around music videos. The young people of the 1990s grew tired of that, and MTV had to retool to include dramas, sitcoms, and other forms of programming. In the 2000s, the MTV audience was flocking to the internet, so MTV developed more than 300 websites.[1]

8.1 Sources of Programs

Programming is not an easy job. Many people want to produce radio programs, movies, and television programs, so there are many sources to choose from. After a programmer has decided what to select, the projects must be nurtured to develop in a manner that will attract an audience, and they must be scheduled so that the audience can find them.

8.1a Self-Produced

Much of programming is produced by the organization that distributes it. A local radio station hires disc jockeys for music shifts. A local TV station produces several news programs a day. A cable system shows live coverage of the city council meeting on its **local origination** channel. CBS produces some its soap operas at its network facilities in Los Angeles. NBC produces its morning show, *Today,* from its New York studios. ABC employees decide what to put on its website. HBO Productions films dramatic series for the network. ESPN sends sports crews to cover the events it cablecasts. A hospital's video department makes a tape about pediatric care to show in waiting rooms.[2]

local programming

8.1b Related and Nonrelated Media

Program departments also get material from each other, especially if they are related. When *Queer Eye for the Straight Guy* (see Exhibit 8.1) became a hit on the NBC-owned cable channel Bravo, NBC put it on its main network. Disney has placed some of the ABC network sitcoms on its ABC Family cable channel in an effort to give that channel a boost. NBC shared the 2008 Summer Olympics with many of its General Electric–owned relatives—MSNBC, CNBC, Bravo, USA, Telemundo, Oxygen, as well as the internet and mobile devices. In some cities

Exhibit 8.1

Queer Eye for the Straight Guy started on Bravo, one of the smaller cable networks in which NBC has an interest. It became a bigger hit than the programmers had predicted, so NBC placed it on its broadcast network and also continued its showings on Bravo.

(Kathryn Indiek/Globe Photos, Inc.)

related sources

where one company owns several TV stations, the stations share news crews and news stories.[3]

In radio, related stations obtain similar program material when Clear Channel undertakes **voice tracking.** The parent company has the announcers for a few stations do intros and "outros" to music for a large number of the company's stations. These intros sound local because the announcers incorporate material about the local community, but they are being piped in from a distant city. Voice tracking saves Clear Channel a great deal of money because the disc jockeys don't sit around while the music plays. In fact, they don't even hear the music; they just record voice for several hours.[4]

Local TV stations **affiliate** with particular networks and obtain much of their programming from them. So although the CBS soap opera might be an original production of the CBS network, it is a program that the CBS-affiliated TV stations throughout the country obtain from the network to help fill their broadcast day. In a similar manner, public TV stations obtain programs through PBS, and Spanish-language stations tap into material from Telemundo. Occasionally, especially with public radio and TV, the stations supply programs to the networks. Cable TV systems and satellite TV systems affiliate with many networks (MTV, USA, ESPN, HBO, etc.) because they provide many channels to the consumer. In a similar vein, XM satellite radio has channels from different radio networks, including one devoted to CNN News in Spanish.[5]

nonrelated sources

Even when programmers are not related through company ties or affiliation agreements, they often receive programming from each other. HBO produces for ABC. NBC has gotten some of its programming from PBS. *Scrubs* was produced by ABC for airing on NBC; then, after seven seasons, when NBC decided not to air it anymore, ABC took it in. If one programmer is producing a show another programmer wants, they sit down and negotiate, and eventually "our lawyer talks to your lawyer."[6]

8.1c Majors and Independents

majors

Another source that programmers tap for material is **major** production companies—the ones that also produce theatrical movies. Again, this often involves "synergy." CBS and Showtime may order series from Paramount because all are part of the Viacom family. NBC owns Universal, from which it obtains, among other things, the many episodes of *Law & Order* that it airs (see Exhibit 8.2). ABC has Disney, and Fox has 20th Century Fox. But it's not all "in the family." For example, *Law & Order* is seen on many broadcast and cable outlets owned by various companies. Some film studios that aren't owned by the same company that owns the network strike deals, called **pod deals,** in which they receive a contract to produce a certain number of series or programs for the network. NBC created such an arrangement with DreamWorks.[7]

These major production companies also sell the movies they produce to programmers. A particular movie might be seen on network TV, local stations, several cable and satellite networks, the internet, and DVDs. Often the majors set up **windows** for their movies. A window is the time between when a film is shown in the theater and when it is shown elsewhere. Usually pay services and home video

Exhibit 8.2

Law & Order illustrates a variety of programming arrangements. The program started in 1990 as an offering from Dick Wolf Productions, a small independent production company. The production company moved onto the Universal Studio lot and utilized the facilities and distribution function of that major studio. Universal sold *Law & Order* to a variety of cable networks such as USA, A&E, and TNT, but its biggest customer was NBC, which regularly programmed the original series and its three spin-offs on that network. In 2004, NBC bought Universal, so it now owns the series, but Dick Wolf retains creative input.

(Courtesy of Academy of Television Arts & Sciences)

can negotiate a shorter window than commercial networks; in other words, a movie might be available for pay-per-view within a month after theatrical distribution but not available to be shown on network TV for a year. In recent years, majors have been eager to get movies into the lucrative DVD market and have shortened or eliminated that window.[8]

In addition to the major production companies, there are smaller **independent** production companies that supply programming. When **fin-syn** was in effect (see Chapter 3), these companies were very important and provided the commercial broadcast networks with most of their prime-time programming. For example, Aaron Spelling Productions supplied *Beverly Hills 90210* and *Love Boat;* Carsey-Werner-Mandabach produced *Cosby* and *Roseanne;* and Bochco Productions provided *NYPD Blue* and *Hill Street Blues* (see Exhibit 8.3). But now that fin-syn has been abolished and networks can "roll their own," small companies are hurting.

independents

Exhibit 8.3

Hill Street Blues was produced by Steve Bochco's independent production company. It went on NBC in 1981 and lasted until 1987. Unlike most previous crime dramas, it had a large cast of interesting characters who sometimes worked on the same case week after week. In fact, some of the cases were never solved. The first year it was on, the show had very low ratings, but NBC renewed it anyway, and then, much to the executives' delight, the series won eight Emmys. After that, it attracted a much larger audience.

(NBC/Globe Photos)

When they do get the ear of a commercial network, they usually have to make financial concessions and give the network part ownership in the programs they produce. Sometimes independents attach themselves to a major production company in an **umbrella deal.** The major gives them office space and some financing and, in return, gets the right to distribute whatever they produce. Some cable TV networks still welcome productions from independent production companies, but they generally do not pay as well as the commercial networks. Independent producers are also active in providing direct-to-video movies and informational programs to video stores, and they hire themselves out for corporate videos.[9]

8.1d Syndicators

Other players in the programming business, called **syndicators,** supply programs to a variety of programming outlets. They are in the business of acquiring programs from others or producing programs themselves and then selling those programs to local radio and TV stations, cable networks, or anyone else who wants to buy them. Unlike many TV networks, which require all affiliated stations to run the programming they supply at the same time, syndicators give their customers total discretion as to when they air the program. So *Jeopardy!,* a show that is syndicated by King World Productions, conceivably could air on Channel 8 in Dallas at 3:00 P.M., Channel 9 in Philadelphia at 7:30 P.M., and the cable Game Show Channel at 10:30 A.M.

types of programming

Syndicators acquire programming in a variety of ways. One way is to buy programs that have already been on major networks. These programs, referred to as **off-net,** have the advantage of a track record. The syndicators can tell stations how the programs have fared in the past, and the programs may have a built-in audience of people who want to see their old favorites again. Now that people can obtain network shows on DVDs, however, the value of off-net has declined. Syndicators also handle old movies, usually obtaining packages of several movies from various movie studios. The movies, too, have a track record, but by the time they get to stations and cable they may have been shown so many places that most interested people have seen them. Syndicators also sell material that has not been seen before because it is made specifically for syndication and is referred to as **first-run syndication.** Examples are *Oprah Winfrey Show, Wheel of Fortune,* and *Judge Judy.*[10]

radio

Syndication is also common in radio, but in radio it is hard to tell a syndicator from a network because generally radio networks do not require affiliates to air their programs at the same time; both networks and syndicators provide short bits

Provider	Services
ABC Radio Networks	News, sports, music, and talk
Business Talkradio	Business news, financial advice, and lifestyle programming
CBS Radio Networks	News, sports, information features, and music
Hispanic Communications Network	Spanish-language news and features
Metro Weather Service	Tailored weather forecasts
Monitor Radio	News and commentary from the Christian Science Monitor
Motor Racing Network	Live auto racing coverage and news and talk about auto racing
National Geographic Explorer	Vignettes featuring interviews and natural sound about the world we live in
North America Network	Talk shows about health, current events, and other subjects
Radio Spirits	Old-time radio shows
Salem Music	Adult contemporary Christian music
Sports Memories	Short stories of great sports moments
Success Radio Network	Interviews with experts on success and personal growth
Westwood One Radio Networks	News, entertainment, business, music, concerts, and sports

Exhibit 8.4

A partial list of radio networks and syndicators.

of material that customers are free to use as they wish (see Exhibit 8.4). Most stations want to sound local, so they do not emphasize that they are playing network or syndicated programming. Some stations affiliate with more than one network and buy from more than one syndicator.[11]

8.1e Others

In summary, the most common ways for programming departments to obtain material are to produce it themselves, obtain it from some other programming department (related or not), or get it from a network, major production company, independent production company, or syndicator. But there are many other sources. Foreign language stations and various cable channels obtain programming from other countries. Radio stations deal extensively with music companies. Sports franchises negotiate with broadcasters to have their games televised. Religious programs are produced by churches. Video game producers obtain rights from movies to make them the basis of a game, and vice versa. Advertisers are getting back into the business of supplying programming. Members of the public contribute to programming on cable access and the internet. Just about anything is possible. That's why the programming department is a creative, exciting place.[12]

miscellaneous sources

8.2 Development

Many TV programs, TV series, and movies that are being developed from scratch (and some that are acquired from other sources) go through a **development** process that takes them from idea to distribution. Even if you are only "programming" your webpage, you probably spend some time thinking about what you are going to include and how you are going to edit any video clips you may have.

ideas

In the network world, development is a complicated process. Ideas don't just spring off the page and become programs. They must be nurtured, tested, and fine-tuned before they are ready for public consumption. The development process starts when someone comes up with an idea. The person may be a network executive who hears a news story on the way to work and decides it would make a good TV movie or watches a short clip on YouTube (see "Zoom In" box) and decides it could be made into a series. He or she calls creative people (independent producers, people within the network, the creator of the YouTube clip) and asks them to flesh out the idea. Or an independent producer may hear that news story or see that clip and set up a meeting to **pitch** the idea to several network executives (see Exhibit 8.5).[13]

ZOOM IN: **Up from the Internet**

Here are three examples of projects that started on the internet and moved to other media:

- Michael Reiss put *Queer Duck* on a little-known website, icebox.com, in 1999. It was a cartoon series about a gay duck who lived with his boyfriend, Openly Gator. A friend, Xeth Feinberg, designed and animated the show out of his home on an iMac computer. Showtime decided it wanted the series to run along with its drama series, *Queer as Folk,* and invested 10 months in legal wrangling to get the series from icebox. After it aired on Showtime in 2002, Michael and others turned the idea into a movie that was distributed by Paramount in 2006. And what is Michael Reiss's day job? He writes for *The Simpsons.*
- Stephanie Klein's media adventures started with her blog, "Greek Tragedy." Among her writings were stories about her life as an overweight child and a chronicle of her return to single life after her first marriage

went sour. Soon the blog ranked among the top 1 percent of all blogs in terms of inbound traffic. It caught the attention of publisher Harper Collins, which gave her book contracts for the two stories that became *Straight Up and Dirty* (2006) and *Moose: A Memoir of Fat Camp* (2007). NBC then approached Stephanie, and she has now started down the development path to make *Straight Up and Dirty* into a comedy series.
- Steven Tsapelas, Angel Acevedo, and Brian Amyot put 11 episodes of *We Need Girlfriends* on YouTube. The series, which is about recent college grads trying to understand the New York dating scene after being dumped by their girlfriends, drew as many as 700,000 views per episode. It also drew the attention of *Sex and the City* creator Darren Star who, in 2007, negotiated a script commitment from CBS. Steven, Angel, and Brian are now writing the script and beginning their journey through development.

If the network executives like the fleshed out idea, they may eventually commission (pay for) a full script. If the idea is for a series, as opposed to a single program, the executives may request a **pilot**—one program as it might appear in a series. Pilots are viewed by network executives and tested on potential viewers (see Chapter 10). If the program idea gets past the pilot stage, the network authorizes production.

Of course, it's not as simple as all that. In fact, development is frequently referred to as "development hell." Often there are rights to be procured, especially if the idea is based on something already published, such as a book, or on something that has already been distributed, such as a YouTube clip or a programming concept that was successful in another country. The executives may want the location or the age of the lead actor changed to take into account a **demographic** they want to attract. Network or movie studio heads may come close to accepting a new idea, then decide it is safer to put their resources into making a **sequel** to a popular movie or a **spin-off** involving characters from a popular TV series. A children's program idea that the creators thought was just right for the Disney Channel may be passed on by Disney but eventually accepted by Nickelodeon. There are bound to be negotiations concerning the budget, and the stars' agents will get involved. And, of course, getting a meeting in the first place to discuss the idea is a daunting task for a newcomer. Then there is the weeding-out process. Broadcast network executives may hear hundreds of pitches, pay for 100 scripts, decide to ask for pilots for 20 shows, then only produce 5. They may not even ask for a pilot because pilots are expensive and often lead nowhere, so some executives make their decisions based on scripts.[14]

Exhibit 8.5

The TV movie *In the Line of Duty: Ambush at Waco*, starring Tim Daly as the Branch Davidian leader David Koresh, is an example of a program that was based on a news event, in this case the February 28, 1993, storming of the compound at Waco, Texas. NBC contacted producer Ken Kaufman on March 1 and asked him to produce a program on Waco for the *In the Line of Duty* series he was supplying to the network. It aired on May 23.

(Courtesy of Patchett Kaufman Entertainment)

pilots

variations

8.3 The News Process

News is handled very differently from other programs and series. Because it is based on whatever is happening in the world, there is no pitch-pilot development process. There are people who decide which stories to put on the air and how much emphasis to give each, but these decisions are made more quickly than the decisions involved with selecting dramas, reality shows, and other forms of programming.

8.3a News Gathering

Gathering news is a complex process. The major providers—CNN, Fox News, NBC, CBS, ABC, local TV stations, all-news radio stations—have their own reporters who go to local, national, and international places where the news is happening to

reporters

Exhibit 8.6

Reporters who were "embedded" with the troops during the beginning of the war in Iraq had sophisticated digital equipment that enabled them to beam pictures that could be telecast live.

(Rhodes/USMC/Globe Photos, Inc.)

gather the facts. Sometimes these reporters find their own stories because they know the locale—if they notice new stop signs at an intersection, they investigate to see why they were put there. Other times they are given stories to cover by an **assignment editor,** a person located at the news headquarters who keeps a list of stories that are likely to become important—the city council will be considering parking problems at 1:00; this is the anniversary of a fire, so there should be a follow-up story to cover what has been done to prevent this type of fire in the future. Other times they are assigned stories by the head of the news organization—for example, the reporters who were **embedded** with the troops in Iraq (see Exhibit 8.6).

Most news organizations also use stories from **stringers**—people who are not on the payroll but who are paid for stories or footage they supply that is actually used. These people are particularly

stringers

valuable for breaking news stories occurring in areas where the news organization does not usually have a reporter.

major providers

Large organizations that program news in a variety of places use the same reporters and same information for various outlets. For example, NBC, CNBC, and MSNBC share resources, not only for their broadcast and cable outlets, but also for their internet sites. Broadcast networks make footage available to their affiliates and often sell it to others. This works in reverse, too. Affiliated stations supply footage to networks, particularly when a major story occurs in their city. Cable News Network makes news available to others besides CNN, including local broadcast stations. Since its beginnings in the 1980s, CNN has been particularly aggressive in selling (or sometimes giving) its news to foreign countries. As a result, CNN receives credit on news clips seen on many countries' local newscasts.

news agencies

Another source of news, particularly for organizations that don't have their own elaborate news-gathering structure, is **news agencies.** Sometimes these groups are referred to as **wire services,** a term that predates electronic news when organizations gathered news for newspapers and sent the information over wires to machines in newspaper offices that then printed out the stories. Today, most news agencies sell not only written copy but also **sound bites** and video footage, which they transmit via satellite. The oldest American-based news agency is the Associated Press. A variety of other news services also supply news, such as Reuters, Accu-Weather, and Agence France-Presse. Some of them provide wide coverage of international news; some provide specialized news, such as weather or sports; and some are local. News agencies have a bevy of reporters stationed at strategic points around the world who gather news the same way other reporters do.

the public

The general public serves as another source of news. People call stations or networks with tips on news stories. Now that many people have **camcorders** or cell phone cameras, ordinary citizens sometimes supply news organizations with footage just because they happen to have their camera with them when a major event occurs. For example, in January 2009, when an airplane made an emergency landing on the Hudson River, the early news photos came mostly from cell phone cameras of people who happened to be on nearby boats. In addition to video and audio files, individuals also put news on the internet, in the form of **blogs.** Some news organizations look at blogs while others are skeptical of their accuracy.

Many organizations prepare written **news releases** or **video news releases** **(VNRs)** that they send to networks and stations, hoping that the material will be included in the news. For example, Congress has a special room where senators and representatives can tape VNRs to send to stations in the districts or states they represent. Sometimes news releases do give news organizations story ideas, but often they are simply self-promotion.

news releases

In addition, news organizations often have satellite equipment to monitor weather conditions. By using **scanners**—devices that monitor police and fire radio communications—they sometimes arrive at the scene of a crime before the police. Some news organizations have their own helicopters, so they can gather news and traffic information from the air. News organizations, also use the general information on the internet to obtain background information for stories.[15]

equipment

8.3b News Compilation

While reporters gather worldwide, national, and local news, other people perform different duties that lead to newscasts. The number of people needed depends on the emphasis given to news. A small radio station may have a news staff of one person who decides what stories to broadcast, rewrites news agency copy, makes a few phone calls to confirm facts or gather material for stories, writes the script for a three-minute news update once an hour, and also reads the news over the air.

At a large news organization, a **news director** has overall responsibility for the news operation, hiring and firing people, and setting the general guidelines for the approach. Each individual newscast is overseen by a **news producer,** who organizes the program and decides what will be included. If a newscast is to include special types of information, such as consumer affairs or entertainment news, **segment producers** may oversee those particular items. Assignment editors, as already mentioned, keep track of the stories that need to be covered and send reporters to cover these stories. Writers rewrite news agency copy and stories sent in by reporters and prepare the introductions and conclusions for the total newscast.

people needed

At most radio stations, the process of selecting the stories that air is ongoing because news is broadcast many times throughout the day. Radio networks generally feed stories to stations as they occur and also provide several news programs throughout the day. Thus, the news producer of a station affiliated with an ABC radio network might decide that the 8:00 A.M. news should consist of five minutes of ABC news as broadcast by the network; a story on a fire in Delhi that came in from Associated Press but was not covered by ABC; an update on the condition of a hospitalized city official that the producer obtained by phoning the hospital; a report on a liquor store robbery gathered by one of the station's reporters; a report of a murder received from the local news agency; and the

radio

weather as received from the weather bureau. The stories of the Delhi fire and liquor store robbery may be included in the 9:00 A.M. news, slightly rewritten and accompanied by two stories excerpted from the 8:00 ABC news broadcast, a report phoned in live by a station reporter covering a school board meeting, an update on the hospitalized official, and the weather. A number of all-news radio stations broadcast news continuously throughout the day. They have many crews, a variety of news sources, and a large staff of people working at the station; this costs money, making an all-news format expensive.

television

Most broadcast television news has a more defined countdown because the major effort is devoted to the evening news. There are exceptions, of course, including late night and early morning newscasts incorporated into both network and station schedules. CNN, Fox News Channel, MSNBC, and other 24-hour news cable services have procedures closely akin to an all-news radio station.

News producers at most local stations and the major broadcast networks spend the day assessing the multitude of news items received to decide which will be included on the evening news and in what order. This is no easy task given that a half-hour newscast (22 minutes without commercials) is equivalent to only three columns in a newspaper. Decision making about what stories to air goes on until close to airtime, when the "final" stories are collected and timed. The anchors get into position, the director and crew prepare for the broadcast, the material to be included in the newscast is put in order, the teleprompter copy is readied, the computer graphics are polished, and the newscast begins. Last-minute changes can still occur, for the news is sometimes changed even as it is being aired if something noteworthy happens while the newscast is in progress.

presentation

websites

Presentation of news is another important area. News for radio is sometimes read by disc jockeys or reporters. Stations, however, vary both content and presentation of news broadcasts in relation to their audience. A rock station has a faster-paced presentation than an oldies station. All-news radio stations generally have anchor teams to present the news (see Exhibit 8.7). Because TV news can be a moneymaker for local stations, they are particularly eager to lead the ratings; therefore, local stations attempt to find newscasters who appeal to viewers. The networks hope to establish trustworthy, congenial newscasters who maintain a loyal audience.

The news organizations that have websites (and most now do) need to update the sites constantly. The people in charge of the website rework the news that is gathered for broadcast or cablecast so that it fits with the style and content of the website. They can include the same video clips used in a newscast, or they can reedit the material. They can also include material that does not make it to the newscast because the web does not have time limitations. As more people are getting their news

Exhibit 8.7
The setup for two news anchors at all-news radio station KFWB in Los Angeles.

from the internet, news organizations are placing more emphasis on the look and accessibility of their websites.[16]

8.4 Formats

Within radio, the main element that is developed by stations is the **format**—the particular type of material that they program. Some formats used by commercial radio stations include adult contemporary, classic rock, top 40, country, alternative, all-news, oldies, and talk. Within the context of the format, stations may develop special features, such as local sports, farm reports, and traffic reports. Radio networks and syndicators also develop original programming, such as business reports, live concerts, and health information. Public radio stations usually develop and program classical or offbeat music and public affairs shows that seek out cultural and thought-provoking elements of the community.[17]

formats and features

Many factors affect what format and features a particular station chooses. One consideration is the programming already available in the station's listening area. For example, a jazz format may be chosen because no other station in town offers that type of music. If the local area has few jazz music fans, however, this format will not draw an adequate audience. Therefore, the composition of the listening audience is another important factor. A rural audience will probably be more interested in detailed farm reports than will a city audience. The interests of the community and the interests of the station management can also affect programming. If the town has a popular football team, the station might broadcast football games. If the station manager is particularly interested in boating, he or she might promote boating reports.

decision factors

Getting material ready to be distributed, whether through the pitch and pilot process, by gathering news, or by establishing a format and features, can be an arduous process. Even when it's finished, the work is not over. Scheduling can be crucial to success or failure.

8.5 Scheduling

Concern with scheduling is most intense on television networks, but other media deal with scheduling also. In radio, the main scheduling element that programmers use is a **clock,** which shows all the segments that appear within an hour's worth of programming (see Exhibit 8.8). Generally, each time segment within an hour resembles that same time at other hours. For example, if news comes on at 9:15, it will also come on at 10:15, 11:15, and so on. In this way, the listener knows what to expect of a station at each segmented time of the hour. Most stations change their clocks to some degree during different **dayparts.** For example, a station may give more frequent traffic reports from 6:00 A.M. to 9:00 A.M. than from 9:00 A.M. to 4:00 P.M. For a station with a music format, the program manager must also decide what music to play, how often, and for what period of time. Even the most popular song wears out after a while, and the program manager must decide when to stop airing it.[18]

radio

The movie business is not as schedule intensive as radio or TV, but it does have a scheduling cycle. Many movies, especially high-budget ones, are released

movies

Exhibit 8.8

A typical radio station clock.

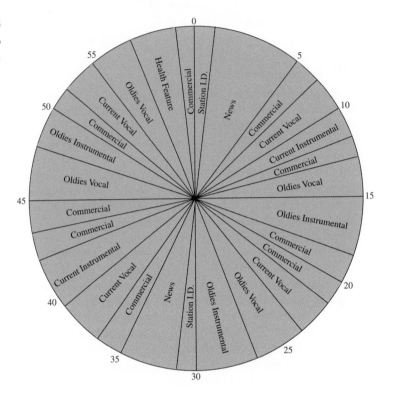

in the summer or during holiday periods when people have leisure time to spend at theaters. Video games are often scheduled to come out near Christmas to maximize sales. The internet, because it is always available, has less to do with scheduling than some media, but websites need to be kept up-to-date, lest surfers lose interest and do not return.[19]

8.5a Scheduling Factors

why schedule?

Programmers in the television business spend a great deal of time deciding how and when to schedule each program. Nowadays, of course, people can totally rearrange the carefully crafted schedule by using their **digital video recorders,** buying DVDs of series, or downloading material from the internet onto an iPod. But scheduling considerations are still important for television programmers because many people use TV to relax. They just want to turn on the power button, surf with their remote control until they find something they like, and then stay put. For that reason, when a program is scheduled can be important to its longevity. If it is against a major hit, it probably will not do well. If it is scheduled at a time when the audience members who would be most interested are not available because they are at work or sleeping, it is unlikely to gain a following.

prime time

For the broadcast networks, prime time is the scheduling period that receives the most attention because that is when most people watch TV. The major scheduling meetings for prime time occur in the spring to set the schedule for the fall season. However, with high program turnover and the recent tendency to start new series during the summer and other parts of the year, prime-time scheduling is considered at other times of the year, too. To determine a schedule, the executives consider not only where to put new programs but also which old programs to cancel or move to a new time. In making these decisions, they consider ratings of present programs, fads, production costs, overall program mix the network hopes to attain, ideas that have worked well or poorly in the past, the kind of audience to which a program will appeal, and types of programs that the other networks may be scheduling.[20]

Cable networks with programs that do well in prime time, such as HBO, engage in serious thinking similar to broadcast networks when they consider where to place their programs. Other cable networks engage in **narrowcasting,** in which they are only trying to attract a particular demographic—sports enthusiasts, children, music aficionados. They usually schedule the best of what they have to offer during prime time. But some channels don't put their best foot forward during that time period. For example, 9:00 P.M. is not a good time to attract children.[21]

    cable

Commercial, public, and cable networks are careful when scheduling their other dayparts, but programming in the morning, afternoon, and late night does not tend to change as much as prime-time programming. Some afternoon soap operas and late-night talk shows have been on for decades in the same time slot. TV stations affiliated with a network only have to worry about scheduling during the periods when the network does not feed them material, usually late morning, late afternoon, and the middle of the night. The most crucial programming they deal with is their locally produced news, which they try to schedule at the best possible time. Other than that, they often schedule syndicated material and (during the overnight period) reruns of programs that are on during the day.

    other dayparts

8.5b Scheduling Strategies

A number of scheduling strategies have arisen over the years. For example, networks often attempt **block programming**—the scheduling of one type of program, such as situation comedies, for an entire evening. This is undertaken in an attempt to ensure **audience flow**—the ability to hold an audience from one program to another. Cable networks that narrowcast essentially have block programming for their entire schedule, but sometimes they break down their already narrow programming so that, for example, they target World War II history buffs for an entire evening. Public broadcasting has also tried block programming, with some evenings devoted to fiction and some to nonfiction.

    blocking

If networks are introducing a new series, they may program it between two successful existing series with the idea that people will stay tuned through the whole block. This is usually referred to as **hammocking.** Another concept, called **tentpoling,** involves scheduling new or weak series before and after a very successful program. **Stripping** is a scheduling strategy that is used mostly by TV stations and cable networks. They show programs from the same series at the same time Monday through Friday, usually during daytime hours. That way audience members can find the show easily (see Exhibit 8.9).[22]

    hammocking

Exhibit 8.9

Programming
scheduling
strategies.

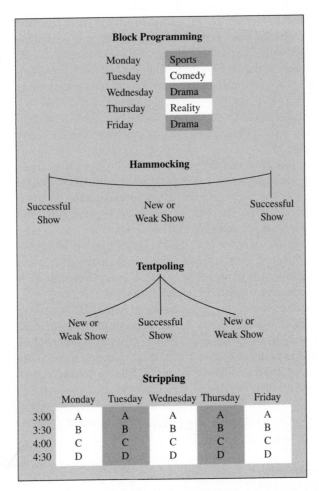

success and
failure

Scheduling is a controversial process. Producers always want their shows scheduled at the best time and resist having to pull some other program along (unless, of course, they have produced that program as well). They are quick to blame poor performance on poor scheduling. Of course, there are times when everything seems right—a show comes from a reputable source, it goes through thorough development, and it is scheduled at an ideal time—and still it fails. Other times a show doesn't seem to work but network executives believe in it and reschedule it until it finally clicks (see "Zoom In" box). Programmers have not come up with a fail-safe method for guaranteeing a hit. In the end, it is usually the public that determines what lives on and what gets canceled.[23]

8.6 Drama

qualities

Programmers are well advised to know the **genres** of programs that have worked in the past and the upsides and downsides of each form. Drama, for example, is a basic, enduring, high-profile form of entertainment that has been part of electronic

ZOOM IN: **Nothing in Life Is Easy**

Getting a big hit like *Seinfeld* on the air wasn't easy; even it went through "development hell" and rescheduling. The idea for the show started because stand-up comedian Jerry Seinfeld was well received during guest appearances on *The Tonight Show* on NBC in the 1980s. A couple of NBC executives supported the idea of Jerry performing in his own situation comedy and gave the go-ahead for a pilot, even though there was no well-defined idea for the series. Jerry and his friend Larry David wrote the script, *The Seinfeld Chronicles,* about "hanging out and doing stuff." It featured Jerry, Kessler (who later became Kramer), and George—but no Elaine. In 1989, the pilot was tested on potential viewers and scored poorly. As one viewer wrote, "You can't get too excited about going to the laundromat."

The idea was shelved, but it refused to die. At one point, NBC had a little extra money for specials, and several Jerry supporters talked the network into using it for four episodes of *Seinfeld.* NBC requested one change from the original concept—add a female. Jerry and Larry thought these "specials" would never see the light of day, so they wrote them for their close friends, not the general TV audience. They did see the light of day and received ratings that were so-so but good enough that NBC ordered 13 more episodes. Network executives still wanted to know what the series was going to be about. And that's how the concept of "the show about nothing" was born. The series went on the air in January 1991, and for several years was moved around the schedule from one spot to another, never doing particularly well. Most sitcoms of the time had two story lines going simultaneously, but since "nothing" was hard to write about, *Seinfeld* often had three to five story lines.

Its big break came in 1993 when the highly successful series *Cheers* voluntarily went off the air. Four months before that show was scheduled to end, *Seinfeld* was moved to the Thursday night spot right after *Cheers,* a move that was highly promoted during the Super Bowl. The two shows clicked as a package, and *Seinfeld* even outrated *Cheers* on occasion. After *Cheers* left the air, *Seinfeld* moved into its coveted Thursday 9:00 P.M. slot.

Many shows (successful and unsuccessful) have a convoluted history. Can you think of any way to make the process simpler? What factors do you think make a TV show a hit? Why do you think *Seinfeld* finally became a hit?

The cast of *Seinfeld.*

(Supplied by Globe Photos)

media since its beginning. Dramatic stories were part of early movies, and early radio had many dramatic series. Today drama is rarely heard on radio. Movies, of course, are still heavy on drama, and many of them are part of television, either as movies made for TV or as movies originally made for the theater that are then shown on various forms of television.

soaps

Drama comes in many forms, one of which is soap operas. Each program is serialized in such a way that it entices the viewer to "tune in tomorrow"; the plot lines trail on for weeks, and adversity is the common thread. A variation on the soap opera, the **telenovela,** has been imported from Latin America. It has continuing stories, but no one telenovela goes on forever; it usually only lasts for several months.

prime-time forms

Prime-time TV dramas have taken many forms over the years. Dramas of the 1950s, usually referred to as **anthology dramas,** were series (*Playhouse 90, Philco Television Playhouse*), but the drama presented each week was an individual story that did not have the same characters as other dramas within the series. Although these plays were popular with the public, they became less acceptable to advertisers, who were trying to sell instant solutions to problems through a pill, toothpaste, or coffee. The sometimes depressing, drawn-out relationships and problems of the dramas were inconsistent with advertiser philosophy, which led to their demise by the 1960s.

docudramas

Anthology dramas were replaced by **episodic serialized dramas,** which had set characters and problems that could be solved within 60 minutes. With series such as *Gunsmoke, Route 66,* and *Marcus Welby, M.D.,* plot dominated character, and adventure, excitement, tension, and resolution became key factors. The nature of episodic dramas has changed over the years. For example, in the 1980s dramas with more complexity, such as *Hill Street Blues* (refer back to Exhibit 8.3), debuted. Audiences had to watch carefully to follow the plot. HBO pushed the envelope in the late 1990s with *The Sopranos,* a highly acclaimed series about the inside workings of a New Jersey mob family. Today's dramas, such as *24* and *CSI,* are highly produced, with production values that often rival those of movies.

Dramas sometimes come in the form of **miniseries** that last for several episodes but do not go on indefinitely. One of the most successful miniseries was the 1977 *Roots* (see Exhibit 8.10), Alex Haley's saga of his slave ancestors, which aired eight straight nights to the largest TV audience up to that time. *Roots* was also one of the first series to be called a **docudrama**—a program that presents material that has a factual base but includes fictionalized events. Both the miniseries and docudrama forms still exist, but they are not aired as often or as successfully as they were during the late 1970s and early 1980s.[24]

Exhibit 8.10

In 1977, *Roots* set viewing records. More than 130 million viewers watched at least part of this David Wolper production, which dealt with author Alex Haley's search for his black ancestors. LeVar Burton, the actor who played Kunta Kinte, the slave brought over from Africa, was a college student appearing in his first TV role.

(Everett Collection)

A downside of dramas is that they are expensive to produce. As media fragment and audiences for any one show become smaller, it is harder to justify the costs. Also, dramas, by nature, contain suspense and conflict, elements that often lead to violence. This violence is highly criticized by politicians and citizen groups, sometimes necessitating changes on the part of program producers (see Chapter 12).[25]

8.7 Comedy

Comedy, like drama, has historically been a mainstay of media. Situation comedies, first on radio and then on TV, aim to make people laugh—not an easy task. It takes strong-penned writers and talented performers to crank out humorous lines. Although sitcoms are the primary form of comedy on TV, other forms do exist. *Saturday Night Live* has had a long run with skits that spoof many facets of society. Stand-up comics appear, particularly on cable channels. Variety shows, which include stand-up comics, skits, musical acts, and the like, used to be more common on TV than they are today, but they are still seen, particularly as specials.

forms

Many of the early situation comedies made an attempt to be believable, but the necessity to crank out programs accelerated a trend toward paper-thin characters and canned laughter. One mainstay became the idiotic father ruling over his patient and understanding wife and children. A breakthrough in comedy series occurred in the 1970s with the debut of *All in the Family*, whose bigot lead, Archie Bunker, harbored a long list of prejudices. This series, unlike any previous comedy series, dealt with contemporary, relevant social and political problems (see Chapter 3). The 1990s had a fair amount of caustic comedy. The Fox network had two hit shows in this category: *Married with Children* dealt irreverently with marriage, and *The Simpsons* glorified the underachiever. Comedy Central's *South Park* showed irreverence toward just about everything. These sitcoms and others were criticized for their negative approach and for their constant references to sex, but they received audience loyalty.

sitcoms

Sitcoms today are somewhat of an endangered species. *Seinfeld, Frasier* (see Exhibit 8.11), and *Friends* went off the air voluntarily, and other sitcoms have not achieved similar popularity. Reality shows seem to be taking away the audience that once belonged to sitcoms.[26]

8.8 Reality

Reality TV came to the fore as its own genre in the 1990s. Prior to that there had been programs featuring real people engaged in real activities, with some, such as *Candid Camera,* dating back to the 1950s. But they did not achieve the popularity that they have in the 1990s and 2000s, and they did not carry the label of "reality TV." One of the initial appeals of modern reality programming to networks was the low cost. The first shows featured primarily heroic and unusual feats by ordinary citizens. There were no actors to pay, no sets to build, and the technical requirements were minimal. Originally the genre was looked down upon as cheap programming whose appeal would be short-lived.

early

Exhibit 8.11

Cast and producers of *Frasier*. This show started in 1993 as a spin-off of the highly successful *Cheers*. Kelsey Grammer played a psychiatrist on both shows, a character that lends itself to many story lines. *Frasier* was the first show to win five consecutive Emmys for Best Comedy Show. It voluntarily went off the air in 2004 after winning 37 Emmys.

(Courtesy of the Academy of Television Arts and Sciences)

Exhibit 8.12

Mark Burnett (*left*) shown here with *Survivor* host Jeff Probst, is often considered the king of reality TV. After the original *Survivor*, he went on to produce many spin-offs of the show and many other reality shows.

(Courtesy of the Academy of Television Arts and Sciences)

Then, in the summer of 2000, reality went in a new direction when Mark Burnett (see Exhibit 8.12) produced *Survivor* for CBS. The program involved a number of ordinary citizens placed on an island, where they undertook unusual tasks and voted each other off the island. The winner of this *Survivor* series received $1 million, a small price to pay for the return that CBS received. An estimated 72 million people watched the final episode, giving CBS the highest ratings it had had in years. Since that time, reality has become a popular genre, with many of the programs eliminating people until only one is left.

The term "reality TV" has been extended to some other forms or programming that were originally considered game shows or music shows—for example, ABC's *Who Wants to Be a Millionaire?* and Fox's *American Idol*. The shows have been criticized for being inane (to the extent that there are reality shows that spoof reality shows) and for endangering people as the grossness and difficulty of the tasks they must perform escalate. But they are still relatively inexpensive, and they garner large audiences, so reality TV has already had a longer life than its detractors predicted.[27]   **criticism**

8.9 Games

Game shows have long been popular on TV. Their formula involves ordinary people using skill or luck to win prizes—and exude excitement. Audience members, both in the studio and at home, empathize and agonize along with the contestants.

Early radio's quiz shows were not as hyped as today's TV game shows. Some of them were quite intellectual, such as *The Quiz Kids,* which presented difficult questions to precocious children. Television took over the quiz-game program idea early in its history. Most of the initial TV shows had modest prizes for the winning contestants, but during the mid-1950s, the stakes began to increase with such programs as *The $64,000 Question.* The 1958 **quiz scandals** (see Chapter 3) gave a temporary blow to game shows. No chance-oriented shows dared touch the airwaves for a while, but gradually, low-stakes programs referred to as game shows emerged during daytime hours. The prize for *The Dating Game,* for example, was an expense-paid date for the contestant and the person he or she selected. The most watched, longest running syndicated game show is *Wheel of Fortune* (see Exhibit 8.13), which started in 1975.[28]

Exhibit 8.13

Host Pat Sajak and card turner Vanna White of *Wheel of Fortune.*

(NBC/Globe Photos)

Games of another sort, **video games,** consume a great deal of recreational time. They are praised for developing eye-hand coordination but criticized for being violent and a waste of time.

8.10 Music

Music is the mainstay of radio, but it is difficult to keep track of the various music formats because music is constantly evolving (see "Zoom In" box). Music that might be referred to as "new wave" doesn't stay "new" for long. When a particular type of music, such as rock, becomes popular, it tends to be subdivided into categories (hard rock, soft rock, classic rock, album-oriented rock) so that stations can have a unique sound.[29]   **examples**

radio

TV

Historically, music had a more minor role in TV than in radio. Dick Clark's *American Bandstand,* a glorified disc jockey program on which teenagers danced to the current hits, had the greatest longevity of any music show, being shown in some form for about 40 years. Public broadcasting and some cable channels frequently air concerts from venues around the country.   **criticism**

The music concept with the most TV success is music videos, started in 1981 by cable's MTV. With MTV's success, music videos began to appear everywhere—on commercial networks, local stations, cable networks, and at dance clubs. They have lost some of their popularity but remain a good way to showcase bands.[30]

Modern music such as rap is criticized for its sexual and violent lyrics, especially those that demean women. As with some other forms of programming, there are periodic inquiries from government and citizen organizations about the negative effects this music has on youth.

ZOOM IN: **Radio Reincarnated**

KNAC was an FM radio station in Long Beach, California, that, starting in 1986, programmed heavy metal and hard rock. Although the station's signal only covered part of the Los Angeles area, it developed a large cult following that extended well beyond its signal reach, in part because of its austere T-shirts and bumper stickers. The station helped popularize Metallica and played the music of many other popular bands of the era—Guns N' Roses, Van Halen, Scorpions, Judas Priest. In 1989, another Los Angeles station, KQLZ (referred to as "Pirate Radio") set out to compete with KNAC. It had a high-powered signal and hired away some of KNAC's on-air personnel, but it had a short life and nowhere near the ardent fan base of KNAC. This was attributed in part to the credibility KNAC earned with its audience, and the perception that the DJs were actually members of the audience, not slick DJs.

About 1992 heavy metal started to lose popularity to various forms of alternative music. The station added some alternative music (Pearl Jam, Nirvana) to its mix, which angered some of the heavy metal fans. Then in 1994, it was announced that KNAC was being sold and the new owner was changing the station to a Spanish-language format. At that point, KNAC returned to its pure heavy metal format and had a long six-month good-bye with its core listeners that lasted until February 15, 1995, when the station switched to the Spanish format. The last song played was Metallica's "Fade to Black."

In 1998, some ex-KNAC personnel revived the station as an Internet radio station, KNAC.COM ("The Loudest Dot Com on the Planet"). Within the first month, the site had half a million unique listeners and eventually grew to more than a million regular visitors.

Some of the principals of KNAC.COM—Diana DeVille, Tony Iommi, Ronnie James Dio, Phil Hulett, and Vinnie Appice.

(Photo courtesy of KNAC.COM)

8.11 News

As previously mentioned, news is currently available in a large number of forms at any time of the day or night. Such has not always been the case. Radio had to fight the newspaper industry in order to provide any news (see Chapter 5), and for many years both radio and TV reported news only at specific times of day. Journalists were referred to as **gatekeepers** because they decided what people needed to know. Usually they selected items that were unusual and timely. Many radio and TV news departments still operate that way, but most also have websites that people can access at any time. In addition, there are stations and networks that program news 24 hours a day.

variety

There are also many more sources for news, so it is easy to find news content and style that are agreeable, whether you are conservative or liberal, young or middle-aged, rich or poor. News junkies can get their fill, while those only interested in

gatekeepers

Election News Primary Sources

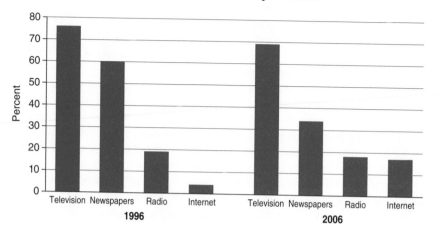

Exhibit 8.14

This chart, which indicates what people consider to be their primary sources of election news, shows changes over a 10-year period, now that the internet has developed.

major headlines that affect their lives are also easily accommodated. In times of disaster, almost everyone depends on the electronic media as the main source of help and information—directing victims to shelters, communicating vital health and safety information, and calming jangled nerves (see Chapter 1).[31]

    The electronic media are particularly important when it comes to delivering political information (see Exhibit 8.14), such as candidate debates, campaigning, coverage of the Democratic and Republican conventions, and election night coverage. This last was particularly dicey in 2000, when the networks declared the results of the presidential election vote count in Florida and then had to change their story several times as the nation waited for weeks to find out whether Al Gore or George W. Bush would become president. The rush to judgment by the news media was highly criticized.[32]

politics

    There are numerous other criticisms of news, many centered on what news the electronic media present and how they present it. Television, particularly, is obsessed with visual stories and sometimes downplays an important story simply because no exciting pictures accompany it. Sometimes stories are so highly visual that the facts become obscured. The weather, for example, can contain so many computer graphics, satellite feeds, and digital effects that viewers are left wondering whether it will rain.

criticisms

    Both radio and TV provide encapsulated news. Five-minute radio updates, revolving 20-minute newscasts of all-news stations, and 30-minute evening news broadcasts cannot cover the day's news in depth, explaining and analyzing it. Yet the news that is chosen is often trivial. Greater coverage is given to a politician eating a taco than to his or her views on crime. If a station or network achieves a scoop, it dwells on that story even though the story is relatively unimportant. In the rush to get a story first, a news organization may get a story wrong.

    Tasteless coverage of victims is also a problem. Some reporters succeed in interviewing people who have just seen a close relative killed or have lost all their possessions in a disastrous flood. Overall, the emphasis is on sensationalism, with murder, rape, and prostitution given more airtime than weighty national and international economic and political issues. The phrase "If it bleeds, it leads" has been used to describe the phenomenon of gaining audience attention with sensational stories. In their sensationalizing, news media are often accused of judging guilt before a suspect has had a chance to receive a court trial. "Trial by the press" is a frequently used ploy of lawyers who think their clients cannot get a fair trial because of adverse radio and TV publicity.

Another criticism is that news is biased. Some claim it is a product of the "liberal Eastern establishment," while others bemoan the conservative slant of Fox news and talk radio. Politicians and businesspeople, who are often targets of news darts, are particularly prone to cry bias. But it is the duty of journalists to act as society's watchdog and report improprieties of public servants and others whose activities have a major effect on the well-being of the citizenry.

Gathering and presenting the news is an expensive proposition, but it can also be a money maker because, of all the programming presented by the electronic media, news is the one that is most necessary to everyday life.[33]

8.12 Sports

The most highly publicized sports programs are those that show America's major sports—football, basketball, baseball—on the broadcast and cable networks. Archery, badminton, tractor pulling, and every other conceivable type of sport, however, can be found somewhere on radio or TV. Throughout the country, local radio, TV, and cable systems have their own sports programs, often in conjunction with a local college or university. A growing number of radio stations have an all-sports format that includes play-by-play and call-in talk shows. Regional cable TV sports networks that show local teams on a number of cable systems are particularly popular. Team and league websites stream sports events, or at least highlights, over the internet.

types

Throughout their short history, sports and electronic media have had an unusual symbiotic relationship. The Dempsey-Carpentier boxing match gave early radio its first big boost (see Chapter 5); early TV had its wrestling matches (see Chapter 3); ESPN was one of the first successful cable networks (see Chapter 4).

history

As sports proved its value to broadcasting, the **rights fees** television paid to air the games skyrocketed. For example, the rights to the National Football League season grew from $4.65 million in 1962 to $50 million in 1970 to $500 million in 1998 to $650 million in 2006. During the 1990s, the networks sometimes lost money on sports events; they had to give money back to advertisers because they could not deliver the audience size they promised. This still happens occasionally, but for the most part the rights fee issue has been solved by compromise. The sports teams slowed the acceleration rate of their rights fee increases, advertisers spent a little more money, and fans remained loyal, especially for major events such as the Super Bowl and the Olympics (see Exhibit 8.15).[34]

rights

Another controversial aspect of sports and broadcasting involves **blackouts.** When TV first started broadcasting sports events, attendance at games plummeted (for example, a 32 percent plunge for baseball between 1948 and 1953). Sports owners realized they were cutting their own throats when they televised all

blackouts

their games. As a result, Congress enacted a law in 1973 that allows major sports games to be blacked out (not aired) up to 90 miles from the origination point unless the game is sold out 72 hours in advance. The biggest critics of this policy are the fans who are occasionally deprived of seeing the game on TV.

There are occasional outcries from the sports world against TV equipment that is so sophisticated it out-umpires umpires and out-referees referees. But the sophistication of modern TV equipment often enhances the viewing with slow motion, split screens, cameras placed within racing vehicles, microphones picking up sounds on the field, and virtual graphics that show such elements as the strike zone or the nationality of the swimmer in each lane of the Olympic pool.

Whatever the problems are with the sports–electronic media relationship, it is not likely that either party will initiate divorce proceedings. The fans would not approve.[35]

Exhibit 8.15

Michael Phelps's successful quest for eight gold medals at the 2008 summer Olympics kept viewers tuned in.

(Photo courtesy of Karen Blaha)

equipment

8.13 Talk Shows

radio

Early radio had talk shows, but not of the kind it has today because the telephone technology that allows viewers to call in was not available. The talk shows were simply in-studio interviews with people in the news. Today's radio talk shows on both commercial and public radio still have many in-studio elements, but they also involve audience members. Some are devoted almost entirely to members of the public who call to express opinions on specific or general subjects. Others feature experts on various subjects who then answer listeners' questions. Talk radio has gained increasing importance in the political world, serving as a sounding board for ideas and disagreements.

public service

The earliest TV talk show was *Meet the Press,* which started in 1947 and can still be seen Sunday mornings on NBC. It and other programs like it feature interviews with people of political and social stature who discuss important issues of the day. Programs with this public-service orientation exist on commercial networks, local stations, public TV, cable networks, and public access. Sometimes they are criticized for being boring, but most of them now have quality production values.

morning

A lighter form of talk show appeared in the 1950s with the advent of *Today* and *The Tonight Show* (see Chapter 3). *Today,* a morning show, covered a broad swath of both news and feature material. It contained interviews, newscasts, weather, and interactions by the host(s). *Today* and its morning competitors still adhere to the same general formula despite many changes in on-air and off-air staff. *The Tonight Show,* with the exception of the opening monologue, was primarily an interview show, but unlike the more serious talk shows, many of its

evening

Exhibit 8.16

Oprah Winfrey hosts a top-rated syndicated talk show.

(Courtesy of Academy of Television Arts & Sciences)

interviewees were entertainment celebrities. It and its many late-night clones capitalize on the average person's desire to know what makes celebrities tick.

The daytime talk shows with hosts such as Geraldo Rivera and Oprah Winfrey (see Exhibit 8.16) came to the fore in the 1970s. These shows tend to feature celebrities or a group of unknown people with unusual stories to tell, all on the same basic theme. Some of these talk shows became controversial because of the increasingly bizarre subjects discussed (sex changes, satanic worship, daughters who date their mothers' boyfriends). After a show that resulted in the death of one of the participants, the shows tamed down, although some of them are still criticized for exploiting people.[36]

The cost of talk shows depends primarily on the quality and demands of the guests and host. Some talk shows are almost free; they are beset with requests from aspiring authors and the like who wish to appear on the program for publicity purposes. Other shows pay top price to obtain "hot properties" of the show business and political worlds.

8.14 Documentaries and Information

daytime

cost

definition

changes

The definition of a documentary has changed over the years. Film theorists who studied and produced documentaries before the days of television believed that documentaries had to deal with controversial subjects and present a point of view. Television's early documentaries, such as those presented by Edward R. Murrow on *See It Now* (see Chapter 3), were in that mold. They presented bold, strong programs on controversial subjects.

Hard-hitting documentaries were still present during the 1960s, but their controversial subject matter caused the networks many headaches in terms of lawsuits and the need to prove that they were handling all sides fairly. As a result, the networks turned to softer documentaries such as *The Louvre* and *The White House Tour with Jacqueline Kennedy*. During the 1970s, **minidocs** became prevalent. Often produced by local TV stations, they consisted of several minutes about a particular subject. Stations sometimes aired information about one subject over a week, covering a different aspect of it for three to five minutes during each day's newscast. Then they would edit all the minidocs into one unified documentary to air as a program by itself.

In 1978, CBS started *60 Minutes* as a magazine program that consisted of a number of minidocs of 5 to 20 minutes each. Purists do not consider *60 Minutes,* or its imitators, to be documentaries because they do not go into enough depth, and some of the subjects are more featurelike than documentary. The success of *60 Minutes,* however, is significant in terms of its influence on society. A number of times, it has effected change by exposing a fraudulent practice or unsavory scheme, and as a result, it, too, has been embroiled in controversy.

Public television has two long-running prime-time documentary series, *Frontline* and *P.O.V.* (*Point of View*), both of which started in the 1980s. *Frontline* is produced by people employed by public TV stations, but *P.O.V.* consists of documentaries produced by independent videomakers who are not part of the established media structure (see Exhibit 8.17).

Cable TV's Discovery Channel, formed in the 1980s, opened the door for more documentary production. Most of its programming is fairly non-controversial, dealing with wildlife, science, and other aspects of nature. Similarly, the National Geographic nature specials, the *Biography* series on A&E, and much of the programming on the History Channel and the Learning Channel qualify as documentaries by today's standards.[37]

There are many other types of programs that provide general information, covering such subjects as health, cooking, home decorating, and farming. C-SPAN covers the House of Representatives and the Senate and many important conferences and meetings. Some of the best instructional programming is produced in the corporate realm, where programs intended to train employees or showcase products and services are well produced and effective.[38]

Exhibit 8.17

This documentary, *Farmingville*, was aired on *P.O.V.* in 2004. It covered a neighborhood on Long Island, New York, that was torn by conflict over immigrant workers.

(A film by Carlos Sandoval and Catherine Tambini, photo courtesy of P.O.V. and Camino Bluff Productions.)

examples

information

8.15 Religion

Religious programming has been part of radio and television since their beginnings. Today some media entities program religious material 24 hours a day; others run religious programs part of the day and another format the rest of the day; and some limit religious programs to Sunday morning.

Although most religious programs do not contain paid advertising, many of them ask people to send in money, buy religious items, or support the particular religion. Some radio and TV stations and networks broadcast programming of one denomination; others are interdenominational, with different religious organizations paying to provide programming throughout the day. Some radio stations have religious music formats that sound similar in style to Top 40; others include a great deal of talk. Some stations produce locally based programs; others air mostly programming that comes through a network, such as Family Stations Inc. (for radio) and the Trinity Broadcasting Network (for TV). Much of the programming is evangelistic in nature, leading to the coining of the term **televangelism** (see Exhibit 8.18).

characteristics

scandals

From time to time, scandals break out in the religious programming area. In the 1980s, Jim and Tammy Bakker of the PTL network were involved in a sex and hush-money scandal that sent Jim to jail. Similar improprieties related to other televangelists caused a temporary crisis in confidence for all religious broadcasting. But the crisis passed, and today religious broadcasting is ubiquitous, with some ministers using satellites, DVDs, and franchise marketing to grow flocks throughout the United States and the world.[39]

8.16 Children's Programming

radio

Children watch and listen to much of the programming that is targeted toward adults, but through the years, radio and TV entities have produced many programs aimed specifically at children. As with other genres, radio first aired children's programs, mostly during Saturday morning and after-school hours. Today, radio programs very little for children, other than Radio Disney, a network that features music and contests for kids 2 to 12.

early TV

Television networks started children's programming early with an emphasis on puppets, such as Howdy Doody. The longest-running kids' show on commercial network TV (1955 to 1982) was *Captain Kangaroo,* starring Bob Keeshan. Children's programming was important on early local TV stations, too. Most programs consisted of a host or hostess whose main job was to introduce cartoons and sell commercial products. During the 1960s, networks overwhelmingly adopted the likes of *Felix the Cat* and *Tom and Jerry*—and therein began a controversy.

For many years, these cartoons dominated Saturday morning TV, making for one of the most profitable areas of network programming. The cartoons were relatively inexpensive to produce, and advertisers had learned that children can be very persuasive in convincing their parents to buy certain cereals, candies, and toys. The result was profits in the neighborhood of $16 million per network just from Saturday morning TV.

The situation changed gradually. Parents who managed to awaken for a cup of coffee by 7:00 A.M. Saturday noticed the boom-bang violent, noneducational content of the shows, along with the obviously cheap mouth-open/mouth-closed animation techniques. A group of Boston parents became upset enough to form an organization called Action for Children's Television (ACT), which began de- **ACT** manding changes in children's programs and commercials. Researchers realized that children under five were watching 23.5 hours of TV a week and that by the time they graduated from high school, they would have spent 15,000 hours in front of the tube. The Children's Television Workshop developed *Sesame Street,* and its successful airing on public TV proved that education and entertainment could mix.

All this led to a long, hard look at children's TV. In 1969, the U.S. Surgeon General appointed an Advisory Committee that selected 12 researchers to investigate the effects of television violence on children. After these 12 had worked for two and a half years, the committee concluded that a modest relationship exists between viewing violence on TV and aggressive tendencies.

Led and cajoled by ACT, a number of organizations demanded reforms in children's TV. They noted that there were some fine children's programs on TV—*The Wonderful World of Disney* and *Mister Rogers' Neighborhood*—but they were after changes in the cartoons, slapstick comedy, and deceptive commercials. In 1974, the FCC issued guidelines for children's television that stated, among other things, that stations would be expected to present a reasonable number of children's programs to educate and inform, not simply entertain. The FCC also stated that broadcasters should use imaginative and exciting ways to further a child's understanding of areas such as history, science, literature, and art.

By the mid-1970s, most stations and networks acquiesced, at least in part, to **1970s** the reform demands. Programs that attempted to teach both information and social values hit the airwaves. These were more expensive (and less watched) than cartoons, so the amount networks and stations spent on children's programs increased. ABC developed after-school specials that dealt dramatically with socially significant problems faced by children, such as divorce and the death of a friend. In cable TV, Nickelodeon was established in 1979 to show solely nonviolent children's programming.

With the 1980s, the tide turned again, and the FCC, in the spirit of deregula- **1980s** tion, dismissed the idea of maintaining or adopting standards for children's programming. The 1974 guidelines were ignored. The networks, eyeing their sinking children's TV profits with fear, canceled many of the expensive education-oriented programs and returned to the world of cartoons. In the mid-1990s, Fox **1990s** began programming the *Mighty Morphin Power Rangers,* which some called the most violent children's cartoon ever—and it became a huge hit. In 1990, Congress passed a law that dealt mostly with restricting commercials within children's TV programs (see Chapter 9). In 1996, Congress passed the **Telecommunications**

SpongeBob SquarePants, one of Nickelodeon's popular programs, was created by former marine science teacher, Steve Hillenburg. He also drew the talking sponge, dopey starfish, and cranky squid who encounter adventures with charming naiveté.

(Henry McGeel/Globe Photos, Inc.)

Act, which specified that **V-chips** be installed in TV sets to help parents shield their children from undesirable programming.[40]

Today, both socially redeeming educational material and "brain-numbing" cartoons for children are abundantly available. Nickelodeon has become the top purveyor of popular children's programs (see Exhibit 8.19), and PBS still has many offerings in this area. The broadcast networks have faded to the background; several have turned their Saturday morning programming over to other companies that rent the network time to broadcast programming they produce or acquire. The internet is laden with sites designed to appeal to children, many of them operated by the same companies that provide children's TV programs.[41]

current

8.17 Issues and the Future

changes

Programming seems to be the one aspect of the media that will go through changes in the near future. What will be the role, if any, of independent producers now that networks are undertaking so much of their own programming? Will radio remain a local-sounding medium, or are large companies making all radio sound the same? Is there really a need for local stations when all they do is pipe through network programming and show the same syndicated fare that is on cable or satellite TV? Will syndicators be needed now that off-network material is available on DVDs or through internet downloads?

internet

The biggest unknown related to programming is what role the internet and portable devices will play in distributing entertainment and information. These technologies have yet to be fully developed in terms of their program potential, and their future course will affect the fortunes of the older media.

choices

As media forms proliferate, so do programming choices. Although this would appear to be a good thing for the consumer, it can lead to choice fatigue. You can spend all evening with the remote control just clicking through the channel options—all these channels and still there is nothing good on TV! It is easy to settle into programs that fit your comfort zone and do not challenge you with opposing viewpoints. This is particularly true of news. It is easy to find programs that correspond to your politics, and you can be lulled into thinking that everyone in the country shares your point of view.

The entertainment area, especially at the commercial network level, is dominated by reality TV—to the detriment of comedy and drama. This makes it particularly difficult for writers to find jobs, and it limits the opportunities for actors. Other program forms, such as soap operas and some public-affairs programs, are in jeopardy because they garner smaller audiences than in the past when there was less choice. Network executives are balking at the cost of pilots. If they eliminate pilots altogether, will the quality of TV suffer?

Criticisms leveled at programs are not likely to go away. Sex and violence will continue to draw fire, but program producers are not likely to give them up because they are elements that sell programs. News will be criticized, especially when journalists must make quick decisions on important topics or when they exhibit viewpoints that are not popular with certain members of the populace. Because children are so impressionable, there will be watchdog organizations demanding improvements in children's TV. Those involved with both entertainment and information programming will make mistakes—they are human beings. But both the media and the people who produce for them will become stagnant if they do not try new ideas, many of which are likely to fail, but some of which will showcase the talents of numerous individuals, including programmers.

criticisms

8.18 Summary

Different types of programs wend their way through the programming process in various ways. The three forms most likely to be programmed during network prime time are dramas, which have historically been a mainstay of TV, in anthology, episodic, docudrama, and soap opera form; sitcoms, which have had their ups and downs; and reality TV, the newest genre. All three forms are sometimes produced by the network and sometimes obtained from majors and indies. They go through a development process that often includes pitches and pilots, and they are carefully scheduled in prime time, often incorporating block programming, hammocking, and tentpoling. Networks deliver these programs to their affiliated stations or cable systems.

Game shows and afternoon talk shows often come from syndicators, who also market off-net and made-for syndication programs. They are likely to be stripped.

Music is predominant on radio, where it is part of a format and where it is usually scheduled with a clock and dayparts in mind. Music videos have also been popular on television, where they started out being narrowcast on MTV.

News is programmed quite differently than other forms of programming because of its immediacy. It comes from reporters, stringers, news agencies, the general public, VNRs, satellites, helicopters, the internet, and many other sources. News producers, directors, assignment editors, website designers, and anchors are involved in getting it to the public through radio, scheduled commercial and public TV news broadcasts, 24-hour cable news networks, and constant-access internet sites.

Sports events are usually produced by the entity that airs them and involve paying rights fees to the sports franchise and dealing with blackouts. Some cable networks, such as ESPN, are devoted totally to sports.

Public-affairs talk shows are usually produced by whatever organization distributes them, ranging from cable local origination to long-term network shows. Daytime and nighttime talk shows have lighter content often involving celebrities. They are syndicated or self-produced.

Documentaries often come from independent producers and are narrowcast on cable TV channels. Religious programming is both station and network oriented, with the bulk of it being of an evangelical nature.

Children's programming is a controversial area that is to some extent governed by laws and influenced by citizens' groups. Several cable networks are devoted to children's programs, and public broadcasting has played a large role in this area. Commercial networks produced children's programs in the early years, but now some networks have farmed out this programming to other companies.

All forms of programming have come under criticism, some of which is generated by the need to try new ideas.

Suggested Websites

www.ap.org (Associated Press, a news agency)

www.cnn.com (news channel Cable News Network)

www.espn.com (sports channel ESPN)

www.mtv.com (music channel MTV)

www.natpe.org (National Association of Television Program Executives, an industry organization that deals with programming)

Notes

1. "I Want My (Web) MTV," *Newsweek,* April 14, 2008, E8-E9; and "Back to Basics," *Hollywood Reporter,* January 26–27, 2008, N1–N8.
2. "Why Cable Originals Are Crucial," *Broadcasting & Cable,* October 13, 2003, 42; "Daytime Drama," *Mix,* September 2001, 82; "For TV News Producers, a Fast-Forward War," *Wall Street Journal,* March 24, 2003, B-1; and "ESPN Turns 25," *TV Technology,* September 8, 2004, 14.
3. "A Season of Seeing Double," *Broadcasting & Cable,* September 30, 2002, 7; "New Media Get Deals," *Los Angeles Times,* July 7, 2008, D5; "Why Local News Is in a Fix," *Broadcasting & Cable,* August 7, 2006, 10; and "NBC's Olympic Proportions," *Broadcasting & Cable,* August 4, 2008, 17.
4. "A Giant Radio Chain Is Perfecting the Art of Seeming Local," *Wall Street Journal,* February 25, 2002, 1.
5. "Can These Marriages Be Saved?" *Broadcasting & Cable,* January 14, 2002, 6; and "Online Education, *Broadcasting & Cable,* April 17, 2006, 9.
6. "HBO to Produce New Programs for ABC," *Los Angeles Times,* August 6, 2002, C-1; "Public Partnering," *Broadcasting & Cable,* January 21, 2002, 32; and "Something Borrowed, Two New," *Hollywood Reporter,* May 14, 2008, 3.
7. "Invasion of the POD Deals," *Electronic Media,* August 19, 2002, 1A; and "Law and Disorder," *Wall Street Journal,* July 12–13, 2008, 1.
8. "Video on Demand Not Yet a Big Movie Player," *Los Angeles Times,* December 1, 2003, C-1; and "DVD Growth Spurs Spending Spurt on Home Video Marketing," *DVD Exclusive,* August 2004, 24.

9. "Table Scraps," *Broadcasting & Cable,* July 26, 2004, 1; "Networks Keeping Comedy Pilots in the Family," *Electronic Media,* February 11, 2002, 4; and "Digital Impasse," *Broadcasting & Cable,* August 28, 2006, 8.

10. "Syndies Turning to Cable," *Broadcasting & Cable,* March 19, 2001, 19; "Superior Court Shows," *Broadcasting & Cable,* March 31, 2008, 26; and "The New Rules of Syndication," *Broadcasting & Cable,* January 23, 2006, 26–27.

11. "The Database," http://www.syndication.net/show_details.php?show=565 (accessed July 12, 2008).

12. "Taking Amateur Video Up a Notch," *Los Angeles Times,* July 16, 2007, C-1; "Pepsi to Put 2 Summer Shows on WB Network," *Los Angeles Times,* April 10, 2003, C-1; "A Telenovela with the Sights, Sounds of L. A.," *Los Angeles Times,* December 1, 2003, C-1.

13. "Star Commits to 'Girlfriends' from YouTube," *Hollywood Reporter,* November 2–4, 2007, 2; "The Web's First Fall Season," *Wall Street Journal,* September 9, 2000, B-1; "Networks Fear the Net as a Copilot," *Los Angeles Times,* March 27, 2007, E-1; and "Quest for TV Magic," *Los Angeles Times,* November 16, 2003, E32–E33.

14. "Zucker: 'It's About Less Waste,'" *Hollywood Reporter,* January 30, 2008, 8; "Ready to Air . . . But Where," *Emmy,* December 2001, 36–39; "Play It Again, Shrek," *Los Angeles Times,* January 3, 2005, E-1; and "Delicate Surgery on 'Grey's Anatomy,'" *Wall Street Journal,* February 21, 2007, B1.

15. "Radio and TV News Services," *Broadcasting and Cable,* August 14, 2000, 44–45; and "News Bulletin: Fox Tops CNN for First Time," *Hollywood Reporter,* January 30, 2002, 1.

16. "New Media, New Newsrooms," *Broadcasting & Cable,* January 14, 2008, 22; "Getting Up to Speed," *TV Technology,* January 23, 2008, 20; and "The All-Digital War," *Wall Street Journal,* March 12, 2003, B1.

17. Michael C. Keith, *The Radio Station* (Boston: Focal Press, 2004), 84–99.

18. Keith, *The Radio Station,* 106–11.

19. "Day and Dateline," *Hollywood Reporter,* August 2006, 19; and "Cable TV, We Need to Talk: I've Started Seeing Other People," *Hollywood Reporter,* June 11, 2008, 16.

20. "Nervous Time for the TV Set," *Los Angeles Times,* May 8, 2002, 1; "CBS' Younger Look," *Broadcasting & Cable,* June 5, 2000, 8; "Dream Time," *Emmy,* Special Issue, 2003, 48–52; and "Fox Plans Sizzling Summer," *Broadcasting & Cable,* March 29, 2004, 3.

21. "Feathering the Niche," *Emmy,* June 2003, 96–101.

22. For more detail on programming, see Philippe Perebinossoff, Brian Gross, and Lynne Gross, *Programming for TV, Radio & the Internet: Practice, Process, and Strategy Demystified* (Boston, Focal Press, 2005).

23. "How 'Seinfeld' Broke the Mold," *Electronic Media,* November 12, 2001, 28.

24. "In the Beginning: The Genesis of the Telefilm," *Emmy,* December 1989, 30–35; "Hot Water for the Soaps," *Hollywood Reporter,* January 22, 2008, 1; Saul N. Scher, "Anthology Drama: TV's Inconsistent Art Form," *Television Quarterly,* Winter 1976–77, 29–34; and "Roots: 25 Years," *Hollywood Reporter,* January 17, 2002, 14.

25. "*24* and More," *Emmy,* June 2003, 114–18; and "Why the Sopranos Sing," *Newsweek,* April 2, 2001, 48–55.

26. "The Way We Laughed," *Newsweek,* February 19, 2007, 54–55; Erica Scharrer, "From Wise to Foolish: The Portrayal of the Sitcom Father, 1950s–1990s," *Journal of Broadcasting & Electronic Media,* Winter 2001, 23–40; and "Frasier Farewell," *Daily Variety,* May 13, 2004, A-1.

27. "72 Million Join 'Survivor' Tribe," *Hollywood Reporter,* August 25–27, 2000, 1; "Mark Burnett: Reality Survivor," *P3,* July 2008, 20–21; "'Idol' Hands Fox Another Ratings Romp," *Hollywood Reporter,* January 30, 2008, 19; and "Reality Binge," *Hollywood Reporter,* July 14, 2008, 9.

28. "Got Game," *Emmy,* Primetime Special 2008, 36–38; "Wheel Keeps on Turnin' with Online Fan Club," *Hollywood Reporter,* September 13, 2004, 17; and "Game Show Frenzy Takes Hold," *Broadcasting & Cable,* November 1, 1999, 22–23.

29. To see how music formats change over a period of years, compare *Broadcasting/Cablecasting Yearbook, 1985* (Washington, DC: Broadcasting Publications, 1985), F-65–F-94; and *Broadcasting & Cable Yearbook, 2000* (New Providence, NJ: R. R. Bowker, 2000), D-646–D647.

30. "Dick Clark Spins Another Record," *Los Angeles Times,* August 26, 1987, B-1; and "VH1 Hits a New High Note," *Broadcasting & Cable,* May 5, 2008, 10.

31. "Newsrooms Go Multiplatform," *Broadcasting & Cable,* March 26, 2007, 38; and "NBC Stations Shift Strategy," *Broadcasting & Cable,* May 12, 2008, 2.

32. "CNN Rolls Out Election Express," *TV Technology,* October 17, 2007, 6; and "Primary Source of Election News," http://www.stateofthemedia.org/2007/chartland.asp?id=539&ct=col&dir=&sort=&col1_box=1 (accessed July 18, 2008).

33. Tien-Tsung Lee, "The Liberal Media Myth Revisited: An Examination of Factors Influencing Perceptions of Media Bias," *Journal of Broadcasting & Electronic Media,* March 2005, 43–64; "TV Stations Reconsider Live Coverage Policies," *Los Angeles Times,* May 2, 1998, A-19; and "Infoganda: The Real Indecency in Broadcast," *TV Technology,* June 23, 2004, 32.

34. Dale L. Cressman and Lisa Swenson, "The Pigskin and the Picture Tube: The National Football League's First Full Season on the CBS Television Network," *Journal of Broadcasting & Electronic Media,* September 2007, 479–97; "Money to Burn," *Broadcasting & Electronic Media,* April 25, 2005, 1; and "Losses Predicted on NFL Deals," *Los Angeles Times,* January 15, 1998, D-1.

35. "Sports Leagues Impose More Rules on Coverage," *Wall Street Journal,* July 16, 2007, B-1; "Games to Fully Enter Digital Age," *Los Angeles Times,* July 7, 2008, D5; "Why TV Needs Sports," *TV Guide,* August 11, 1990, 3–14; and "NFL on Television," http://en.wikipedia.org/wiki/NFL_on_television (accessed July 17, 2008).

36. Carroll J. Glynn, et. al., "When Oprah Intervenes: Political Correlates of Talk Show Viewing," *Journal of Broadcasting & Electronic Media,* June 2007, 228–44; Jonathan Tankel, "Reconceptualizing Call-in Talk Radio as Listening," *Journal of Radio Studies,* Winter 1998, 36–48; "A Morning TV Wake-up Call," *Newsweek,* March 20, 1995, 68; "'Jenny Jones' Guest Killed after Taping," *Electronic Media,* March 13, 1995, 3; "Russert: Old School Rules," *Broadcasting & Cable,* September 6, 2004, 8; and "The Guys behind the Desk," *Emmy,* March/April 2005, 32–34.

37. Jill Godnilow, "Kill the Documentary as We Know It," *Journal of Film and Video,* Summer/Fall 2002, 3–10; "Documentary TV: Genre Heats as Demand Grows," *Television Week,* June 6, 2005, 22; and "Investigative Journalism Under Fire," *Broadcasting & Cable,* June 23, 2008, 10.

38. "C-SPAN Branches Out," *Broadcasting & Cable,* January 1, 1996, 48; "Room at the Top," *Emmy,* July/August 2007, 96–99; and "'60 Minutes' Keeps Ticking," *Broadcasting & Cable,* September 22, 2008, 14.

39. "Inspired by Starbucks," *Wall Street Journal,* June 13, 2008, W1; Kevin Howley, "Prey TV: Televangelism and Interpellation," *Journal of Film and Video,* Summer/Fall 2002, 23–37; "It's Pitch and Pray and Wish the Scandal Away," *TV Guide,* August 15, 1987, 5–9; and "TV's Religious Revival," *Broadcasting & Cable,* April 25, 2005, 26.

40. "Updated Disney.com Offers Networking for Kids," *Wall Street Journal,* January 2, 2007, B-1; George C. Woolery, *Children's Television: The First Thirty-Five Years, 1946–1981: Part II. Live, Film and Tape Series* (Metuchen, NJ: Scarecrow Press, 1985); "'Morphing' Karate Teens Provoke Parental Parry of Television Violence," *Christian Science Monitor,* November 16, 1994, 1; "TV Networks Find Ways to Stretch Educational Rules," *Los Angeles Times,* February 23, 2002, 1.

41. "The N Finds Its Niche," *Broadcasting & Cable,* September 19, 2005, 17; "4Kids Makes a Bold Move," *Electronic Media,* March 18, 2002, 11; and "Kids TV, Now with More Vitamins," *Broadcasting & Cable,* November 14, 2005, 25.

Chapter 9

SALES AND ADVERTISING

Just about everyone who works in electronic media is engaged in sales. A talent agent sells a producer on using a particular writer or actor in a production. A producer who pitches a show to a network executive is selling an idea. Recent college graduates need to sell themselves to directors in order to be hired as crew members. But the types of activities that are usually referred to as sales are those that occur when media companies try to sell products or services to customers or when media companies sell to each other. Within the media business, advertising constitutes the largest element of sales.

Money is the root of all television.

Anonymous

9.1 Media-to-Customer Sales

Many forms of electronic media sell directly to the customer. For example, movie theaters sell tickets, and perhaps even more important, they sell refreshments to the people who come to watch the movie. Although concessions account for only 25 percent of revenue for most theater chains, they provide a profit margin that averages about 85 percent, with popcorn and soda being particularly profitable (see Exhibit 9.1). Refreshments are so important to movie theaters that some chains have ushers with pushcarts to sell to people who don't want to wait in line at the concession stand.[1]

movie theaters

Profit of Movie Concession Stand Products			
Item	**Price**	**Margin**	**Profit**
Small popcorn	$3.25	96%	$3.11
Large popcorn	$5.00	94%	$4.71
Small soda pop	$2.60	91%	$2.37
Large soda pop	$3.35	86%	$2.88

Exhibit 9.1

This shows the typical profit on high-margin concession stand products, although it does not take into account such costs as electricity and pay for the sales clerks.

competitive services

public broadcasting

Internet service providers, satellite radio, satellite TV, cable TV systems, and phone companies sell services directly to customers. **Internet service providers,** such as Yahoo and AOL, sell access to the internet, usually for a monthly fee between $10 and $20. **Satellite radio** sells packages of audio channels ranging in cost from about $7 to $14 a month, depending on the number of channels you choose. **Cable television, satellite TV,** and phone services compete with each other, especially in television offerings, but also in terms of **broadband** internet and telephone services (see Exhibit 9.2). In order to remain competitive, they constantly market "deals." They may, for example, offer internet service free for two months if you sign up by next Thursday. Or they may give you three free **pay-per-view** choices if you sign up for the **DVR** and **high-definition TV.** Or they may give you a free month of call waiting if you sign up for caller ID. The marketing departments at the various companies continually eye what the competition is doing and adjust their offerings accordingly.[2]

Public radio and TV have a particularly difficult sales job in relation to their consumers. They must convince audience members to help pay for public broadcasting even though there are no restrictions on people receiving the programming for free. The stations do this through on-air fundraising campaigns that stress the value of their programming and their need for money. However, the on-air campaigns, although usually successful, often irritate listeners and viewers because they take away from the very programming the stations want the audience members to support. Public broadcasting is constantly looking for new ways to convince people to donate money, from printed mailers to wine tasting parties.[3]

Exhibit 9.2

These are typical offerings of a cable TV system. Usually there are discounts for subscribing to several services.

Generic Cable TV System Packages	
Product	**Cost per Month**
Digital cable—45 channels	$35.00
Digital cable—150 channels	$45.00
Digital cable—200+ channels	$50.00
High definition	$15.00
Each additional HDTV receiver	$10.00
Digital video recorder	$7.00
Digital video recorder service	$10.00
High-speed internet	$35.00
Phone service	$45.00
Pay-per-view	Depends on event
Video-on-demand	Depends on event

Many media items are sold to customers in the same manner as other consumer goods—through a wholesale and retail distribution network. Film companies and others distribute **DVD**s for sale or rent through various stores. **Video game** consoles, **cell phones, iPods,** and software are often sold the same way. But there is a difference between selling consumer products such as food and furniture and selling many media products because, unlike eggs or a coffee table, media products such as music, video games, and movies can be delivered through computer networks. Most sales that are internet-assisted involve a two-step process: the consumer sees a product on the website and then goes to a store to buy it or requests that it be delivered by United Parcel Service or a similar service. With some media products, the consumer can view and receive all in one step. This means the website must be a particularly effective sales instrument.[4]

through computer networks

9.2 Media-to-Media Sales

Media entities frequently sell to each other. For example, a TV station may rent out its studio to an independent production company that wants to use it to record a **corporate video.** A movie studio may sell rights to a video game company so that the company can create a game based on the movie, or a book publisher may sell rights to a studio so it can produce a movie based on the book.[5]

renting facilities

Syndicators sell programming to media outlets such as radio and TV stations and cable networks. They charge their customers one of three ways. The simplest is **cash.** The station or network writes a check to cover the cost of the programming it wants from the syndicator. Another method is **barter.** The station pays nothing for the program, but the syndicator gets to sell and keep the money from most of the commercials aired within the programming. A few of the commercials are left for the buyer to sell to cover its cost of transmitting the program—electricity, handling, and so on. The third method is **cash plus barter.** The buyer pays the syndicator some money and, in return, sells a larger portion of the commercials than is the case with barter. TV syndication used to be dominated by small, independently owned companies that made most of their sales at a yearly convention held by the National Association of Television Programming Executives (NATPE). But, as with many other aspects of the business, big conglomerates have bought the syndication companies or have built departments to syndicate the material they produce, thus diminishing the importance of the NATPE convention.[6]

syndication

Another media-to-media sales process involves cable networks and local cable TV systems. Cable networks such as USA, CNN, and Lifetime depend on local cable systems, such as Comcast and Time Warner Cable, to get their programming seen. If cable systems, especially the big conglomerates, do not include a particular network in their offerings to customers, the network will not have viewers and, therefore, will not be able to sell advertisements. As a result, cable networks have affiliate representatives who convince cable systems to start carrying or keep carrying the network. Cable networks charge the systems for carrying the networks based on how many subscribers the system has (see Exhibit 9.3). The price is determined basically by the degree to which the systems value the network (that's why ESPN can charge more than Hallmark); if the price is not high enough, the network will go out of business.[7]

cable affiliates

Cable Network Charges	
ESPN	$3.65
TNT	.93
USA	.52
CNN	.47
MTV	.32
Lifetime	.25
Hallmark	.06

Exhibit 9.3

The figures shown here are what selected cable networks charge cable systems per subscriber per month. ESPN is frequently criticized for its high rates, but cable systems continue to pay.

broadcast affiliates

movie business

Interestingly, the process involving broadcast TV networks (NBC, CBS, etc.) and their affiliated stations is quite different. The networks pay the stations to show the programs. This may seem backward because the stations are the ones obtaining the goods, but the networks retain the rights to sell many of the commercials within their programs and thus make the lion's share of the money. In recent years, as network fortunes have fallen, the networks have tried, occasionally successfully, to cut back on their payments to affiliates or even get stations to help pay for the programming.[8]

The movie business has a complicated sales process. Movie theaters obtain the films they show from distributors, which either work independently or are part of the studio structure. The theater owners usually decide what films they want based on hearsay, sales pitches, and **trailers**—not on actually seeing the movies. This is called **blind bidding.** Most exhibitors are large companies with multiplexes in many cities, so they buy for a large group of theaters. Generally, the theater owners keep 30 percent of the money brought in through ticket sales during the first week of a movie's run, and the studios receive 70 percent. As the weeks progress, the studios get less, and the theaters get more, until the percentages reach about 90/10. These percentages are only general; each movie has its own contract between the distributor (studio or independent) and exhibitor (theater), and all the terms are negotiable. For example, a studio might insist on more than 70 percent of the first-week ticket sales for a super blockbuster, arguing that it needs a higher percentage to recover the production costs and, besides, all the people who will flock to the movie will buy lots of popcorn, candy, and soda, adding to the theater's profits. However, movies earn most of their money during the first week and then burn out, so theaters would not be happy with this arrangement.[9]

Syndicators, networks, and movie studios also sell their products overseas (see Chapter 14). Usually they have a foreign sales department that is separate from the domestic one. For people who like to travel, handling foreign sales is an interesting job (see "Zoom In" box).

9.3 The Nature of Advertising

indirect

Most advertising does not involve direct sales to the customer. Ads are placed on radio, television, and the internet and are shown at the beginning of DVDs or before a movie starts in hopes that the ads will cause the consumer to buy whatever is advertised. But there is no guarantee that any particular consumer will watch the commercial or that it will affect behavior. A decision to buy something may be the result of hearing an ad on radio, reading a magazine story about the product, searching for it on the internet, glimpsing an ad while fast-forwarding through a TV program, and hearing about the product from a friend. However, there have been many instances when sales have gone from very low to high at the same time a product is heavily advertised. For that reason, advertising is considered to be a very effective way of encouraging people to buy.

ZOOM IN: How to Close a Deal

Usually the hardest part of media selling (or any selling for that matter) is closing the deal. You do have to get the ear of potential customers (usually by making an appointment to see them) and then get them interested (because you have a good product or service), but if you have targeted your customers well, those two steps are usually not as difficult as closing the deal. Here are a few tips.

1. **Emphasize the customer's needs.** By the time you are ready to close the deal, you will have talked to the customer and learned what he or she needs. Right before you plan to close, reemphasize a particular need that your product can fill: "The content of our syndicated program will fill the hole you have at 3:00, serving as a bridge from soap operas to children's programming."
2. **Validate your offer.** Save some powerful sales tool until right before you want to close. This might be a testimonial from a satisfied customer, a favorable newspaper

review, or a short, snazzy video about your service. Show it to the customer and then place the contract on the table.
3. **Offer options.** Say, "Which would you prefer, 150 channels or 200 channels?" In other words, don't make "nothing" an option. When they select one of the options, sign them up.
4. **Move quickly.** Give them a reason to close the deal *now*. "The offer of a free DVR will expire tomorrow." "I will be offering this movie to your competition tomorrow morning." Many people are on the verge of signing, but they want to take their time. Try to keep that from happening because giving them time means they can have second thoughts or contact your competition.

Of course, you must also be prepared not to close. Sometimes the circumstances and the timing are not right. If you think the person will be interested in the future, make sure you leave the door open to contact him or her again.

People who sell advertising are really selling potential eyes and ears, not a particular product. They convince someone who has a product or service to sell that using their medium is the best way to move the product off the shelf. They cite ratings (see Chapter 10) as evidence for the number of people who are listening to or looking at material that surrounds the commercial and try, to the extent they can, to describe who those people are and why their ears will perk up at the mention of a particular product. Some media entities, such as small radio stations, don't have very good ratings, so they engage in **concept selling;** they try to sell advertisers on unique elements of their content that will theoretically attract the type of people the advertiser wishes to reach. But concept selling is difficult; it is much easier to approach an advertiser armed with statistics on how many and what type of people will be exposed to a commercial or web ad.

ratings

Selling at the local level is fairly straightforward. Local radio and TV stations and cable systems have sales reps who contact local businesses and try to sell them commercial time (often referred to as **inventory**). At small-town radio stations, the disc jockeys may also be salespeople, spending four hours a day announcing and playing music and four hours going to the local hardware stores,

sales reps

specialty shops, and car dealers selling ads. Often the general manager of a small station doubles as a salesperson. Larger radio and TV stations and cable systems use a distinct sales force. These people are generally paid on a salary-plus-commission basis. Sometimes they specialize according to product line—one person sells to restaurants, another to real estate agents. Often they sell ads for the on-air facility and for its website. Generally, the people who sell local ads are expected to service them as well so that advertisers repeat their business. Servicing ads includes making sure the ads are run at the appropriate times and facilitating any copy changes that may be needed.

A station may also have an on-staff salesperson to handle ads from nationally based companies, but now that many stations are owned by conglomerates, most of dealings related to national ads are handled by headquarters staff for all their media properties. Even stations that are not part of large conglomerates often farm out their national sales. It would be too expensive to send a salesperson to deal with each potential advertiser for only one station or website, so instead they hire a

station reps

station representative. Station representatives are national sales organizations that operate as an extension of the stations with which they deal. Usually these reps handle national sales for dozens of stations around the country. Often they try to line up similar stations (e.g., a number of country music radio stations) so that they can sell ads to companies wishing to attract that demographic group. A station rep earns income from commissions on the sales obtained for the station.

networks

A large portion of advertising dollars goes to the commercial and cable TV networks. These networks maintain significant sales forces that are devoted to selling and servicing major national ads. Network commercial inventory is a fairly limited commodity. Only a certain number of spots are available, and these cannot be expanded as easily as a newspaper can add pages when ad demand is heavy. A network has the same amount of time available during the holidays, a heavy season, as it does the rest of the year, so the sales force must be careful to sell in such a way that meets the overall needs.

Generally, network management is deeply involved with sales and makes sure that a knowledgeable team presents ideas. For many years, commercial TV network advertising sold out rather easily at high prices. Usually most of it was gone **upfront** (before the fall season actually started). As other forms of media, such as cable and satellite TV, have eroded the audience, however, network sales have gone soft, and sometimes spots go unsold.[10]

All of the media representatives that sell commercial time on a national basis—conglomerate sales staffs, station representatives, network salespeople—are likely to deal with advertising agencies because they are the organizations that usually make media decisions for large companies that place many ads.

9.4 The Role of Advertising Agencies

duties

Most major **advertising agencies** are termed **full-service agencies.** They establish advertising objectives for companies with products or services to sell and try to position their clients so that there is something unique about them that consumers will remember. They place their clients' ads in all forms of media—newspapers, direct mail, radio, TV, internet—and often design campaigns that can be executed internationally as well as nationally. They conduct research to test the advertising concept and then attempt to obtain the best buy for the money available. Full-service agencies design the advertisement, oversee its production, and determine when a company should initiate a campaign and how long it should continue. After ads have run, agencies handle postcampaign evaluations. Throughout the process, the ad agency keeps the company executives well informed of the successes and failures of the advertising campaign.

For its efforts, the advertising agency generally receives 15 percent of the **pay** billings. In other words, a billing of $1,000 to its client for commercial time on a TV station means that $850 goes to the station and $150 is kept by the ad agency to cover its expenses. Because dollar amounts are fairly constant from agency to agency, the main "product" an ad agency has to sell to its customers is service. Advertisers sometimes change agencies simply because one agency has run out of creative ideas for plugging its product, and a new agency can initiate fresh ideas.

Some of the largest ad agencies (see Exhibit 9.4), like many other compa- **organization** nies, have undergone mergers and consolidation. Advertising has also become an important international business, so ad agencies have offices in many countries. Most agencies are divided into departments, some of which deal with administrative responsibilities such as accounting, while others deal with creative aspects such as writing and art design, and still others handle media and marketing. A team works on the ad campaign of any one company. Not all agencies are large, however. Some one-person operations service only a few clients who like the attention they receive from the person directly responsible.

Not all major advertisers employ an ad agency. Some maintain their own **in-house** services, which amount to an ad agency within the company. Other **in-house and** companies do not hire a full-service agency but hire what are sometimes referred **boutique**

Leading U.S. Advertising Agencies		
Name	**Founding**	**Current Owner**
McCann Erickson	1902	Interpublic
BBDO (Barton, Batten, Durstine & Osborn)	1928	Omnicom
JWT (J. Walter Thompson)	1877	WPP
Y&R (Young & Rubicam)	1923	WPP
DDB (Doyle Dane Bernbach)	1949	Omnicom
Ogilvy & Mather	1948	WPP
Grey	1917	WPP

Exhibit 9.4

Some of the top advertising agency names in the United States. They have all been around for quite a while, but although they operate somewhat independently, they are now all part of one of the four holding companies that control most of the worldwide advertising business—Omnicom (American), WPP (British), Interpublic (American), and Publicis (French).

to as **boutique agencies.** These agencies handle only the specific things a company asks for—perhaps the designing of an ad, the postadvertising campaign research, or advice on reaching the Spanish-speaking audience. Usually this type of work is paid for by a negotiated fee rather than a 15 percent commission.[11]

9.5 Traditional Advertising

There are many changes afoot in advertising, but the type of advertising that is still used most frequently and that receives the lion's share of revenue (at least for the present) is the traditional advertising that has been used by radio and TV for decades.

spot

At the root of this advertising is **spot buying.** The advertiser (through its ad agency) buys 15, 30, or 60 seconds of airtime for certain times of the day or for insertion within or adjacent to certain programs. Hence, a disc jockey whose show runs from 6:00 A.M. to 9:00 A.M. may present ads from several dozen advertisers, all falling under the umbrella of spot advertising. In radio, spots can be aired at almost any time because program segments are short—three or four minutes for a musical selection or two or three minutes for a group of news items. In television (both broadcast and cable), spots are aired at more definite times because the programming has natural breaks where ads are inserted.

program

Occasionally advertisers engage in **program buying,** in which they pay the total cost for producing and distributing a program. This is rare because production costs are high. More frequent is the practice of several noncompeting companies paying for the cost of the program. Then their ads, and usually only their ads, are inserted within the program.

network

Another way to categorize traditional advertising is to divide it into network, national, regional, and local. When an advertiser makes a **network buy**—either broadcast or cable—its ads go everywhere that the network program goes. In the case of a commercial radio or TV network, such as CBS or ABC, the ads play on all the stations affiliated with the network. For a cable network, the ads appear on all the cable and satellite systems that carry that network. Network ads are heard or seen on all stations or systems at the same time, just as network programs are. Most network ads are for products that are bought by many people throughout the country, such as automobiles, toothpaste, or cereal.

national

National buying is also for products with wide appeal, but it involves buying time on individual stations. The advertiser places the ad on many stations, usually enough to cover the whole country, but the ads do not play at the same time on all stations. An advertiser may take this route because its product needs a certain demographic audience. Maybe an automobile manufacturer wants to promote a car that it thinks will particularly appeal to women with small children. It can select stations throughout the country that target that audience.

regional

Regional buying covers a particular area of the country and is most often used for products that do not have a national appeal but can be useful in a number of places. A tractor manufacturer, for example, might want to advertise throughout the rural areas of the Midwest. Regional buys can be made by purchasing time on a number of different individual stations in the region, or they can be made by purchasing time on a regional network. The latter are most common with radio and with cable-TV sports programming.

Local buying involves placing commercials in a limited geographic area. It is used primarily by merchants who service only a local area—used car dealers, restaurants. A local advertiser often places the same ad on many different local radio and TV stations and cable local channels. In this way, the ad can reach a maximum number of people in a particular area.

local

A company may decide to undertake network, national, regional, and local advertising and even use the same commercial spot for all. More often, however, there will be at least slight variations in the commercial to take into account different audience composition.

This traditional approach to advertising worked quite well during the early days of radio and television when programs had large relatively captive audiences who had little recourse but to sit through commercials. But now people are exposed to approximately 5,000 advertising messages a day; they are so inundated they have developed mental mechanisms to block commercials. In addition, through channel surfing, mute buttons, and digital video recorders, they have much more control over what they receive.

traditional model

The old model—in which networks and stations obtained money from advertisers to produce and transmit programs, so the programs could be seen or heard by audience members who also noticed commercials—isn't working very well anymore. It is still used, but advertisers are constantly searching for new and better ways to reach potential customers.[12]

9.6 Product Placement

Product placement is actually a fairly old practice, particularly within movies, where companies pay to have their product shown in the film. The Coke can that the actress drinks from is not accidentally a Coke, and the Chevy Impala the actor drives is in the movie because Chevrolet paid to have it there. But product placements have increased in intensity and number as a way to get viewers to notice products, now that they have so many ways to avoid commercials.

product placement

Television reality programs, especially ones that feature makeovers, lead the count of product placements, but soap opera actresses dab branded perfume and music video performers carefully show off the brand label on their jeans. Sometimes the products aren't actually placed in the program but are seen as **virtual ads.** This is common in sports programs where the billboards seen in the background aren't actually on the walls of the stadium but are inserted digitally. Video games have become an excellent place for product placement; players actually seem to enjoy driving past billboards that advertise audio equipment or down a road past a well-known fitness center. There is even product placement on radio (see "Zoom In" box).[13]

virtual ads

An extension of product placement is **product integration.** In this case, the product is not just shown; it is an integral part of the story—the teenager who has to find a way to earn money so he can buy Reebok shoes to impress his girlfriend, a group of employees who talk Office Depot into sponsoring a talent show. Product

product integration

ZOOM IN: How Do You Do Product Placement on Radio?

Several radio stations, in an attempt to cut down on commercial clutter, have started weaving mentions of sponsors into disc jockey chatter or even other commercials. For example, a DJ or talk-show host might ask listeners to call in on the AT&T request line. Or an ad for a music concert might include a mention of a restaurant where people can eat before the concert. On a rainy day, the weather report might include a comment about London Fog raincoats. In listing local events coming up for the weekend, the announcer might offhandedly add that (you name the brand) potato chips would go well with a particular outing.

Do you think this type of "product placement" will be effective? Do you see any possi-

ble ethical problems? Do you think it would be appropriate to include product mentions within the news?

integration is bringing advertising agencies back into the TV production business to create shows that will highlight their clients.

infomercials

Both the **infomercial** and **home shopping** (see Chapter 4) are another step beyond product placement. In both cases, the entire program relates to a product (or a number of products) that someone wants to sell. Because infomercial producers pay stations and networks to air the infomercials, these programs sometimes bring in more money than could be made from commercials, especially in the wee hours of the morning. Home shopping networks make their money by keeping a percentage of what viewers pay for the products.

merchandising

Another variation on product placement is **merchandising,** in which movies, TV programs, or video games purposely include props or characters that can be turned into products—the cartoon bear that can be sold as a cuddly stuffed animal, the detective's coffee cup that is later included as part of a dish set.[14]

9.7 Interactive Internet Advertising

keyword search

The interactive capabilities of the internet have brought forth new concepts in advertising, the most successful of which has been **keyword search.** Although credit for the original concept is usually given to goto.com (later part of Yahoo!), it has been utilized most effectively by Google. Whenever you undertake a search on Google, on the right side of the results page you see a list of sponsored links to products and services related to your search (see Exhibit 9.5). Companies pay Google every time someone clicks on their site link, a process known as **pay-per-click (PPC).** The order in which the sites are listed is determined by a number of factors, including the relevance of the company to the search query and the

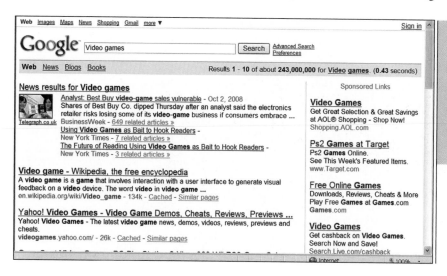

amount the company bids for each particular concept. Companies (often with the help of consultants) select the search words they want to be identified with and design their sites in such a way that they appear to be obviously relevant to those terms. In addition, companies that want to be at the top of the list are wise to bid high for the terms they really want. The sites on the left of the results page are also listed by relevance and other algorithms, but those sites do not generally pay Google.

Google has also made money selling its technology and expertise to other search engines and internet-based activities. The concept is effective because it gives consumers information they want. Most traditional advertising aims to interest people in something they may never have thought of and may not need or want until they see the ad. Keyword search gives information, in a low-key manner, for something that they have thought about and may even be looking to buy.[15]

selling technology

Other types of interactive advertising on the internet have had some degree of success, albeit it less than keyword search. For example, many sites sell **banner ads** that they place at the top of their site that may or may not have any relevance to the particular site (see Exhibit 9.6). Usually the sites charge a set fee for these rather than pay-per-click, but the traffic they generate is not high. There are **pop-up ads** that cover up some of the webpage and disappear only when someone clicks them off. However, there are also pop-up blockers to screen those out because they are generally irritating rather than effective. Companies often place graphic or full-video ads on their own webpages in order to advertise their products; sometimes the video ads are the same ones that appear on TV.

other ads

Exhibit 9.6

A banner ad.

(Powells.com)

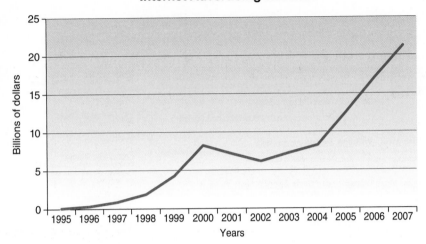

Internet Advertising Revenue

Internet interactive advertising has become very popular, as can be seen by
the amount of money advertisers spend on it. The amount has increased every
year except during the dot-com bust (see Chapter 2), and in recent years has risen
at an increasing rate (see Exhibit 9.7).[16]

9.8 Underwriting

**public
broadcasting**

Underwriting is a method of generating income used mainly by public radio and
TV stations and networks. Because public broadcasting is nonprofit and noncommercial, its relationship with commercials has always been on a slippery slope.
The Public Broadcasting Act of 1967 allowed stations to obtain funds from corporate sponsors and then acknowledge that "This program was made possible by
a grant from XYZ Corporation." However, public broadcasters have always had
trouble making ends meet, and underwriters weren't particularly happy with the
unadorned credit they received for helping NPR or PBS.

changes

In 1981, Congress sweetened the pie for corporations by allowing public stations to air company logos. This was referred to as **enhanced underwriting.**
Over the years, the enhanced underwriting guidelines have become more liberal,
allowing for corporate slogans, locations, toll-free numbers, celebrity endorse-
ments, and descriptions of product lines and services. The length of these
announcements has also grown from a maximum of 15 seconds to a maximum of
30 seconds. Public broadcasting underwriting announcements are still not allowed
to have qualitative statements, comparisons, price information, calls to action, or
inducements to buy, but underwriting announcements now look very much like
commercials. In fact, the FCC has decided that public television stations can
place genuine commercials on some of their new digital channels.

other money

Underwriting is not the only form of income for public broadcasting, but it is
the one that most closely resembles advertising. Public broadcasting has become

more dependent on corporations and subscribers because other forms of income, such as money from
governments and educational institutions, has been
shrinking. In 2004, NPR received a windfall contribution of $235 million from the estate of Joan Kroc, widow
of Ray Kroc, the founder of McDonald's. The network
is using the money, in part, to diversify and grow its audience, which was already on a growth curve, having
more than doubled to 22 million listeners between 1999
and 2004 (see Exhibit 9.8).[17]

9.9 Advertising Cost Factors

Media entities want to charge advertisers as much as
they can to air commercials or place ads on the internet.
Advertisers, of course, want to pay as little as possible
for the ads. Economic forces of supply and demand and
what the market will bear affect the interplay of the advertising world, but in both good times and bad there are
a number of factors that the media and advertisers keep
in mind.

Exhibit 9.8

The bulk of the Kroc money went into a
contingency fund to provide NPR with funding
beyond revenue sources that can be impacted
by the economy and other outside factors; but
some is being used to further the activities of
this new NPR facility in Los Angeles, which
was started in 2002, in part to increase
diversity of programming from that created in
Washington, DC.

One of these is **cost per thousand** or **CPM** (M from
the Latin *mille* for thousand). This is the amount of
money the media company charges the advertiser to
have an advertisement in front of 1,000 people (see
Chapter 10). Theoretically, advertisers should want the
same CPM for each ad they place, whether it is in an
NBC prime-time show or a History Channel morning
show. The price for the ad on the History Channel would
be much lower because it would probably not have as many thousands of people   **CPM**
watching, but the CPM would be the same. But such is not always the case. Advertisers believe the major broadcast networks still have more glitter than cable or
syndication, so prime-take broadcasting is able to charge a higher CPM (about
$12) than daytime cable (about $2). Also, certain programs can charge a higher
CPM than others (see "Zoom In" box).[18]

Some media companies, particularly radio and TV stations, local cable chan-
nels, and websites, prepare a **rate card,** a listing of the prices they charge for dif-   **rate cards**
ferent types of ads (see Exhibit 9.9). Although they usually do not put the CPM in
writing, they tie their rate card prices to the number of people who will see or hear
the ad. Even if a media company does not have a rate card, company executives
go through the process of deciding how much they need to charge for ads in order
to cover costs and reap some profit.[19]

Many variables affect the prices of advertisements on radio and TV (and to
some extent the ads that show in movie theaters). One is the number of commercials the advertiser wishes to air. Electronic media facilities rarely accept only
one ad at a time—it would be too expensive in relation to the time taken by the

ZOOM IN: **Super Commercials**

The Super Bowl is an outstanding example of selling ads on the basis of what the market will bear—and the market bears a lot. A 30-second Super Bowl ad can cost up to $3 million. Even with 100 million people watching, that is a CPM of $30, about 2.5 times the CPM of prime-time fare. But advertisers love the Super Bowl. It brings a sought-after young male demographic, and it's a time when people actually make a point of watching ads because they have such a great reputation. In fact, the commercials are as likely to be talked about around the water cooler the next day as the game itself. Some of the memorable ones throughout the years are:

- Joe Namath for Noxzema in 1969
- A young boy giving "Mean" Joe Green his Coke in 1980
- Apple's classic 1984 futuristic ad in which a female athlete destroys the Orwellian "Big Brother," symbolically freeing workers from office tedium
- Bugs Bunny and Michael Jordan interacting for Nike in 1992
- Former governors Mario Cuomo of New York and Ann Richards of Texas for Doritos in 1995
- Numerous Budweiser commercials including the "Bud-Wise-Er" frogs
- E-Trade's hillbillies and monkey in 2000
- Britney Spears going through various decades with Pepsi in 2002

- Carrier pigeons trying (unsuccessfully) to compete with Federal Express in 2008

The E-Trade commercial was particularly entertaining and prophetic because it poked fun at all the money dot-com companies were spending on Super Bowl ads. Two hillbillies clapped and a monkey danced while a narrator said, "We just wasted $2 million on a commercial. What are you doing with your money?"

Do you think companies are justified in spending so much for commercials on the Super Bowl? Can you think of any other programming in which commercials might take on the aura that they have with the Super Bowl? What is your favorite Super Bowl commercial?

Britney Spears in Pepsi's Super Bowl commercial.

(Reuters/TimePix)

---

frequency

salesperson to sell the ad. Usually a media company requires that an ad be aired at least 12 times and tries to induce the advertiser to buy even more airtime by offering **frequency discounts**—lower prices per ad as the number of ads increases. For example, a company that places an ad 12 times on a radio station might be required to pay $500 per ad, or a total of $6,000. If it places the ad 24 times, it would pay $450 per ad, or a total of $10,800.

length

Another variable is the length of the ad—usually 15, 30, or 60 seconds. Obviously, a one-minute commercial costs more than a 30-second commercial, but

Rate Card
Generic Radio Station

Day	Time Slot	Rate
M/F	6:00 A.M. – 10:00 A.M.	$150
M/F	10:00 A.M. – 3:00 P.M.	$120
M/F	3:00 P.M. – 7:00 P.M.	$145
M/F	7:00 P.M. – 12:00 A.M.	$80
Sa/Su	6:00 A.M. – 10:00 A.M.	$115
Sa/Su	10:00 A.M. – 3:00 P.M.	$100
Sa/Su	3:00 P.M. – 7:00 P.M.	$115
Sa/Su	7:00 P.M. – 12:00 A.M.	$75
All days	12:00 A.M. – 6:00 A.M.	$70

30-second spots are 75% of 60-second spots
15-second spots are 45% of 60-second spots

Exhibit 9.9

An example of the type of rate card that a small radio station chain might use.

usually not twice as much. For example, 12 one-minute ads might cost $550 each, and 12 30-second ads might cost $500 each. Station costs, such as handling, are more expensive for two separately produced 30-second commercials than for one 60-second commercial.

Audiences are of varying sizes at different times of the day (different **dayparts**), so this factor also becomes a variable on rate cards. Radio stations and networks usually have the largest audiences in the morning and early evening, when people are in their cars. The audience is lower during the midmorning and evening, and lower still in the middle of the night. Television is a different story, with most viewers congregating between 7:00 P.M. and 10:00 P.M. The other parts of the day can vary from channel to channel, depending on the programming. Some stations have a different rate for **adjacencies**—commercials that are aired right after or right before a popular program. The advertisers pay extra to be the first or last commercial, rather than one that may get lost in the middle of a set of commercials.

daypart

Some stations create a **grid rate card.** This includes prices for different dayparts but also different prices for the same daypart depending on whether the station has a large number of spots available or has only a few **availabilities (avails).** Obviously, the rates are higher when there are only a few avails.

Television facilities sometimes sell ads based on particular programs rather than for particular times. An ad bought for 9:00 P.M. Tuesday during *The Tuesday Night Movie* might cost more than an ad purchased at 9:00 P.M. Wednesday during *This Week's Report.* Product placements are negotiated on an individual basis, but take into account the popularity of the TV program or the movie actors. Because ads are always available on the internet, the rate structure is different, with the length of time the ad stays up and its physical size being important determinants. In short, there are many variations in the factors that determine the cost of a particular ad.

programs

9.10 Advertising Practices

In addition, media companies create special "deals" to assist and entice advertisers to buy their advertising time. Some of these are:

trade-out: This is particularly popular in radio and involves trading advertising airtime for some service the station needs. Perhaps the station owns a car that needs occasional service; to receive this service free from Joe's Garage, the station broadcasts 12 ads a week for Joe's Garage with no money changing hands. Similar arrangements are negotiated with restaurants, stationery suppliers, gas stations, audio equipment stores, and the like.

run-of-schedule (ROS): The salesperson and advertiser decide how many ads should be run and sometimes make up a package that consists of ads in different dayparts. The seller decides on the specific times the ads should run based on the availability of commercial time. For this privilege, the advertiser receives a discount and does not pay the going rate for the times selected.

fixed buy: The advertiser states the exact time each ad should run and pays a premium price. This often involves some negotiating. If two advertisers want 9:00 A.M. Monday, someone must compromise. In this circumstance, some media companies employ the **bump system,** in which the advertiser who offers the most money will get the spot even though some other advertiser was promised it first.

local discount: A local advertiser pays less than a national advertiser for the same airtime. The rationale is that the national advertiser can profit from reaching all the people in a station's coverage area while a local advertiser might not. For example, a national automobile manufacturer has the potential for selling cars to people in an entire city that a station covers, but the people in the southern part of the city are not likely to respond to an ad for a car dealer in the northern part of the city. For this reason, part of the station's coverage area is practically useless to the local car dealer advertiser.

orbiting: This practice appeals to advertisers who want many different people to see or hear their ads because ads play at slightly different times each day. For example, all the ads might be aired during morning drive time but on Monday they would be at 6:10, 7:15, and 8:50 while on Tuesday they would play at 6:30, 7:45, and 9:15.

annual flight schedule: Companies that sell leisure products often like to advertise around holiday times, and for them, stations and networks provide an annual flight schedule. The ads run only before major holidays and not at other times of the year.

volume power index (VPI): An advertiser decides who its target audience is and pays only for the number of listeners or viewers who fit that pattern. For example, a cosmetic company may designate women 18 to 34 for a particular product and base the CPM on only those people.

make-goods: Networks promise an advertiser a certain size audience; if the program does not deliver that large an audience, the broadcaster returns money to the advertiser or offers the advertiser free ads to make up for the shortfall.

rate protection: This is usually only available to longtime customers who advertise a great deal. The media organization guarantees that even if the advertising prices increase, these good customers can still buy ads at the old prices.

And on and on. When economic times are bad, media companies tend to offer more enticements to advertisers (and their agencies), trying to find something that will close the deal.[20]

9.11 Advertisement Production

Production of commercials is a very important facet of advertising. Companies rarely produce their own commercials because they simply do not have the equipment or know-how to do so. They are in the bread-making, car-selling, or widget-manufacturing business, not the advertising business.

radio

Small businesses that place ads on small radio stations generally have the stations produce the ads. The people who sell the time to the merchant write and produce the commercial at no extra cost. If they are good salespeople, they spend time talking to the merchant to determine what he or she would like to emphasize in the ad, and they check the ad with the merchant before it is aired. Often the commercial is prerecorded with music or sound effects, and occasionally the merchant talks on the commercial. Large companies that advertise on radio usually have much more elaborately produced ads that include jingles, special effects, and top-rated talent. The production costs are generally in addition to the cost of buying time.

TV

Television commercials are still more elaborate. Usually they are slick, costly productions that the advertiser pays for in addition to the cost of commercial airtime. Local stations sometimes produce commercials for local advertisers and charge them production costs—the cost of equipment, supplies, and personnel needed to make the commercial.

agencies

Much of local TV advertising and most network commercial production is handled through advertising agencies. As part of its service to the client, the agency decides the basic content of the commercial, perhaps by coming up with a catchy slogan. Usually a **storyboard** (see Exhibit 9.10)—a series of drawings indicating each shot of the commercial—is made. The advertising agency then puts the commercial up for bid. Various independent production companies state the price at which they are willing to produce the commercial and give ideas as to how they plan to undertake production. The ad agency then selects one of these companies and turns the commercial production over to it. Some production companies maintain their own studios and equipment, whereas others rent studio facilities from stations, networks, or companies that maintain facilities.

Exhibit 9.10

Software used to generate storyboards from PowerProduction Software in California.

(StoryBoard Artist Software © copyright PowerProduction Software www.powerproduction.com)

Exhibit 9.11

This commercial for the Olympics, featuring Billy Crystal, is being shot on Broadway in New York City.

(Rick Mackler/Globe Photos, Inc.)

The time, energy, pain, and money that are invested in producing a commercial often rival what goes into TV programs (see Exhibit 9.11). One commercial often takes weeks to complete, and many use the very latest in visual effects. Once the commercial is "in the can," it may air hundreds of times over hundreds of stations and networks—hence, the obsession with perfection.

internet

Ads for the internet can be the same commercials that are produced for radio or TV, but they are more likely to emphasize graphics (see Exhibit 9.12). The jury is still out on how fancy these graphics should be. Some people think eye-catching graphics are necessary to grab the attention of web surfers; others think they are negatively distracting and that simple is better.[21]

9.12 Advertising to Children

ads and programs

Commercials aimed at children generate particularly strong controversies. On early radio and TV, the hosts and stars of children's programs presented many of the ads. For example, in his early programs of the 1950s and 1960s, Captain Kangaroo, after doing a stint about growing vegetables, would walk over to a set featuring boxes of cereal and tell the children how good the cereal was. This practice, called **host selling,** came under fire when research showed that children have difficulty distinguishing between program content and commercial message.

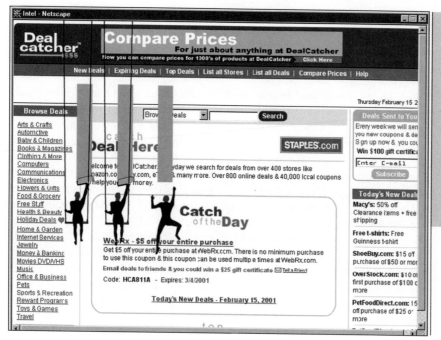

Parents, particularly the group of Boston parents who formed Action for Children's Television (ACT), also complained about the ads (see Chapter 8).

FCC guidelines

As a result of the research and the complaints, the FCC in 1974 issued guidelines stating that programs' hosts and stars should not sell products and that the display of brand names and products should be confined to commercial segments. In other words, a box of Kellogg's Corn Flakes should not appear within the program itself. In 2005, the FCC extended the "host selling" guidelines further to prevent programs from directing children to related commercial websites where they could buy products. The 1974 guidelines also stated that special measures should be taken to provide auditory and/or visual separation between program material and commercials. This led to **islands** at the end of most children's commercials. These were still pictures with no audio followed by a definite fade to black. This was to help children understand that the commercial had ended and the program was about to begin again. Starting about 1983, when the FCC was in a deregulatory mood, these guidelines were no longer enforced, but some commercial producers abided by them anyway.

time for ads

The amount of time devoted to commercials within children's programming has also varied in controversial ways over the years. For several decades, the National Association of Broadcasters had a self-regulatory code that stated the maximum number of minutes per hour that should be devoted to commercials within children's programs. During the 1960s and early 1970s, this number was 16. In 1973, the code was amended (largely at the insistence of ACT) to reduce the number to 12, and soon after, it was further reduced to 9.5 minutes per hour. Although this code was not a law (it was just an advisory standard decided by people within the television industry), most stations abided by it.

toy-based programming

The code was outlawed by the Justice Department in 1982, however, and the FCC, in its deregulatory mood, said there should be no limits on the amount of advertising on any programming. This led to increases in the number of ads in children's programs and led to a short-lived phenomenon known as **toy-based programming.** In these programs, a particular toy was the "star" of the show. Often the toy was created before the show with the idea that the program would sell the toy. This did not sit well with ACT and other parents who thought their children were being exposed to commercials that were 30 minutes long.

1990 provisions

In 1990, commercials during children's programs were once again restricted, this time by Congress. It passed a bill limiting commercials to 12 minutes per hour on weekdays and 10.5 minutes on weekends. Some stations do not abide by the rule, and the FCC periodically fines them. This 1990 bill essentially outlawed toy-based programs, although they had already fizzled as a programming form because not enough children were watching them.

content

In addition, the content of children's commercials has been criticized. For example, all the ads for sugary foods have been related to childhood obesity. The production techniques of children's ads are also questioned, especially those that appear to be deceptive, such as camera lenses that make a toy car look as if it travels much faster than it does and sound effects that add to the excitement of a board game but are not part of the game.[22]

9.13 Other Controversial Advertising

deceptions

Not only children's ads are accused of being deceptive. Ads aimed at adults are also less than what they seem to be. One classic case occurred when Campbell's put marbles at the bottom of a soup bowl so the vegetables would rise to the top and make for a richer-looking soup. The Federal Trade Commission made Campbell's stop airing the commercial because it was deceptive—no one puts marbles in soup. In a more recent case, the FTC barred Slim Down Solution from making bogus weight-loss claims. Even when demonstrations are valid, they may not be applicable. For example, a watch put through a hot and cold temperature test may operate perfectly at 110 degrees and –20 degrees, but not be satisfactory at normal temperatures.

testimonials

Testimonials have received their share of criticism because the well-known stars or sports figures used for the commercials may not know whether what they are reading from the cue cards is correct or not. Stars actually have to use the product they are advertising (this arose after football star Joe Namath advertised panty hose), but they cannot be expected to know all of its assets and liabilities. Sometimes the implications conveyed by or about the stars are misleading— "John Q. Superstar runs 10 miles a day and eats Crunchies for breakfast." The implication is that the Crunchies enable him to run 10 miles—in all probability, a false premise.

types of products

Another area of controversy involving commercials concerns the types of products advertised. Congress outlawed cigarette advertising on the airwaves in 1972 after a Surgeon General's report linked cigarettes to cancer, but that is the only time a category of product has been removed. Beer, wine, and personal sanitary products are considered by some groups to be inappropriate TV fare, but advertising them has not been outlawed. The liquor industry, which for many years had a self-imposed ban on advertising hard liquor on radio and TV, lifted its prohibition in 1996, and several stations started accepting liquor ads. Critics claim that advertising nonprescription medicines over the air and making them available through the internet encourages people to use them excessively. In the 1980s, a debate raged over whether condoms should be advertised. Stations and networks aired steamy love scenes within the programs without mention of any type of sexual protection but would not accept ads for condoms. Eventually many stations did accept the ads, primarily because of AIDS.[23]

Political advertising is also criticized for both its content and its presentation. Part of the hostility is directed at politicians who smear each other on minor personal points (or even the quality of each other's commercials) rather than dealing with the issues. Sometimes "special interest groups" buy ad time and do the smearing, keeping the candidate above the fray. But this, too, is highly controversial, and the Federal Election Commission and others have changed the rules governing such organizations a number of times. Some believe that, given the power of the media, all candidates should be given free time. Politicians favor this idea because it would greatly reduce campaign expenses, eliminating many of the perils inherent in raising large amounts of money to gain office, but broadcasters are opposed to the idea. At present, radio and TV stations are supposed to charge candidates their **lowest unit charge**—that is, the lowest rate they would charge an advertiser wishing to advertise at the same time. Lowest unit charges are hard to determine, however, because of the many deals that are made between advertisers and stations. This has led stations to falsify their prices, further fueling the arguments for free time for all candidates.[24]

political ads

9.14 Issues and the Future

One thing that is certain about the future of sales and advertising is that it will see both experimentation and change. Salespeople, advertising agencies, corporations, broadcasters, internet site owners—all are trying to find more effective ways to engage consumers so that they are more likely to buy advertised goods.

One method for doing this is to **target** ads more tightly. For example, technology allows cable and satellite TV companies to send different commercials to different homes based on the demographics of the occupants. So the rich senior citizens living in an affluent neighborhood might see a Cadillac commercial while their poorer nephews will be delivered a Chevy commercial, even though they are watching the same program. Internet ads can be targeted according to what websites a person visits. For example, those who surf boating websites can be shown ads related to water recreation products while those who access fashion information might see ads for cosmetics and shoes. Of course, obtaining the information about a person's interests in order to target the ads is fraught with privacy concerns.[25]

targeting

interactive

Another area of experimentation involves sophisticated **interactive ads.** For example, there is technology that allows you to click on a skirt that a TV drama actress is wearing and trigger a screen that gives information about the skirt and how you can buy one just like it. This information can appear at the bottom of the TV screen while the drama continues, or you might be switched to a different screen with the purchasing information and then when you have finished you can switch back to where you left off in the drama. Also, if you are watching TV on your computer screen, you can, in some instances, click on an ad to print out a discount coupon. The interactive anonymity of the internet poses problems, however. Someone can place an ad for a bogus product, collect money electronically, and disappear.[26]

ad skipping

In order to counter the ad skipping possible with digital video recorders, media owners are toying with the idea of placing ads on the TV screen as the program is playing. These ads could be in the lower third of the screen, where there already are promos for upcoming programs, or they could be along the side of the screen, essentially keeping the screen the same size it has been with standard definition TV and using the extra width of HDTV for ads. Conversely, while ads are playing, a part of the screen could be showing a "behind the scenes" aspect of the program so that the viewer stays with the ad and doesn't fast-forward or switch channels.[27]

agencies

Advertising agencies are becoming more aggressive about providing programming to networks that can benefit their advertisers. Some have even designed programs with plots that can accommodate all their client's products in an intense product placement style. Some programmers are skeptical about accepting programming from agencies, remembering that the quiz show scandals of the 1950s were caused, at least in part, because advertisers were supplying the programs.[28]

new media

Advertisers are also ascertaining how best to exploit new media opportunities in ways that enhance viewer engagement. They want to beam ads to cell phones and get people on social networking sites to send ads to each other. They would like to tag along with computer downloads and present ads on hotel video channels. They are trying to find ways to capitalize on texting, such as some sort of pay-to-text scheme, and they are constantly trying to increase the return on internet keyword search. Most salespeople now sell ads for **multiple platforms.** In other words, once a salesperson convinces a company to advertise on one property, such as a broadcast network, he or she then tries to get the advertiser to place ads on various other related media properties such as websites, cable networks, and radio stations for "a small extra fee."[29]

control

Another issue related to advertising is control. Critics point out that "he who pays the piper picks the tune" and think that heads of large companies and advertising agencies control program content because they provide most of the money that supports it. They point to the fact that automobile dealerships pulled advertising from CBS stations after *60 Minutes* aired a segment on financing practices of car dealerships. Not only are there specific instances of advertiser influence, but also overall censorship occurs simply because advertisers refuse to sponsor certain types of programs, such as controversial documentaries. On a more subtle level, advertisers affect program content because they demand certain (young) demographics, and that drives the content of programming that networks and stations show.[30]

Another issue, one that has been around for a while, is **clutter**—the amount of nonprogram material (mostly commercials) that occurs within an hour. As stations and networks see their dollars shrink, they expand the number of minutes devoted to ads. In the early 1980s, the commercial networks were airing no more than 9.5 minutes per hour of ads during prime time, but by 2005 some networks were up to 19 minutes. In addition to being annoying to the viewers, this increase in clutter bothers advertisers because their messages get lost. Although clutter is not declining, it is not getting worse because advertisers and media personnel realize its liabilities.[31]

    clutter

Much broadcast advertising criticism is aimed at the content of commercials, accusing them of being misleading, insulting, abrasive, loud, and uninformative. Critics complain that commercials spend too much time plying emotions and not enough giving information about the product. In fact, some commercials don't talk about the product at all. Research has demonstrated that ads instilling negative feelings in people often lead to sales of the product because people remember the brand name. Because 15 seconds, 30 seconds, or even a minute is not enough time to explain the assets of a product in an intelligent manner, the intent of commercials is to gain the audience members' attention, so they will remember the name without knowing much about the product.[32]

    content

Some people see advertising as having an overall negative effect on the structure of our society. It encourages a society dominated by style, fashion, and "keeping up with the Joneses," and it retards savings and thrift. It fosters materialistic attitudes that stress inconsequential values and leads to waste of resources and pollution of the environment. Advertising also fosters monopoly because the big companies that can afford to advertise can convince people that only their brands have merit. The counterargument to all this is that advertising drives the economy. If people were not enticed into buying things, the economy of the country (and world) would stagnate.[33]

    effect on society

Although advertising is severely criticized from many quarters, it has its positive elements. It defrays costs, reducing what the public has to pay directly, and it informs people about products available to them. Advertisers need the media, the media need advertisers, and consumers need both. Advertising is not likely to go away.[34]

    positive elements

9.15 Summary

Sales and advertising practices of electronic media differ, but generalizations about what happens related to the internet, television, radio, and movies summarize the various facets of advertising.

For the internet, ISPs sell directly to customers, and the internet in general helps sell goods directly to consumers. Google sells its advertising know-how to other media organizations. The internet is the newest player in advertising, with sales rising regularly. Keyword search is the most effective advertising method, especially combined with pay-per-click. Other types of ads include banner ads and pop-ups. Production of internet ads involves intensive use of graphics. Internet ads can be targeted and interactive, but privacy is a problem.

Cable TV systems do direct-to-consumer selling by offering video, phone, and internet services. Public television must try to get viewers to contribute money. On the media-to-media front, TV facilities sometimes rent out facilities to each other. Syndicators sell to TV stations and cable networks using barter, cash, and cash plus barter. Cable networks must convince affiliated cable systems to carry their services and pay for them, whereas broadcast networks pay their affiliates.

Advertising is the lifeblood of broadcast TV and basic cable. Stations and networks have sales forces, and broadcast TV networks, in particular, are concerned with upfronts. Television entities deal heavily with advertising agencies—full service, in-house, and boutique. Networks are likely to be involved in network buys while stations deal more with national and local buys. Some cable systems are well suited for regional buys. Product placement, product integration, and virtual ads are becoming more common on all forms of TV.

Commercial broadcast networks have the highest CPM. Public television has enhanced underwriting, which leans toward advertising. TV stations are likely to offer local discounts, orbiting, annual flight schedules, and volume power indexes. Sometimes networks must offer make-goods. Production of television ads is elaborate and handled by advertising agencies. Commercial stations have been subject to many regulations related to advertising on children's TV, and many other forms of television ads are controversial, including political ads. Television organizations are particularly interested in countering ad skipping and clutter, and they want to keep control of their programming and not allow advertisers to interfere.

Satellite radio sells its product direct to the consumer. In media-to-media selling, both syndicators and networks sell to stations. Radio stations have small sales forces, use a station rep, or sell from headquarters. Most of the ads are of a spot nature, but there is some program selling. Radio networks are most likely to be part of a national buy, while stations emphasize local buys. Radio stations are likely to have rate cards that take into account frequency, length, and daypart. They engage in trade-out, run-of-schedule, fixed buy, rate protection, and other sales practices. Often a station will produce an ad for a client. Radio advertising is less controversial than TV advertising, but it has also dealt with controversial products such as cigarettes and condoms.

Movie theaters sell tickets directly to customers and also sell refreshments, which are quite profitable. The theaters display movies by dealing with studios and splitting the income under what are often complicated formulas. Commercials often run before a movie or at the beginning of a DVD, but in general, movies are less concerned with advertising than the internet, television, and radio.

Suggested Websites

www.adweek.com (*Adweek* magazine, an industry publication)

www.aaaa.org (American Association of Advertising Agencies, a trade group)

www.omnicomgroup.com (Omnicom, a major advertising agency holding company)

www.rab.com (Radio Advertising Bureau, an organization to help radio organizations advertise effectively)

www.tvb.org (Television Bureau of Advertising, an organization to help television entities advertise effectively)

Notes

1. "Popcorn Is Salt of the Boxoffice," *Hollywood Reporter,* December 10–12, 2004, 14.
2. For examples, see http://www.sirius.com; https://www.directv.com; http://www.comcast.com; and http://www22.verizon.com/content/ConsumerFios.
3. "As Sponsorship Sales Blossom, Public Radio Walks a Fine Line," *Wall Street Journal,* March 17, 2006, 1.
4. "Super Circuits," *Hollywood Reporter,* February 12–18, 2002, 12–13; "The Brave New World of TV," *Broadcasting & Cable,* November 14, 2005, 6; and "Store Aims to Plug People into iPhones," *Los Angeles Times,* July 10, 2008, C-1.
5. "TV Turns to Retail," *Broadcasting & Cable,* November 28, 2005, 14; and "Unreality TV: Soap Opera Plots Spawn Lucrative Products," *Wall Street Journal,* June 27, 2005, B1.
6. "Syndies Sing Happy Ad Tune," *Hollywood Reporter,* August 25, 2008, 2; and "Picture Getting Smaller for Syndication," *Electronic Media,* February 12, 2001, 1.
7. "Pot of Gold," *Broadcasting & Cable,* July 7, 2008, 10.
8. "Can These Marriages Be Saved?" *Broadcasting & Cable,* January 14, 2002, 6.
9. "An Ancient Legend Spans the Film Formats," *TV Technology,* November 7, 2007, 18.
10. "U.S. Ad Market Growing, Especially Cable," *Broadcasting & Cable,* September 1, 2003, 2; "Upfront's Upside," *Broadcasting & Cable,* June 23, 2008, 17; and "Working Smart with Your Rep," *Sound Management,* April 1988, 18.
11. "Advertising Agencies," *P3,* August 2005, 16; and "The Leading US Agencies in 2007 by US Revenues," http://www.adbrands.net/us/index_agencies.html (accessed July 26, 2008).
12. "Local Focus of Spot Runner Has Appeal to Global Investors," *Los Angeles Times,* May 7, 2008, C-1; "DVR, for Divide, Vex, and Rankle," *Los Angeles Times,* May 14, 2007, E-1; and "Spending on Cinema Ads Spikes," *Hollywood Reporter,* June 16, 2008, 10.
13. "Vid Game Score Will Soar," *Hollywood Reporter,* June 18, 2008, 1; "FTC: No Need to Label Products," *Hollywood Reporter,* February 11–13, 2005, 3; and "Texas Station Interrupts Commercial Interruptions," *Wall Street Journal,* April 23, 2007, B1.
14. "And Now, A Story from Our Sponsor," *Hollywood Reporter,* March 18, 2008, S-4; and "Infomercials: A Mega-Marketing Tool," *Emmy,* June 1995, 114.
15. "The Internet Allows Consumers to Trim Wasteful Purchases," *Wall Street Journal,* November 29, 2006, B1; "FIM Links to Google in Ad Pact," *Hollywood Reporter,* August 8–14, 2006, 1; and "Online Ads Are Google's Strength," *Los Angeles Times,* May 1, 2004, 1.
16. "Internet Advertising Revenues Top $21 Billion in '07, Reaching Record High," http://www.iab.net/insights_research/iab_news_article/299656 (accessed July 27, 2008).
17. "Who Pays for Public Broadcasting?" http://www.cpb.org/pubcast/#who_pays (accessed September 14, 2004); and "Dishing Out the Hamburger Money," *Los Angeles Times,* May 16, 2004, E-1.

18. "Cheaper by the Thousand," *Broadcasting & Cable,* February 4, 2002, 20–22; "The Road to Super Bowl XL," *Wall Street Journal,* February 3, 2006, 1; and "Super Rates," *Los Angeles Times,* January 28, 2000, C-1.

19. "Loyalty Factor," *Broadcasting & Cable,* May 5, 2004, 42. The rate card is based on one for the Ozark Radio Network eight-station combo found at http://www.ozarkradionetwork.com/rates2.html (accessed July 31, 2008).

20. "Marketing ROI: Every Brand's Opportunity," *ANA/The Advertiser,* October 2001, 30–34; and "Stations Add Incentives and Generate Revenue," *Television Week,* August 15, 2005, 24.

21. "How a Gecko Shook Up Insurance Ads," *Wall Street Journal,* January 3, 2007, B1; and Larry Elin and Alan Lapides, *Designing and Producing the Television Commercial* (Boston: Allyn and Bacon, 2003).

22. "FTC Commissioner Tackles Ads for Kids," *Wall Street Journal,* August 20, 2008, B7; "Ad Violations Cost Viacom, Disney," *Hollywood Reporter,* October 22–24, 2004, 4; "Senate Unanimously Okays Blurb-Curbing Kidvid Bill," *Daily Variety,* July 20, 1990, 1; and Barbara J. Wilson and Audrey J. Weiss, "Developmental Differences in Children's Reactions to a Toy Advertisement Linked to a Toy-Based Cartoon," *Journal of Broadcasting & Electronic Media,* Fall 1992, 371–94.

23. "Distillers Reverse Ban on Radio, TV Ads," *Broadcasting & Cable,* November 11, 1996, 7; "The Ups and Downs of Ad Rejection," *Broadcasting,* September 16, 1991, 40; and "Radio Executives Open to Condom Spots," *Broadcasting,* November 25, 1991, 46.

24. "FCC Seeking Way Out of 'Lowest Unit Charge' Confusion," *Broadcasting,* June 17, 1991, 21; and "Campaigns Get Personal," *Newsweek,* April 28, 2008, 20.

25. "Targeted Ads Raise Privacy Concerns," *Wall Street Journal,* July 8, 2008, B1; and "Regulatory Battle Rages over Web Ads," *Broadcasting & Cable,* August 4, 2008, 10.

26. "NBCU, DISH Offer Ad 'Triggers,'" *Broadcasting & Cable,* May 26, 2008, 3; and "Cash on the Wirehead," *Byte,* June 1995, 71–74.

27. "Beware the Space Invaders," *Broadcasting & Cable,* August 11, 2008, 14; and "Ways to Ensure They Don't Touch That Dial," *Hollywood Reporter,* July 2, 2008, 1.

28. "The Pitch That You Won't See Coming," *Los Angeles Times,* August 22, 2004, E-1.

29. "R U Texting?" *Broadcasting & Cable,* September 25, 2006, 10; "New Ways to Drive Home the Message," *Newsweek,* May 30, 2005, 56; "Strike Up the Brand," *Hollywood Reporter,* April 4–6, 2008, S-10; and "Trying YouTube's Hottest Trick," *Broadcasting & Cable,* February 26, 2007, 10.

30. "*60 Minutes* Story Drives Away Ads," http://www.broadcastingcable.com/article/CA411062 (accessed April 19, 2004); and "The Tyranny of Eighteen to Forty-Nine," *Emmy,* Issue No. 6, 2003, 60–63.

31. "Television Networks Fatten Commercial Calf," *Broadcasting,* June 18, 1990, 21; "Ad Clutter Keeps Climbing," *Broadcasting & Cable,* December 22, 2003, 13; "Kelley: Primetime Is Short," *Hollywood Reporter,* July 27, 2005, 3; and "Study: Ad Clutter No Worse in '06," *Hollywood Reporter,* April 24–30, 2007, 6.

32. "License to Shill," *Los Angeles Times,* June 5, 2008, E1; and "Thirty and Counting," *Emmy,* May/June 2006, 198–99; and "A Word from Our Sponsor," *Newsweek,* August 4, 2008, 58–59.

33. Arthur Asa Berger, *Manufacturing Desire* (New Brunswick, NJ: Transaction Publishers, 1996).

34. "Survey Says: Ads Support Best," *Hollywood Reporter,* May 5, 2008, 9.

PROMOTION AND AUDIENCE FEEDBACK

P romotion and audience feedback are, in a manner of speaking, two sides of the same coin. Within the electronic media business, the main purpose of promotion is to attract an audience for a particular product, such as a radio or TV program, a website, a movie, or a video game, in hopes that the feedback from that audience will be positive. The main reason for desiring positive feedback is financial. For the most part, positive feedback related to video games, DVDs, and movies shown in theaters results in instant financial gratification because consumers pay directly for these products. With much of the internet, television, and radio, the results of the feedback are more indirect.

In current TV the blood and gore
Hardly ever ceases to flow;
Especially in the executive suites,
At times when ratings are low.
Edward F. Dempsey

These media forms sell advertisements, and it is the advertisers who provide (or withhold) financial success. It is important for media forms such as radio and TV to have valid proof of their audience feedback so that they can convince potential advertisers to buy commercial time. Advertisers know that they are not going to sell products to every person exposed to an ad, but they want to make valid judgments so that they receive maximum return for their advertising dollars. As John Wannamaker once said, "I know I waste half my money on advertising. I just don't know which half."[1]

Advertising as a whole is a $450 billion industry, with more than $100 billion going to electronic media. Advertisers want information that will help them waste as little money as possible. But they cannot simply depend on the promotion and statistics from members of the electronic media because these practitioners have a vested interest. For that reason, an independent industry has arisen for the purpose of measuring audience feedback.[2]

Promotion (and sales) people use the information from these independent audience research companies to plan their strategies to attract an audience, and thus attract advertisers eager to connect with that audience.

10.1 Promotion

time frame

As mentioned in Chapter 7, **promotion** has become very important to media entities because potential viewers and listeners have so many choices. Even getting someone to know about a TV program, let alone sample it, is difficult. A great deal of promotion occurs when something new happens—a network starts airing a new show, a new movie comes out, a radio station starts a new contest, a new video game hits the stores. It used to be that TV promotion was heaviest in the fall when all the networks started airing their new series. But now new shows are introduced throughout the year, so TV promotion, like other promotion, is year-round.

promos

Exhibit 10.1

A shot from an animated promo for TalkSPORT Radio.

(Supplied by Alpha/Globe Photos, Inc.)

One of the easiest and best ways for radio and TV to promote is to place **promotional spots (promos)** on their own stations or networks. People have loyalty to particular media outlets because they like what they see or hear, so they are a potentially receptive group. In addition, companies that own many media forms often **cross-promote.** A promo for HBO might be cablecast on CNN because Time Warner owns both networks. Promos look like commercials except they are about future programming (see Exhibit 10.1). Promos often contain short scenes (of a teasing nature) from the program, along with information about when it will air. Because promos are like commercials, viewers often skip them in the same way they skip commercials. As a result, stations and networks have started using **snipes.** These are pop-ups that appear during

Ad Spending by Major Movie Distributors

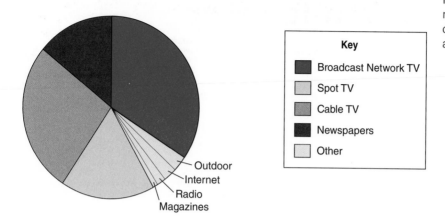

Key
- Broadcast Network TV
- Spot TV
- Cable TV
- Newspapers
- Other

Outdoor
Internet
Radio
Magazines

Exhibit 10.2

How the major movie distribution companies split their advertising in 2007.

a program, usually in a corner of the screen. Viewers cannot fast-forward through them because they are part of the program material the viewer is watching.[3]

Stations and networks are not the only media forms subject to promos. Many DVDs start with promos for other home video products produced by the same company, and **trailers** have been a promotional vehicle for movies for years. An interesting aspect of media promotion is that media forms use each other for promotion. A TV show advertises on radio; an internet site is part of the on-screen ads the come up in a theater before the movie starts; a movie advertises on TV. Exhibit 10.2 shows how the major movie distributors split their advertising dollars.[4]

movies

Promotion includes other forms of advertising such as signs on billboards or the sides of buses, listings on internet search sites (see Exhibit 10.3), and ads in newspapers. Most promotion is geared toward the potential audience for the media product, but some promotion is geared toward members of the industry. For example, at Oscar voting time, the movie studios promote heavily to members of the Academy of Motion Picture Arts and Sciences, whose votes determine the awards.

other promotion

The best promotion is word of mouth. Friends telling each other about movies, TV programs, or music they really like have much more of an impact than anything a media entity can do. Bloggers have become an important part of promotion; however, they often express negative opinions that can hurt the promotion process.

Promotion is closely tied to public relations and publicity, with the three often being in the same department. **Public relations** builds general goodwill that can enhance sales and audience numbers. Typical public relations activities include giving station tours, answering viewers' emails, and engaging in public-service activities, such as conducting campaigns to stop pollution (see Exhibit 10.4). Public relations can also include the airing of **public service announcements (PSAs),** which promote nonprofit organizations or causes. Stations and networks do not charge to air these, so they are often put at undesirable times. But effectively produced PSAs are noticed and often lead to increases in desirable actions, such as more women seeking mammograms or people contributing money to AIDS research.[5]

public relations

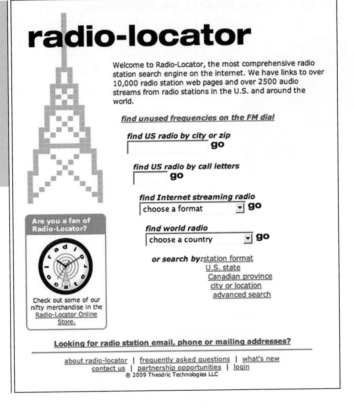

Exhibit 10.4

A sample of public relations activities undertaken by networks and stations.

Sample Public Relations Activities

HBO Helped aspiring filmmakers by sponsoring film festivals and providing money for winners of script contests to finish their films.

History Channel Encouraged civic groups to become active in historic preservation, including conducting a contest (in conjunction with Bank of America) to award $10,000 for the top local preservation project.

CBS Participated in an initiative to fight HIV/AIDS.

KEYE, Austin, Texas Helped more than100 animals find homes through its Annual Pet Marathon.

KTVX, Salt Lake City, Utah Sponsored an Immunization Care-a-Van that offered free inoculations.

KDZA, Pueblo, Colorado Collected 23,000 pounds of donated food for a local food bank during its Stuff the Bus campaign.

WTMJ, Milwaukee, Wisconsin Got its listeners to donate 10,000 teddy bears to be given to children in crisis situations.

WMC, Memphis, Tennessee Raised more than $200,000 to support families of fallen police officers and firefighters.

Publicity is similar to promotion except that it is free. For example, an advertisement in a local newspaper urging people to tune in to a particular disc jockey would be promotion because a station would pay for the ad. A review of the disc jockey's show in the same newspaper would be publicity because the station would not pay for it. Publicity is preferred over promotion because it is free, but the problem with publicity is that it may be negative. A positive review of the disc jockey's show is helpful, but a negative one is often worse than no review.

publicity

Promotion, public relations, and publicity are all subject to failure. Sometimes networks spend millions of dollars promoting a series only to yank it off the air after two weeks because it hasn't met expectations. Sometimes promotion campaigns go awry because they don't attract their targeted audience. For example, if a cable channel is trying to promote itself to college students, its promotion materials shouldn't feature grade school children. Promotion needs creative minds and energetic bodies that can respond quickly to the changing needs of the media.[6]

cautions

10.2 Forms of Audience Feedback

People give feedback to the media in many ways. One of the earliest was fan mail. During the 1920s, research showed that 1 person in 17 who enjoyed a program wrote to make his or her feelings known. During the decades that followed, fan mail was occasionally responsible for keeping a show on the air or resuscitating a canceled show.[7]

fan mail

Today, what was formerly fan mail is more likely to come as emails or text messages. In fact, the internet has opened up many more avenues for interaction between fans and their shows. Most shows have websites that encourage comments and chatrooms. Some producers lean heavily on what goes on in the chatrooms to develop their shows. They find out, for example, which subplots and actors their fans like the best and emphasize those in upcoming scripts.

internet

The internet, by its structure, can have built-in feedback. Those who operate online services or sport webpages know how many people use their services because their computers can keep track of the number of **hits.** Even these services, however, would like to know more about the type of users they attract and what new services these people would like.

Sometimes people give feedback unknowingly. For example, radio stations use the number of CD purchases as reported in *Billboard* magazine as a basis for deciding what music to play on the air. They also find out what songs people are exchanging, legally and illegally, on the internet and emphasize those on their playlists. As mentioned previously, people give feedback with their pocketbooks by buying tickets, DVDs, or products from websites.[8]

unwitting feedback

But the type of feedback that is used most often involves formal research with samples of people who supply information about their media habits that is then statistically analyzed and reported. The best-known form of audience research is ratings, which indicate how many people watch or listen to or access media products. However, other research goes beyond tallies to look at what types of people like certain products and why. In addition, companies solicit opinions from people before completing a project so that they can make it better.

research

10.3 Audience Research Companies

Nielsen

The name most readily associated with audience research is Nielsen, which is presently owned by a Dutch company, VNU, that operates not only in the United States but also in many other countries. Nielsen was established in the United States in 1923 by A. C. Nielsen, Sr., and conducted market research primarily for drugstores. Nielsen had a sample of drugstores save their invoices; the company then analyzed these and sold the information to drug manufacturers, so they could predict national sales. In 1942, Nielsen launched a National Radio Index, a report that indicated how many people listened to various programs. In 1950, Nielsen added reports about the television audience. In 1964, because of economic considerations and the changing nature of radio, Nielsen dropped its radio research and concentrated on television.[9]

updating

Nielsen has undertaken many changes over the years. When cable channels proliferated in the 1980s and satellite TV came to the fore soon after, Nielsen was faced with providing information about many more channels than in the days when three networks dominated. One challenge was determining what program was on what channel. Originally this was done through network lineups and phone calls to local stations to determine what they aired, but that became unwieldy as channels multiplied. In 1982, Nielsen introduced the **Automated Measurement of Lineups (AMOL),** in which equipment coded each network and syndicated show in a manner similar to product bar codes. As each program aired, the Nielsen equipment picked up the code and verified whether the program was what was expected. If not, Nielsen employees still made phone calls, but AMOL greatly simplified the verification problem.

Then along came the VCR, and Nielsen had to decide what to do about programs that were taped. Should they be ignored? Should they be counted when they were taped? What if no one ever played them back? Nielsen decided, amidst controversy, to count them as being viewed if they were taped. When the DVR came along, it was easier, technologically, to know if someone played the tape at a later date, so Nielsen started counting recordings as viewed if they were played back within a week. The internet, although it can monitor its own hits, is nevertheless a ripe area for validated measurement. Nielsen, not wanting to be left out of that area, bought an internet measurement company, NetRatings, and now publishes reports that include data about which sites are accessed most often.

Nielsen has had to adjust its technology as the population has grown and as more people watch TV outside the home. Advertisers want to know how many people watch their ads, not just the programs that surround the ads, and Nielsen has obliged with a new ratings service called C3 that gives minute-by-minute results of commercial watching. Because product placement has grown in importance, Nielsen now has employees who watch hours and hours of programming, noting each time a product appears on-screen. Program producers want figures on how many total people watch a particular program, even if it is shown on different broadcast, cable, and internet platforms at many different times. For this, Nielsen has devised Unduplicated Average Audience Ratings. It has also added ratings for portable devices such as cell phones and iPods.[10]

The other well-known name in audience measurement is Arbitron Inc. Formed in 1948, at one time it measured both local television and radio audiences. In 1994, it dropped the TV ratings because they were unprofitable and now concentrates on commercial radio stations and networks. It, too, has had to keep up with new technologies, such as satellite radio and internet radio, and it must accommodate listeners who are constantly on the move.[11]

Arbitron

Many companies (for example, AGP, The Pulse, and Percy & Co) have tried to compete with Nielsen and Arbitron in the past and have even received financial support from programmers who would like to see more competition in the audience measurement field. But gathering ratings is expensive, and most of the companies have failed because they simply could not make money. However, various companies, including Google, are working on internet counting systems, and cable systems are starting to use their digital set-top boxes to facilitate audience measurement.

competing companies

Some companies provide boutique services related to audience measurement. For example, Marketing Evaluation Inc. researches attitudes and opinions about specific television programs and personalities; companies such as ASI, MarketCast, and OTX Research pretest movies before they are released; another company undertakes face-to-face interviews regarding Hispanic viewing in homes where only Spanish is spoken; still another examines lifestyle and shopping patterns of media consumers.

other companies

Numerous companies also take ratings data that a station or network obtains from Nielsen or Arbitron and analyze it to find particular strengths that can be pitched to advertisers. They might find, for example, that a station's 4:00 P.M. to 6:00 P.M. slot has the second-highest ratings by combining women and children. Often these companies also offer (for a fee) to make suggestions as to how a particular entity's ratings can be improved. Nielsen and Arbitron also provide these for-fee services.

Because ratings are "secretive," in that individual participants are not identified and the ratings services do not divulge much of their methodology, it is difficult to verify the validity of what the audience measurement companies do. For this reason, broadcasters in the 1960s, at the behest of Congress, formed the Media Rating Council to accredit the various rating companies. This council must approve new equipment and techniques before they can be used as part of audience measurement.[12]

Media Rating Council

10.4 Sampling

Companies that undertake research need people they can survey or study in order to compile their data. If a company such as Nielsen wants to know how many people in the entire United States watch TV programs, it tries to reflect in its sample the demographics of the entire United States. This involves using census data to identify a core of counties throughout the country that represent the main U.S. demographics. Nielsen then identifies specific households through **random sampling.** Random sampling involves chance; in essence, the addresses of all the homes that qualify are put in a box and some are drawn out. Of course, a box is not actually used. Most random sampling is now done by computers programmed to select every third (or some other number) household.

random samples

households and people

Nielsen's base for measuring used to be households; that is, it reported what percentage of households watched a particular program. But as viewing became more individualized, Nielsen changed its base to people. It still selects households for its sample, but it considers what each person over 12 watches when reporting its statistics. A great deal of TV viewing occurs in bars, dorm rooms, hotel rooms, workplace cafeterias, and the like. For many years, the ratings services ignored this viewing, but Nielsen now includes out-of-home viewers in its sample (see "Zoom In" box).[13]

 ZOOM IN: **The Power of College Students**

Advertisers consider college students to be a very important demographic. They are starting to build brand loyalties, and they are fairly easy to convince to try different brands. They are tomorrow's great consumers. But for many years, Nielsen did not survey college students living away from home. In 2007, after running a three-year test, the company decided to include these students.

The test showed that the addition of college students boosted the ratings for some shows dramatically. For example, ABC's *Grey's Anatomy* increased 53 percent among women 18 to 24 if college students were included. CW's *America's Next Top Model* increased 39 percent among the same group. Knowing that college students are now included in the survey, some programmers initiated promotions to target them. For example, CW threw campus costume parties to promote the premiere of a new reality show, *The Next Pussycat Doll.* ABC arranged college viewing parties for *Jimmy Kimmel Live.* College students, if they can be lured away from their cell phones and video games, now have a greater influence on television programming.

College students who live on campus are now included in Nielsen's surveys.

Number of Households

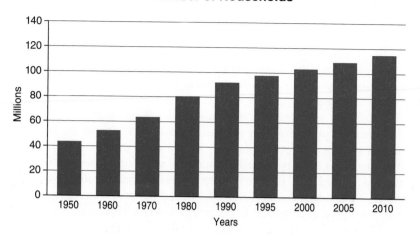

Exhibit 10.5
Census data,
beginning when TV
was first introduced,
show how
households have
grown and will
continue to grow.

Nielsen has had to increase its sample size as the number of households in‐ **size**
creased and the number of TV options proliferated. The original sample was
1,000, but now for national ratings, Nielsen samples 14,000 homes (about 36,000
people), which is about .012 percent of the approximate 110 million American
households or 300 million people (see Exhibit 10.5). The size of the sample for
the local areas that both Nielsen and Arbitron canvas is determined by the popula‐
tion of the particular area.[14]

Not all audience research projects want or need a sample of the entire U.S. **focused samples**
population, or even the entire population of a particular city. If a research com‐
pany wants to know, for example, how many Spanish-speaking people watch the
various Spanish-language networks, it will draw a sample only from Spanish-
speaking households. A radio station that wants opinions about minor changes in
its music would skew its research if it used a broad sample because it's only inter‐
ested in people who listen to the station and like its music. Much better would be
a sample taken from people who call the station to try to win contests. Companies
do have to worry about self-selected samples, however. If a sports TV network
were to put a poll on its website asking people to rate the importance of sports in
overall programming, they would get a biased reply. They could not (or should
not) generalize with a claim such as "90 percent of people in the United States
think there should be more sports on TV."

Once a research company selects its sample, representatives try to solicit the
cooperation of the households selected. They succeed about 65 percent of the
time. When they are not successful, they select alternate homes, which suppos‐ **replacements**
edly have similar demographic characteristics. The participants are paid a mini‐
mal amount of cash, such as $10, or are given gifts to encourage them to
cooperate and do what is needed to obtain the ratings data. The demographics of
the country change fairly rapidly; for example, the divorce rate goes up, or aver‐
age income takes a dip. For this reason, audience measurement companies con‐
stantly change their samples. Typically, a household remains in a sample no
longer than two years.[15]

10.5 Collecting Data

peoplemeter

LPM

Once the sample is solidified, the audience measurement company must gather the needed information from the participants. For national or local ratings, this usually involves some sort of equipment—equipment that has evolved over the years (see "Zoom In" box). The equipment that Nielsen currently uses is the **peoplemeter** and its outgrowth, the **local peoplemeter (LPM).** Nielsen introduced the peoplemeter in 1987 (see Exhibit 10.6) for its national ratings. The machine includes a handheld keypad that looks something like a remote control. Each member of the family is assigned a particular button on the keypad; extra buttons are provided for visitors. They are to turn on this button when they begin watching TV, push it periodically while they are watching, and turn it off when they are not watching. The peoplemeter gathers information about which channel the set is tuned to and who is watching what programs. This information is sent over phone lines to computers in Nielsen's Florida headquarters.

The local peoplemeter is a device about the size of a pager that the survey participant is to wear all the time. The LPM is capable of monitoring all audio occurring in the vicinity of the device wearer, including audio codes from radio and TV programs and internet sites. The eventual plan is to use LPMs to make calculations for a variety of media, but Nielsen is currently using them for local TV markets.[16]

PPM

Arbitron has developed a **portable peoplemeter (PPM)** similar to Nielsen's LPM. In fact, when Arbitron began work on its meter in the 1990s, it wanted to work jointly with Nielsen, but the two companies eventually went their own ways. Arbitron's meter has been approved by the Media Rating Council for both radio and TV, but the company is only using it for radio.[17]

Both Nielsen and Arbitron are still using older methods that involve diaries in some smaller markets. They plan to convert all markets to meters, but the

Exhibit 10.6

A Nielsen peoplemeter.

(Courtesy of A. C. Nielsen Company)

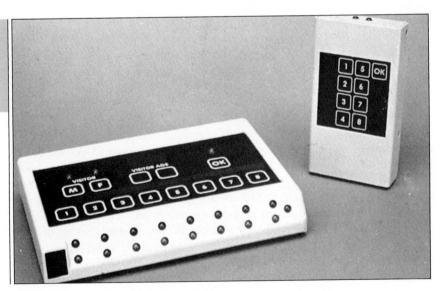

ZOOM IN: **Early Technological Ingenuity**

Nielsen's first machine (a), purchased in 1936 from two MIT professors, was called an audimeter. It was used to launch the 1942 National Radio Index based on 1,000 homes. The machine provided a link between a radio and a moving roll of punched paper tape in such a way that the paper could make a record of what station the radio was tuned to. Specially trained technicians visited the homes at least once a month to take off the old punched tape and put on a new tape—a very involved process.

The sophistication of audimeters increased, and by the 1970s (b) they could be connected directly from TV sets to phone lines that sent the information to a central computer in Florida in the middle of the night. The audimeters could indicate only whether the set was on and what channel it was tuned to, not whether anyone was viewing.

The next step was a different kind of device called a recordimeter that was used in conjunction with a diary (c). This machine was used mainly to make sure the diary entries weren't too different from when the set was actually on. The recordimeter operated somewhat like a mileage counter in that the digits turned over every six minutes that the set was in use. People keeping the diary indicated the "speedometer" reading at the beginning of each day and listed the programs they watched. They also indicated demographic factors about members of the family so that these could be matched with the programs.

a.

b.

c.

Meters Nielsen has used over the years: (a) a 1936 audimeter utilizing punch tape; (b) an audimeter from the 1970s; (c) a recordimeter and diary.

(Courtesy of A. C. Nielsen Company)

Exhibit 10.7

A sample page from an Arbitron diary.

(Courtesy of © Arbitron Ratings Company)

2	**TUESDAY**									
TIME			**STATION**			**PLACE**				
			Fill in station "call letters" (If you don't know them, fill in program name or dial setting)	Check One (✔)		Check One (✔)				
						At Home	Away From Home			
	From	To		AM	FM		In a Car	Some Other Place		
1 ⇨ **Early Morning** (5AM to 10AM)										
2 ⇨ **Midday** (10AM to 3PM)										
3 ⇨ **Late Afternoon** (3PM to 7PM)										
4 ⇨ **Night** (7PM to 5AM)										

IF YOU DID NOT LISTEN TO RADIO TODAY PLEASE CHECK ☑ HERE ➡ ☐

Please review each day's listening to be sure all your entries are complete.

diaries

conversion process is expensive and complicated, so they are rolling out meters city by city. For the **diary** method, survey participants are mailed packets. Each person in the household over 12 is to fill out a daily diary (see Exhibit 10.7) and then mail the diaries to Nielsen or Arbitron. Getting people to fill out these diaries with any degree of accuracy has become quite difficult and is a major reason both Nielsen and Arbitron are switching to meters.[18]

discrepancies

Any time a rating company changes its methodology, however, there is a hue and cry from those rated because the ratings change. For example, when Nielsen first introduced the peoplemeter, sports programming increased and children's programming decreased. The speculation was that women, who were the main ones filling out family diaries, knew what their children were watching but were less concerned about reporting what their husbands were watching. With the peoplemeter, men watching a championship sports event felt compelled to push the button often, fearful they might lose reception of the game. When Arbitron introduces its PPM, radio stations usually find they have a larger number of people listening than was reported by the diaries but that these people spend less time listening than previously thought. The complaints are loudest when ratings go down and costs go up. Audience measurement companies usually increase their sample size when they switch from diaries to meters and pass the added cost on to the users who subscribe to the service.[19]

Some types of audience research utilize trained researchers rather than equipment to obtain opinions from people in large- or small-group settings. Earlier forms of research involved researchers visiting survey participants in their homes or calling them on the phone. These forms of research have largely disappeared because people are more cautious about letting strangers into their homes and are irritated by phone calls.[20]

personal contact

Equipment used for audience research will no doubt continue to evolve. Certainly computers have untapped capabilities for gathering data, and so do set-top boxes, DVRs, and portable devices. As both radio and TV transition to digital, the possibilities for tapping into listening and viewing patterns increase, but this must, of course, be tempered by privacy rights.[21]

10.6 Analyzing Quantitative Data

Research that can be quantified is usually analyzed by computers. In the case of audience measurement, most of the meter data obtained from survey participants goes directly into a computer while diary entries must be retyped or scanned into the computer by humans. The computers are programmed to count how many people watch during various time periods and extrapolate those data into meaningful statistics.

rating

Historically, the main statistics audience research companies reported were **rating** and **share.** A rating is simply the percentage of households (or people) watching a particular TV program or listening to a radio station. Rating percentages are based on the total number of households with TV sets or radios. Assume that the pie in Exhibit 10.8 represents a sample of 1,000 television households drawn from 100,000 households in the market being surveyed. The rating is the percentage of the total sample. Thus, the rating for station WAAA is 160/1000, or 16 percent; the rating for WAAB is 100/1000, or 10 percent; and the rating for WAAC is 140/1000, or 14 percent. Usually when ratings are reported, the percentage sign is eliminated; thus, WAAA has a rating of 16. Ratings may be reported

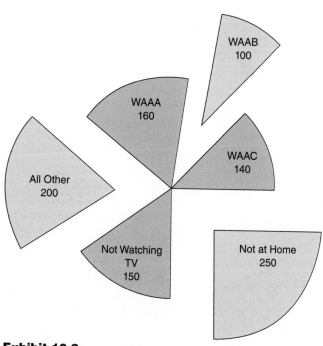

Exhibit 10.8

A ratings pie.

for certain stations or for certain programs. If WAAA aired network evening news at the particular time of this rating pie, then this news would have a rating of 16 in this particular city. National ratings, or course, are drawn from a sample of more than just one market.

share

A share is also a percentage, but it is based on the number of households (or people) with the TV set (or radio) turned on. In the pie shown in Exhibit 10.8, 600 households had their sets on—160 to WAAA, 100 to WAAB, 140 to WAAC, and 200 to all others. The other 400 households either had no one at home or had the TV off, so they did not count in the share of audience total. Therefore, WAAA's share of audience would be 160/600, or 26.7; WAAB's share would be 100/600, or 16.7; and WAAC's share would be 140/600, or 23.3. A share-of-audience calculation is always higher than a rating, unless 100 percent of the people are watching TV—an unlikely phenomenon.

Shares and ratings worked well in the early days of television when there were three networks and a few independent stations to comprise "all other." But times have changed. There are now so many program choices that both ratings and shares for some individual programs or time slots are so small that they are not statistically significant. Also, the original ratings and shares were based on the assumption of one TV set or radio shared by everyone in the household.

other statistics

As a result, a number of other concepts are now used to convey audience numbers. Frequently the raw numbers, rather than the percentages, are reported. Saying that a program has 3.5 million viewers sounds more impressive than saying it has a 1.5 rating or a 2.8 share. Syndicators like to use **gross average audience (GAA),** also called **impressions.** This measurement takes into account the audience numbers if a particular program is shown more than once in a market—something that often happens with syndicated fare. Another popular statistic is the **average quarter-hour (AQH).** This calculation is based on the average number of persons listening to a particular station (or network) for at least 5 minutes during a 15-minute period. The AQH is particularly popular with radio and some cable networks that have small groups of audience members who tune in and out.

Radio and cable also like the **cume**—the number of different persons who tune into a station or network over a period of time. This is valuable for letting advertisers know how many different people hear their message if it airs at different times, say 9:05, 10:05, and 11:05. The cume is important for differentiating services. For example, a radio station with a rating of 1.4 is usually not statistically different from a radio station with a rating of 1.5, but a cume can indicate a wider difference because it draws from a larger time spread. **Reach** and **frequency** are related to cume. Reach is the number of different people or households that are exposed to a particular commercial, usually over the period of a week. Frequency is the average number of times a person sees or hears a particular commercial over a specific time period.

Two other measurement concepts are **length of tune (LOT)** and **frequency of tune (FOT).** LOT is how long people watch shows on a particular network, and FOT is how often viewers return to a network. The two concepts are particularly valuable to cable TV because they give some idea of viewer loyalty to a particular network and also say something about the network. News and sports networks usually rank low in LOT but high in FOT; movie networks register the opposite. Several other useful terms are **people using television (PUT), people using radio (PUR),** and **households using TV (HUT).** These are the percentages of people or households who have the set tuned to anything. In the ratings pie example, the HUT figure is 60; 600 out of 1,000 people had the sets tuned to something.

Internet audience measurement is still in a nascent stage. Two of the main measurements are **clicks** and **unique visitors (UV).** Clicks are noted anytime anyone goes to a site, whereas unique visitors are only people who have not visited the site before.[22]

Even though **quantitative** audience measurement statistics are derived by counting and applying formulas, there is room for controversy and human intervention. Sometimes companies do not get a return of data that represents the sample. For example, a company such as Nielsen may give 51 percent of its diaries to women and 49 percent to men, but perhaps 30 percent of the women return the diaries and only 20 percent of the men do. In cases like this, the company weights the sample so that it appears that they are in the proper proportion. In other words, in this case Nielsen would give more weight to each man's returned diary than to each woman's diary.[23]

counting

There are many clever people in the industry who try to game the system. When folks at NBC knew that ABC's *Good Morning America* was having a particularly interesting guest, they aired the segment of the *Today* show that was opposite that *GMA* segment without any commercials. They knew Nielsen did not rate segments that had no commercials, so they did this to avoid getting a low rating.

gaming the system

During periods when ratings are particularly important, stations and networks often hype their programming, sensationalize their news, have star talent guest on a show, or give away huge prizes. On occasion stations have tried to contact members of the sample group either directly or in subtle ways such as airing a promo about ratings. People at the measurement companies have punished such acts by taking the stations out of the ratings report so that they have no ratings at all to tout to advertisers.

hyping

On the internet side, some website authors have learned to weave the word "sex" into their sites so they get more clicks, even though the site doesn't deal with sex. Also, there is some disagreement over how to count unique visitors. Should the count be according to computers, so that if two people use the same computer, only one of them will count? Or should it be by individuals, with some sort of safeguard in place to keep people with several computers from counting more than once?[24]

internet

10.7 Qualitative Research

Ratings, shares, and their kinfolk are considered to be quantitative research because they deal with concrete numbers related to how many people are watching or listening at particular times. Many advertisers and programmers want to know about the quality of the audience. They want to know that they are reaching the right kind of people so that they maximize their **return on investment (ROI).** To take an extreme example, an automobile insurance company may be happy to hear that its commercial reached 15 million people, but if 90 percent of those people were under the age of 16, the company's ROI is not going to be high.

Nielsen and Arbitron break down their data into various **demographics** so that clients at least know the age and sex of their audience, but that is usually still considered quantitative information. What clients often want to know is the **psychographics**—people's lifestyle characteristics and how these relate to media preferences and buying habits. What are the characteristics of people who are most likely to buy a BlackBerry? Are people who consider themselves upwardly mobile more likely to shop for organic products than those who are satisfied with their status quo? What shows are watched by people who are most likely to attend movie theaters? Research on subjects such as these is considered **qualitative** because it deals with factors that are hard to assign number values, even though in the end there may be statistical tests run to attempt to validate a particular concept. People have been divided into many categories—early adopters, brand loyalists, extroverts, imitators, and so on. Advertisers want to learn the psychographics of people who are interested in their products and then place their ads on programs or media forms that also attract those people.

engagement

Engagement is another qualitative characteristic that has risen to prominence. It looks at the degree to which the creative content of a program or media form results in meaningful communication regarding a brand. In other words, does the content of a program affect the viewer in such a way that he or she wants to buy the product associated with the content? Advertisers are also interested in brand recall—whether people remember a brand after they have finished watching a program. One study found, for example, that brand recall for commercials run during the 2008 Olympics was 130 percent greater than for other prime-time shows and 33 percent higher than for spots run during the Academy Awards.[25]

10.8 Pretesting

Sometimes electronic media organizations want to know what potential audience members think of something before they put it on the air. For example, radio stations want to play music that people will listen to so that they do not switch to another station. To help decide which music they should play, stations undertake **music preference research.** Sometimes researchers phone selected people who are likely to enjoy the type of format being tested, play a short section of music for them, and ask their opinion by following a prepared questionnaire. A variation on this approach is to have listeners call a toll-free number where, using a special ID, they can rate songs. If they don't finish in one session, they can call back at their convenience. Instead of using the phone, companies sometimes send iPods loaded with songs to a certain location, such as a community hall, and invite people to go there and listen to the music.

Another music preference research method is to invite a group of people to a minitheater and have them listen to music and give reactions, usually on a questionnaire form. Advantages of **minitheater testing** are that the fidelity can be better than it is over the phone and more songs can be played. This form of testing is more expensive than phone testing, however. Another research method involves inviting

psychographics

music preference

minitheater and focus groups

small groups of people to participate in **focus groups.** Again, the music can be played, but instead of reacting on questionnaires, the people in a group discuss their likes and dislikes. Focus groups can also tackle other issues, such as whether a disc jockey should be removed or whether a format change is in order. Focus groups give more in-depth information, but they are the most expensive method.[26]

Minitheater testing and focus groups are often used to pretest movies, TV programs, and commercials (see "Zoom In" box). Producers like to know that a product will be successful before they invest heavily in distributing it. Organizations such as Audience Studies Institutes (ASI) gather people in a room and show them versions of media productions. They elicit audience reactions through questionnaires and various button-pushing techniques. If the **pretesting** is done with a focus group, it involves more discussion.

Most TV program pilots are pretested as part of a weeding-out process; those **improvements** that test poorly don't go into production. But sometimes pretesting points to simple, fixable problems, so the program (or movie or commercial) is returned to the editing room, and the reedited version is pretested again. Or pretesting may result in such things as cast changes, the insertion of more trendy material, or a different marketing plan.[27]

ZOOM IN: Pretesting in Vegas

If you are interested in having a say as to what TV programs get on the air, visit Las Vegas. There Viacom and CBS have a 5,000-square-foot facility at the MGM Grand used to pretest programs for CBS, MTV, Nickelodeon, Showtime, and other programming outlets owned by the conglomerate. The theory is that Las Vegas tourists represent the wide demographics that are desired for program testing—midwesterners, southerners, rich, poor, male, female, young, old.

The facility includes two small focus group rooms and two large screening rooms, each holding up to 250 viewers. Every seat has a touch screen so that network executives located in any part of the country can track participant responses. Or, if they prefer to be on-site, they can watch and hear the focus groups through one-way glass.

But that's not all there is to this Las Vegas attraction. A retail store sells souvenirs of products related to the various channels and programs. Individuals who participate in the testing are given coupons for items in the store. You can't miss the facility, which has 50 TV monitors

featuring CBS and Viacom programming. The concept has been so successful that NBC now has a testing center in the Venetian Hotel.

Would you be interested in pretesting programming in this facility? Is this too garish an environment? Should pretesting and product selling be mixed?

The Las Vegas testing facility at the MGM Grand.

(Courtesy of Donny Sianturi)

10.9 Other Forms of Research

Q scores

Another form of research delves into the degree to which people are aware of particular shows or talent and how much they like the shows or personalities. Marketing Evaluations, Inc., conducts this research, and their results are called **Q scores.** Each month the company mails random people a questionnaire with a list of programs and names. The respondents indicate whether they have heard about each and, if so, the degree to which they like or dislike the performers or shows. The **performer Q** score of each person or **TVQ** score for each program—a number that indicates popularity—is then calculated. Having a low Q can mean unemployment lines for talent and cancellation for shows.[28]

educational

Corporate and educational organizations often undertake effectiveness research. They want to know if the programming they have produced has accomplished its goal in terms of teaching certain information or establishing certain attitudes. Often this research involves giving the people who are going to be watching a particular program a pretest to determine what they know or how they feel. Then, after viewing the program, participants are given the same test again to see if there was a significant change.

pay channels

The pay cable channels, such as HBO and Showtime, are concerned about people disconnecting from the service. Therefore, they sometimes undertake research directed toward finding the right combination of movies and other programs to retain the largest possible number of subscribers.

websites

Sometimes website designers have a number of people test their sites before they formally post them. With this help, the authors can find design elements that people don't like and problems with the site navigation. There are many other forms of audience research, and many companies willing to undertake them—for a fee, of course.

10.10 Reporting Results

Usually after conducting research, the organization that has undertaken the procedures issues reports to the clients who pay for the information. These reports can range from a simple chart to a multipage document. If, for example, the network executives were present when a program was pretested, there may be little need for a detailed report. If the material is distributed widely, as in the case of Nielsen and Arbitron, the reports are more detailed.

Nielsen

The way Nielsen and Arbitron report their results has changed over the years. At first everything was analyzed by hand, and it took months for the reports to be issued after the data were collected. With the advent of computers, this changed. Data sent over phone lines from meters could be analyzed almost instantly. This led to Nielsen's **overnight reports** on the previous night's programming that could be delivered to executives by noon the next day, and later to **fast nationals** that could be delivered by midmorning. The data from both Nielsen and Arbitron diaries took longer to analyze, but it too was aided by the speed of computers and could be used to issue reports within weeks of the time the information was collected. As clients obtained their own computer systems, they could interconnect their computers with those of the measurement companies and obtain the information on screens in their own offices without waiting for hard-copy reports to be delivered.

Nielsen's reports on programs shown nationally emphasize the commercial networks, but they also include data on cable TV networks, public broadcasting, and syndicated shows. They give overall numbers on the number of people who watched the various programs, and they also break down these data by various subcategories, including age, sex, and part of the country.

national

For reports on local stations, Nielsen divides the country into 210 nonoverlapping **designated market areas (DMAs).** All areas of the United States are surveyed at least four times a year—February, May, July, and November, periods that are sometimes referred to as **sweeps.** When portable peoplemeters are instituted, markets are surveyed on a constant basis.

local

Nielsen issues many other reports (see Exhibit 10.9). It does cable TV overnights and reports about syndicated programming. It has special reports on Hispanic viewing, although it also includes Spanish-language networks in its main national report. Its reports related to DVRs include data about people who view the programs live, those who view them within one day of recording them, and those who view them within the first week after recording. As previously mentioned, it now issues reports that cover only commercials within programs, and it issues reports on how people use the internet.

other reports

Nielsen also has a specialized service that shows minute-by-minute ratings in graph form on top of part of the picture of a program or commercial. In this way, executives can see such things as how people respond to a particular talk-show guest or how they respond to the clutter of numerous commercials. In addition, Nielsen does specialized reports for just about any entity that wishes to have the data programmed in a particular way to uncover particular trends or traits. These specialized services cost above and beyond the payment for the regular reports.[29]

Arbitron has reports for local radio stations and for radio network programming. For each local market, Arbitron measures two areas, **metro survey area (MSA)** and **total survey area (TSA)** (see Exhibit 10.10). The TSAs are geographically larger than the MSAs and do not receive the low-power stations as clearly as the MSAs. That is why Arbitron has two areas—the low-power stations do not

Arbitron

MSA and TSA

Selected Nielsen Reports	
Report	**Frequency**
Cable Network Weekly Ranking Report	Weekly
DMA Market and Demographic Rank Report	Annually
Household & Person Cost Per 1000 Report	Monthly
Households Using Television (HUT) Summary Report	Monthly
Local Market Sports Report	Monthly
Market Daypart Summaries	Quarterly
Market Primetime Feature Film Directory	Annually
NHIH Spanish Language Cable Network Data Files	Weekly
Report on Syndicated Programs	Quarterly
Total Viewing Sources	Monthly

Exhibit 10.9

Some of the reports that Nielsen publishes, along with how frequently they are offered.

Exhibit 10.10

Map showing an MSA and a TSA.

(Courtesy of © Arbitron Ratings Company)

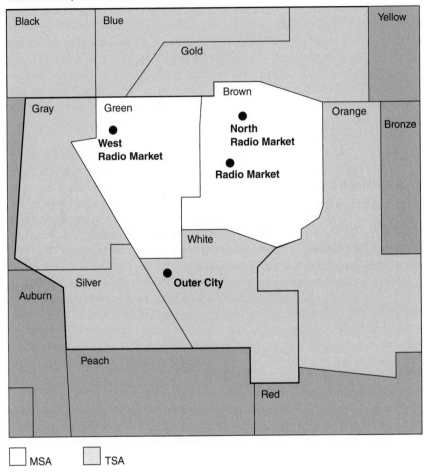

Radio Market, USA

MSA   TSA

compare favorably with the powerhouses in the TSAs but can hold their own in the MSAs. The station data are analyzed according to dayparts, and the reports give overall statistics and a breakdown of various demographics (see Exhibit 10.11). The network data are referred to as Radio's All Dimension Audience Research

RADAR
(RADAR). Arbitron checks to make sure the network commercials actually aired when they were supposed to and then calculates audience size.[30]

charges
Making improvements that keep up with the evolving industry and an ever-changing society can be very expensive. This cost is paid by the customers who subscribe to audience measurement services—networks, advertising agencies, stations, cable TV services, and others concerned with advertising. The services raise their prices from time to time, but must be sensitive to what the market will bear. They also charge differing amounts to different entities. While TV networks may pay millions of dollars per year for the reports they desire, stations in small markets may only pay thousands per year.[31]

Specific Audience
MONDAY-FRIDAY 6AM-10AM

	Persons 12+	Men 18+	Men 18-24	Men 25-34	Men 35-44	Men 45-54	Men 55-64	Women 18+	Women 18-24	Women 25-34	Women 35-44	Women 45-54	Women 55-64	Teens 12-17
WAAA WRRR METRO	8.6	7.2	1.2	10.1	13.1	5.4	4.3	9.8	11.2	10.4	11.1	3.1	14.9	8.6
WBBB METRO	.9	.7			3.6			.8	1.0				6.4	2.9
WCCC METRO	6.4	5.0		.8	6.0	5.4	14.9	8.5	3.1	3.2	4.0	12.5	12.8	
WDDD METRO	.4							.6	1.0	1.6				1.4
WDDD-FM METRO	8.7	9.8	23.8	12.4	4.8	1.8		6.2	14.3	10.4	1.0	3.1		18.6
TOTAL METRO	9.1	9.8	23.8	12.4	4.8	1.8		6.9	15.3	12.0	1.0	3.1		20.0
WEEE METRO	6.8	5.7	17.9	3.9	3.6		2.1	5.0	10.2	7.2	2.0		6.4	25.7
WFFF METRO	6.6	9.1	4.8	3.9	8.3	16.1	19.1	5.4	1.0	3.2	7.1	7.8	8.5	
WGGG METRO	1.3	2.1		2.3		3.6	6.4	.8		.8	2.0		2.1	
WGGG-FM METRO	20.6	23.6	19.0	25.6	21.4	37.5	21.3	20.6	6.1	19.2	28.3	37.5	25.5	2.9
TOTAL METRO	22.0	25.8	19.0	27.9	21.4	41.1	27.7	21.4	6.1	20.0	30.3	37.5	27.7	2.9
WHHH METRO	3.9	3.8	2.4	8.5			6.4	4.0	8.2	4.8	3.0	3.1		4.3
WIII METRO	11.3	10.7	26.2	8.5	6.0	12.5		10.2	21.4	12.8	4.0	4.7	4.3	22.9
WJJJ METRO	4.3	3.8	2.4	3.9	4.8	1.8	8.5	4.4	4.1	2.4	6.1	3.1	4.3	7.1
WKKK METRO	2.4	3.8	1.2	3.1	11.9	1.8		1.5		3.2	3.0			
WLLL METRO	7.3	4.3		2.3	7.1	5.4	4.3	10.8	7.1	10.4	15.2	9.4	4.3	1.4
WMMM METRO	2.7	.5					2.1	4.8	6.1	5.6	4.0	3.1	6.4	1.4
WZZZ METRO	.5	.7		2.3				.4	1.0	.8				
TOTALS AQH RTG	26.3	26.2	23.5	28.2	27.5	29.0	29.9	28.3	29.0	28.3	31.4	31.5	26.1	17.8

Footnote Symbols: * Audience estimates adjusted for actual broadcast schedule. + Station(s) reported with different call letters in prior surveys - see Page 5B.
Both of the previous footnotes apply.

ARBITRON RATINGS

Exhibit 10.11

A sample of the type of report Arbitron furnishes to subscribing radio stations.

(Courtesy of © Arbitron Ratings Company)

10.11 How Audience Measurement Is Used

selling time

The most important use for audience measurement calculations revolves around selling advertising time (see Chapter 9). The higher the rating (or share, AQH, cume, etc.) for a particular program or a particular time period, the more the station or network can charge for placing a commercial or product placement within that program or time period. Nielsen estimates that each of its rating points for prime-time broadcast programming is worth about $30,000. That means a program with a rating of 10 can charge about $300,000 for a commercial, whereas a program with a rating of 8 can charge only $240,000. It is no wonder that networks and stations strive for high ratings.

determining CPM
and CPP

Advertisers use ratings to make sure they are getting a good return on investment. They need the ratings data to determine **cost per thousand,** or **CPM** (see Chapter 9). This indicates to the advertiser how much it costs to reach 1,000 people (or households, if that is what is being measured). For example, if an advertiser pays $30 for a radio spot and the ratings show that spot is reaching 5,000 people, then the advertiser is paying $6 for each 1,000 people. A related statistic is **cost per point,** or **CPP.** The amount of money the advertiser pays for a week's advertising is divided by the number of ratings points the commercial accrues during a week to determine the cost for each rating point.

industry health

Audience measurement is also used to determine the overall health of the electronic media industry. When the PUT, PUR, or HUT levels fall, the industry becomes nervous. Ratings and shares also show the relative health of different segments of the electronic media. The three commercial TV networks used to garner about 90 percent of the share of audience. Network programs with ratings of less than 20 usually did not last long. Now, with the three-network share below 40 percent, programs with single-digit ratings are considered acceptable.

comparison

Measurement is also used as a basis of comparison. For example, the share tells a station or network how it is faring in relation to its competition. Although the PUT and ratings may be going down, a station can consider itself successful if its share of the viewing public is increasing. A 3 might sound like a low rating, but it is acceptable for most radio stations because radio is so diverse and specialized that a 3 is about the most many of the stations in a large market can hope for. A station or network can also use ratings to compare against its past. If a station's rating (or share, AQH, cume, etc.) decreases from month to month over the course of a year, the station has reason to worry.

program
decisions

Audience measurement is also used extensively to make programming decisions. Many programs that flunk pretesting don't get on the air. Programs with high ratings are kept on the air and allowed to build up numerous episodes that can then be sold for syndication (see Chapter 8). Programs with low ratings are often erased quickly, without having the chance to grow or develop an audience slowly. Similarly, people often disappear from the airwaves as a result of audience measurement. If a radio station's AQH slips during a certain time of day, the disc jockey on at that time can expect a pink slip. Television station news anchors are often removed because they have low Q scores.

product
placement

Companies use the product placement research to make sure their products got on the TV or movie screen in an acceptable form, one that maximizes their return on investment. They also use it to keep track of where their competitor's products are being placed.

Sometimes media organizations use research to learn new things about their audience. For example, NBC placed the 2008 Olympics on many different media platforms and wanted to know which of them garnered the most success. It did its own massive research project to figure out how many people consumed the Olympics through traditional methods and how many utilized new media forms. The network used many methods, including diaries, in-depth interviews, data from Nielsen, online surveys, and focus groups.[32]

new input

Measurement calculations are varied in terms of types and uses because of the needs of different clientele. One advertiser might be more interested in the cumulative number of people hearing an ad than in the number hearing at any particular time; another advertiser might be interested primarily in finding a station that delivers the largest reach.

10.12 Issues and the Future

Promotion is not as widely criticized as audience measurement, but it does present some issues. Many viewers find snipes to be annoying. Sometimes the promotion department and the sales department are at odds. The sales department wants prime airtime so it can sell commercials for the highest rate, while the promotion department wants that same time for its promos. Promotion's rationale is that ratings will decline if people don't know about the programs, and if ratings go down, sales will decrease. But on the whole, promotion is accepted as a necessity.

promotion

The customers who subscribe to audience measurement services have many gripes. For starters, the accuracy is frequently questioned because the results do not make common sense. For example, the Nielsen Hispanic TV report, which should be a subset of its overall data, often has radically different numbers from the Spanish population considered in the national reports. Nielsen's overnights from machines often differ from its later reports. Sometimes the report of the number of people watching a program while they record it on a DVR plus the number who watch it within a week is lower than the report that only includes the people watching as they record. In 2003, the prized young male demographic suddenly took a dip in the Nielsens only to be back again several months later—a highly unlikely occurrence.

different results

Another factor that affects the validity of ratings is the size and composition of the sample. As media forms proliferate, the number of people watching or listening to any particular program or service shrinks, and the sample size must be increased in order for statistics to be significant. Many contend that a sample that is only it .012 percent of the population cannot contain all the different types of people found in society. Although rating companies try to make their samples representative of the entire population, with the same percentages of households

samples

with particular demographic characteristics in the sample group as there are in the U.S. population, this presents problems. For example, a rating sample might attempt to have the same percentage of households headed by a 30-year-old, high-school-educated African American woman with three children living in the city and earning $30,000 per year as in the population of the whole United States. Which of these characteristics are important to audience measurement can be debated. Is a 25-year-old, high-school-educated African American woman with no children who earns $25,000 a year a valid substitute?

noncooperation

Even if they can determine the ideal sample, companies are faced with the problem of uncooperative potential samples. About 35 percent of the people contacted refuse to have machines installed on their sets, and about 45 percent do not want to keep diaries. Substitutions must be made for uncooperative people, which may bias the sample in two ways—the substitutes may have characteristics that unbalance the sample, and uncooperative people, as a group, may have particular traits that bias the sample. Particularly troubling is the concept of weighting one demographic group's data more heavily than another because not as many of them responded.

meters

Problems with ratings methodology do not cease once the size and composition of the sample are decided and the people selected. There is also the problem of receiving accurate information from these people. Any system that uses meters is dealing with a technology that can fail. A certain number of machines develop mechanical malfunctions, further reducing the sample size in an unscientific manner. Individuals using peoplemeters can watch TV and not bother to push their button. This deflates the ratings.

diaries

Diaries are also subject to human deceit. People can lie about what they actually watched to appear more intellectual or to help some program they would like to see high in the ratings. The multitude of programming choices taxes people's ability to remember what they saw or heard. Fewer than half the diaries sent out are returned in a form that can be used for analysis.[33]

cost

Because of these flaws and many more, clients want the rating companies to develop better measurement techniques—larger samples, more reliable machines, more minute-by-minute analysis, more qualitative research to predict buying patterns, more analysis of internet and cell phone viewing and listening. Audience research companies can do much of this, but it costs money, and when they do improve methodology or add reports, they find that the clients who demanded the improvements are not willing to pay for them. Companies have dropped their subscriptions, sometimes because they are unhappy with the scope and reliability of the ratings and sometimes because the costs are too high. They usually return, however, because the data the researchers provide are valuable for operating their businesses.[34]

interpretation

Although rating company methodology is often criticized, management interpretation of the data is often faulted to an even greater degree. Rating companies publish results and cannot be held responsible for how they are used. This area of error is the domain of media and advertising executives. The main criticism is that too much emphasis is placed on ratings. Even though rating companies acknowledge that their sampling techniques and methodology yield imperfect results, programs are sometimes removed from the air when they slip one or two rating points. Actors and actresses whose careers are stunted by such action harbor resentment. Television history is full of programs that scored poor ratings initially but were left on the air and went on to gain large, loyal audiences (e.g., *Hill Street Blues, Cheers*). Trade journals sometimes headline the ratings lead of one network over the others when that lead, for all programs totaled, may be only half a rating point. Ratings should be an indication of comparative size and nothing more, but in reality their shadow extends much further.

The overdependence on ratings often leads to programming concepts deplored by the critics. In a popularity contest designed to gain the highest numbers, programming tends to become similar, geared toward the audience that will deliver the largest numbers. Programmers emphasize viewer quantity, often to the neglect of creativity, availability to the community, and services to advertisers.

Similar complaints arise against pretesting. Many programs, such as *All in the Family, Batman, The Sopranos,* and *Everybody Loves Raymond* (see Exhibit 10.12), tested poorly but went on to successful runs. Programs that are unique are most likely to test poorly because the audience doesn't know what to make of them. But if executives have the courage to try them anyhow, they can sometimes have a hit on their hands.[35]

overdependence

Exhibit 10.12

Everybody Loves Raymond, starring Ray Romano (above), tested poorly in 1996. People thought the stories were too thin—just one big mother-in-law joke. They also thought the cast was weak and lacked charisma. But the show (with some changes) remained on the air for nine years and garnered multiple Emmys.

(Courtesy of the Academy of Television Arts & Sciences)

Looking at it another way, however, ratings' effects on program content are simply the result of the democratic process. Audience members get what they vote for. If programmers were to use another criterion, say creativity, as the basis for advertising rates, the situation would be far more unjust than is the present quantitative rating system. Creativity is an abstract concept that cannot really be defined, let alone counted.

democracy

Audience measurement technology can stand improvement. One way for it to generate better numbers would be for all radios, TVs, and computers to be equipped so that all viewing and listening are reported to a central location. Pay-per-view programming has this capacity because people must make a phone call or push a button to obtain the programming. Some digital video

future possibilities

recorders also have this capability. However, incorporating this technique for all programming could be cumbersome and smacks of "Big Brother." However, there would be no need for sampling because everyone would be included. The newer media are bound to add techniques (and complications) to the audience measurement that already exists.

10.13 Summary

A major function of promotion, publicity, and public relations is to get people to partake of programs or media forms. Promotion departments use such elements as promos, PSAs, snipes, cross-promotion, trailers, billboards, and newspaper articles. Audience feedback is one way to tell if promotion has been effective.

Feedback includes fan mail, website hits, and formalized audience research. A look at Nielsen, Arbitron, and ASI shows characteristics of audience research.

Nielsen uses random sampling to undertake TV audience measurement for the whole country and for local markets that it divides into DMAs. It has increased its sample size over the years, but many people refuse to be part of the sample. To gather data, it uses peoplemeters and LPMs and still a few diaries. It uses computers to calculate ratings and shares, but with both of these declining for individual networks, other measurements such as number of viewers, GAA, LOT, and FOT have come into vogue. PUTs and HUTs are also important to the industry as a whole. Nielsen's measurements are quantitative, and sometimes those being rated try to game the system. The company issues many reports, including the overnights and the fast nationals. Its ratings are used to determine advertising costs, including the CPM and CPP. They are also used, some say too extensively, to make programming decisions. Nielsen has had to keep up with the times by instituting AMOL, figuring out ways to count VCR and DVR recording, adding out-of-home viewing, initiating C3, and rating portable devices and the internet. Clients resist paying for the added services that they want.

Arbitron uses different-sized samples to measure radio nationally and in local markets. It uses the PPM and still some diaries and encounters resistance when it switches from one data-gathering form to another. Ratings and shares are even less important to radio because the numbers are so low. Popular radio statistics are AQH, cume, reach, frequency, LOT, FOT, and PUR. Arbitron reports demographics and issues a national report, RADAR, and local reports that include MSAs and TSAs. Its ratings are used, like Nielsen's, for advertising and programming decisions.

ASI undertakes pretesting and aims for diverse samples representing the entire country. Some of its sister pretesting services, such as music preference, want people who know and like the particular format being tested. The methods used for pretesting involve personal interaction, usually through minitheaters or focus groups. The results sometimes involve psychographics as well as demographics. Pretesting is used to make decisions before something is put on the air.

There are many other forms of research, such as Q scores, retention, and engagement. The internet and other new media will definitely affect audience feedback of the future.

Suggested Websites

www.arbitron.com (audience research firm Arbitron)

www.marketcastonline.com/Default.asp (a movie pretesting organization)

www.mediaratingcouncil.org (the organization that accredits rating companies and methodology)

www.nielsenmedia.com (audience research firm Nielsen)

www.qscores.com (site for Marketing Evaluation, Inc., suppliers of Q scores)

Notes

1. "Made to Measure," *Fortune,* March 3, 2008, 70.
2. "The Next Big Thing," *The Economist,* October 4, 2007, 1; "Out-of-Home, Sweet Out-of-Home," *Broadcasting & Cable,* July 18, 1995, 23; and "Internet Advertising Revenues Top $21 Billion in '07, Reaching Record High," http://www.iab.net/insights_research/iab_news_article/299656 (accessed July 27, 2008).
3. "Prime Time's New Reality," *Broadcasting & Cable,* August 23, 2004, 8; and "Pop-Up Promos Drive Industry Crazy," *Broadcasting & Cable,* December 20, 2004, 20.
4. "Slim Pickings," *Hollywood Reporter,* May 30, 2008, 31–32.
5. "Helping Out, from Coast to Coast," *Broadcasting & Cable,* June 23, 2008, 12; "Helping Hands," *Broadcasting & Cable,* June 28, 2004, Supplement; and "Thinking Outside the Cable Box," *Hollywood Reporter,* July 16–18, 2004, 22–23.
6. "All or Nothing," *Emmy,* January/February 2007, 44–49; and James T. Walker and Susan Tyler Eastman, "On-Air Promotion Effectiveness for Programs of Different Genres, Familiarity, and Audience Demographics," *Journal of Broadcasting & Electronic Media,* December 2000, 618–37.
7. "Saved! How a Groundbreaking Series Came Back from the Brink," *Emmy,* January/February, 2008, 50–53.
8. "Pirated Music Helps Radio Develop Playlists," *Wall Street Journal,* July 12, 2007, B1; and "New Audience Metrics Aim to Leapfrog Web Buzz," *Broadcasting & Cable,* June 9, 2008, 14.
9. Nielsen's website is www.nielsenmedia.com.
10. "Nielsen on Track of Product Placements," *Media and Technology,* May 31, 2004, 14; "Cable Takes the Measure of C3," *Broadcasting & Cable,* November 12, 2007, 22; "Measurement Tools Spell Good News," *Television Week,* August 13, 2007, 10; and "Nielsen Ratings Digitized," *Hollywood Reporter,* June 15, 2006, 1.
11. Arbitron's website is www.arbitron.com; "Arbitron to Exit TV Ratings; Cites Sagging Profit," *Los Angeles Times,* October 19, 1993, D-2.
12. "SMART Bombs," *Broadcasting & Cable,* May 31, 1999, 11; and "How to Count Eyeballs," *Newsweek,* November 27, 2006, 42.
13. "TV Networks Launch Big Campus Push," *Wall Street Journal,* March 5, 2007, B3; and "Nielsen Extends Its Reach," *Broadcasting & Cable,* April 17, 2006, 14.
14. "Nielsen's Population Boom," *Hollywood Reporter,* September 27, 2007, 3; and Bureau of the Census, "Projecting the Number of Households and Families in the United States: 1995 to 2010," http://www.census.gov/prod/1/pop/p25-1129.pdf (accessed September 12, 2008).

15. "Nielsen Adapts Its Methods as TV Evolves," *Wall Street Journal,* September 29, 2003, B-1.

16. "Portable People Meters to Be Tested," *Electronic Media,* June 5, 2000, 35; "New Nielsen System Is Turning Heads," *Broadcasting,* May 18, 1992, 8; and Paul F. Lazarfield and Frank N. Stanton, *Radio Research* (New York: Duell, Sloan, and Pearce, 1942).

17. "Nielsen: 'Portfolio' over PPM," *Hollywood Reporter,* March 2, 2006, 3.

18. "Dear Diary, Farewell," *Broadcasting & Cable,* July 17, 2006, 26.

19. "Cleveland Rocked by People Meters," *Broadcasting & Cable,* September 8, 2008, 8; "LPMs Shock Ratings," *Broadcasting & Cable,* December 20, 2004, 9; and "Programmers, Nielsen Disagree over Claim Peoplemeters Underreport Kids," *Broadcasting,* December 21, 1992, 10.

20. Willaim J. Boxton and Charles R. Acland, "Interview with Dr. Frank Stanton: Radio Research Pioneer," *Journal of Radio Studies,* Summer 2001, 191–229.

21. "Who's Really Watching?" *Broadcasting & Cable,* May 16, 2005, 14; and "Nielsen Taps into TiVo Info," *Broadcasting & Cable,* August 12, 2002, 15.

22. "Share and Share Alike? Hardly," *Electronic Media,* March 15, 1999, 1A; "Networks Like When Viewers Watch a LOT," *Broadcasting & Cable,* July 21, 2003, 26; and "TV Households, Viewers Increase in U.S.," *Hollywood Reporter,* August 24–26, 2007, 51.

23. "Nielsen Adds Weight to U.S. Sample," *Broadcasting & Cable,* September 8, 2003, 4.

24. "Spot-less Performance by 'Today' Irks 'GMA.'" *Broadcasting & Cable,* June 4, 2007, 8; "Duel over Delisting," *Broadcasting & Cable,* December 15, 2003, 16; and "Here's How to Game the System," *Los Angeles Times,* May 7, 2008, E-1.

25. "Measuring Engagement," *Broadcasting & Cable,* April 28, 2008, 12; "It's an Olympic Triumph for Advertisers," *Hollywood Reporter,* August 20, 2008, 6; and Thomas R. Lindlof, "The Qualitative Study of Media Audiences," *Journal of Broadcasting & Electronic Media,* Winter 1991, 23–42.

26. Michael C. Keith, *Radio Programming* (Boston: Focal Press, 2004), 108.

27. "Test Pattern," *Hollywood Reporter,* July 25–31, 2006, 18–19; Libby Slate and William J. Adams, "How People Watch Television as Investigated Using Focus Group Techniques," *Journal of Broadcasting & Electronic Media,* Winter 2000, 78–93; and "How CBS Plays in Vegas," *Broadcasting & Cable,* April 23, 2001, 31.

28. Marketing Evaluations, Inc. "The Q Score Company," http://www.qscores.com/pages/Template1/site11/30/default.aspx (accessed September 13, 2008).

29. "Nielsen Bows to Latino Viewers," *Los Angeles Times,* December 20, 2005, C-1; "TV Braces for DVR Ratings," *Broadcasting & Cable,* December 5, 2005, 17; and Nielsen Media Research, "Reports A to Z," http://www.nielsenmedia.com/nc/portal/site/Public/menuitem.9b2b8e2c645235df211ba0a347a062a0/?vgnextoid=5a8afeded47d3010VgnVCM100000880a260aRCRD (accessed September 14, 2008).

30. "Arbitron Picks Up RADAR Service in $25 Million Deal," *Hollywood Reporter,* July 3–9, 2001, 43; and Karen S. Buzzard, "James W. Seiler of the American Research Bureau," *Journal of Radio Studies,* December 2003, 186–201.

31. "Nielsen Adjusts to New Universe," *Hollywood Reporter,* August 24, 2006, 4.

32. "Tracking Embedded Ads," *Los Angeles Times,* July 21, 2008, C-1; "NBC Develops Olympian Research Effort," *Hollywood Reporter,* July 7, 2008, 2; and "TV Milestones," *Hollywood Reporter,* September 2004, 20–108.

33. "Nielsen Ratings Come Under Fire," *Wall Street Journal,* November 17, 2003, B-10; and "The Battle for Channel Surfers," *Fortune,* September 1, 2008, 32.

34. "Nielsen Gets Dumped," *Broadcasting & Cable,* June 13, 2005, 16; "Low-Rated Whine," *Broadcasting & Cable,* July 28, 2008, 4; and "Who's Watching Those Webisodes?" *Wall Street Journal,* October 11, 2006, B4.

35. "When Not to Trust the Feedback," *Los Angeles Times,* July 9, 2002, F-1.

Chapter 11

LAWS AND REGULATIONS

Many intermingling governmental entities regulate the electronic media industry. In the U.S. system of government, the legislative branch writes the laws, the executive branch administers them, and the courts adjudicate them. In this way, the branches have checks and balances on one another. This basic process affects media regulation. An electronic media organization that believes it has been wronged by a decision of the Federal Communications Commission can appeal that decision to the courts and also lobby Congress to change the offending law.

> A function of free speech under our system of government is to invite dispute. It may indeed best serve its highest purpose when it induces a condition of unrest, creates dissatisfaction with conditions as they are, or even stirs people to anger.
>
> **William O. Douglas, former Supreme Court justice**

Laws and regulations dealing with a wide variety of subjects affect telecommunications entities differently. Broadcast stations are very much affected by licensing provisions, whereas the home video industry, which is not licensed, keeps a close eye on copyright legislation. Cable networks that program news must be more concerned about libel than those that deal only with entertainment.

Different entities are involved in telecommunications regulation to varying degrees. The Federal Aviation Administration is involved only when improperly lit antenna towers may be a hazard to airplanes. The Federal Communications Commission's main function is regulation of the airwaves. The next few sections deal with the primary regulatory bodies; the rest of the chapter covers the main laws and regulations these bodies create that affect the electronic media industry.

11.1 The Federal Communications Commission

origins

The Federal Communications Commission (FCC) is an independent regulatory body that Congress created because of the mass confusion and interference that arose when early radio stations broadcast on unregulated frequencies. Congress passed the **Radio Act of 1927,** which created the Federal Radio Commission (FRC), and seven years later, it passed the **Communications Act of 1934** (see Chapter 5), which formally established the FCC with powers similar to its predecessor, the FRC. Much later, Congress passed the **Telecommunications Act of 1996** (see Chapters 4 and 5), which did not change the basic structure of the FCC but did change some aspects of what it regulates. Over the years, the FCC's responsibilities have grown. Starting primarily with telephones and radio, its realm now encompasses not only broadcast television but also aspects of cable TV, the internet, and portable devices.

description

The commission maintains central offices in Washington (see Exhibit 11.1) and field offices throughout the country. It is composed of five commissioners appointed for five-year terms by the president, with the advice and consent of the Senate. The president designates one commissioner to be chairperson, but generally no president has the opportunity to appoint many commissioners because their five-year terms are staggered. The FCC used to be funded solely from taxes, but now the entities that the FCC regulates must pay a yearly fee to help pay for the FCC's services. This charge can range from several hundred dollars for a small-market radio station to tens of thousands of dollars for a TV station in a large market.

organization

Policy determinations are made by all the commissioners, with the chairperson responsible for the general administration of the commission's affairs. Most of the day-to-day work, such as handling interference complaints, public inquiries, and frequency applications, is undertaken by the staff (see Exhibit 11.2).

Exhibit 11.1
The FCC headquarters in Washington, D.C.

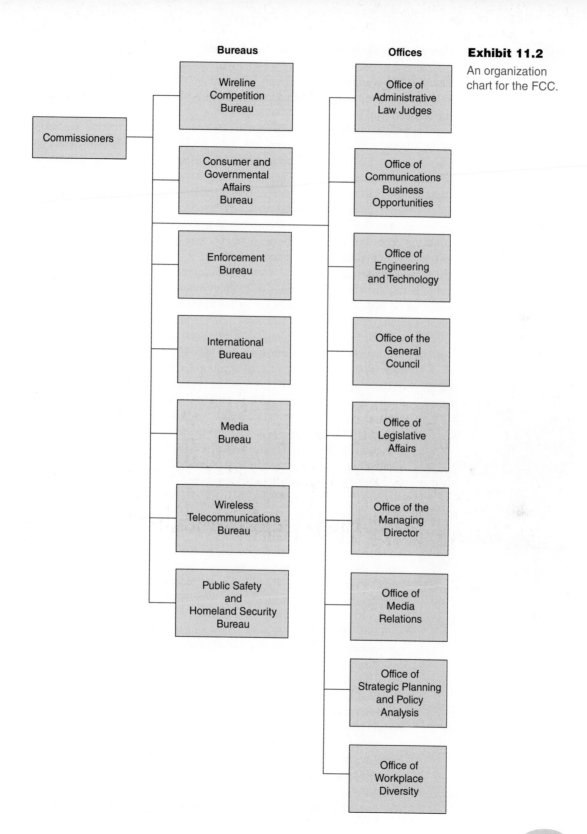

Bureaus

- Wireline Competition Bureau
- Consumer and Governmental Affairs Bureau
- Enforcement Bureau
- International Bureau
- Media Bureau
- Wireless Telecommunications Bureau
- Public Safety and Homeland Security Bureau

Offices

- Office of Administrative Law Judges
- Office of Communications Business Opportunities
- Office of Engineering and Technology
- Office of the General Council
- Office of Legislative Affairs
- Office of the Managing Director
- Office of Media Relations
- Office of Strategic Planning and Policy Analysis
- Office of Workplace Diversity

Commissioners

Exhibit 11.2

An organization chart for the FCC.

The FCC has myriad functions, many of which are not related to electronic mass media. For example, it has jurisdiction over ship-to-shore radio and police and fire communications.[1]

rules and regulations

The FCC establishes rules and regulations that relate to the general operation of the telecommunications industry. FCC policies have addressed issues such as must-carry (see Chapter 4), financial interest–domestic syndication (see Chapter 3), and children's programs (see Chapter 8). Suggestions for these rules can come from media practitioners, the public, Congress, or from within the FCC itself. Once a viable suggestion is received, it is referred to the appropriate FCC office or bureau, and the commission then issues a notice asking interested parties to comment. The staff evaluates these comments and reports on them to the commissioners. Sometimes formal hearings are held so the commissioners can hear from specific individuals or groups and ask pertinent questions. The commissioners then decide whether to issue a new order, amend an old one, ask for more study, or do nothing. If some member of the electronic media industry does not like the result, it can petition for reconsideration, and the process starts all over.

frequency allocations

EAS

president

Justice Department

Many of the FCC functions are technical. It assigns **frequencies** to individual stations, determines the power each can use, and regulates the time of day each may operate. The FCC also controls the general allocation of frequencies, deciding which frequencies go to satellite radio, which to cellular phones, which to wireless internet, and so forth. It also has the jurisdiction to set technical standards, such as those for color television or digital TV. The Emergency Alert System (EAS) is also under the jurisdiction of the FCC. This national service ties together all radio and TV stations so that information can be broadcast from the government to the citizenry during a national emergency.[2]

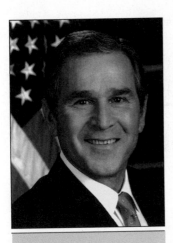

Exhibit 11.3

President George W. Bush established an Office of Global Communications. The purpose of this office was to advise the president and heads of executive departments and agencies on the most effective means for the United States to ensure consistency, prevent misunderstanding, build support, and inform international audiences.

11.2 The Executive Branch

The president influences the media, both formally and informally. Formally, the president can suspend broadcasting operations in time of war and call into action the EAS. However, most of the interaction between the president and the media comes informally, with the president seeking positive coverage from radio and TV. Most presidents devise their own way of handling the media through press secretaries or others who give advice on media relations (see Exhibit 11.3).

Various executive departments also interact with the media. For example, the Justice Department oversees antitrust and as such was involved in the breakup of AT&T (see Chapter 4). The Department of State has a bureau that advises on media issues with possible international consequences.

Other agencies that are either part of the executive branch or are independent affect broadcasting. The Federal Trade Commission (FTC) deals with fraudulent advertising. If the FTC believes that a company's ads are untruthful, it has the power to order the company to stop broadcasting the ads. The FTC also issues guidelines on potentially controversial subjects such as children's advertising and internet advertising to fend off problems before they occur. In addition, the Federal Aviation Administration monitors antenna towers so that planes will not crash into them. The Food and Drug Administration occasionally stops the mislabeling of advertised products. The Surgeon General's Office participated in the 1972 ban of cigarette advertising on radio and TV and in reports related to the effects of TV on society (see Chapters 9 and 12).[3]

FTC

other agencies

11.3 The Legislative Branch

Congress generally sets broad policies that are handled on a day-to-day basis by other agencies. These broad policies are influential in setting direction for the electronic media. The whole basis for the government's intervention in the electronic media business was, and occasionally still is, debated in Congress (see Exhibit 11.4). Ordinarily, the First Amendment would prohibit the government from infringing on the rights of citizens to communicate by whatever means they wish. Congress invaded this area, however, because of the **scarcity theory.** Not everyone who wants to can broadcast through radio frequencies because this would cause uncontrollable interference. As a result, Congress needed to intervene and determine a mechanism for making decisions regarding who could and could not use the airwaves. Today, with many avenues of distribution available, the scarcity theory doesn't hold much water—hence the controversies.[4]

congressional laws

state and local laws

In addition to overall regulatory decision making that is tied to the 1927, 1934, and 1996 communications acts, Congress passes other media-related laws, such as copyright laws, cable TV laws, and the law that set up the Corporation for Public Broadcasting (see Chapter 3). Congress monitors the FCC through both the House and Senate subcommittees on communications. These and other committees often conduct special investigations into such aspects of telecommunications as the quiz scandals, rating practices, or TV violence.

State legislative bodies can also affect media. Under the Constitution, federal laws such as the Communications Act take precedence over state laws. But occasionally, state laws dealing with privacy, advertising, or state taxes affect elements of the industry. City councils, with local legislative control, have jurisdiction over most cable TV franchising.

11.4 The Judicial Branch

Because FCC decisions can be appealed, the courts of the United States have a significant impact on electronic media. Courts of appeal can confirm or reverse

Exhibit 11.4

Many important laws regarding telecommunications are debated and passed by Congress in the Capitol.

the commission's decisions or send them back to the FCC for further consideration. From any of these courts, final appeals can be taken to the Supreme Court,
which can confirm or reject a lower court's decision.

court decisions

Courts also deal with issues other than those arising from FCC decisions.
Rulings on such subjects as freedom of speech, obscenity, libel, copyright, equal
time, media ownership, and even access to the courts determine the direction of
electronic media.

11.5 The First Amendment

**freedom of
speech and
press**

The **First Amendment** to the U.S. Constitution is basic to much of what occurs
in media. This amendment states that "Congress shall make no law . . . abridging
the freedom of speech, or of the press." Issues involving the First Amendment are
raised under many banners, one of them being **censorship.** A breach of freedom
of the press occurs when the government tries to censor or withhold information.
However, the government may contend that giving out the information would be

**clear and
present danger**

a **clear and present danger** to the country.

Clear and present danger also extends to individuals. The Supreme Court has
ruled that freedom of speech does not give someone the right to falsely yell "Fire"
in a crowded theater because this would create a clear and present danger to those
in the theater. But governmental agencies are reluctant to label words as dangerous if freedom of speech is involved. During the 1960s, the NAACP wanted the
FCC to censor speeches aired on Georgia TV by a candidate for the Senate on the
grounds that the speeches contained racially inflammatory remarks that were a

prior restraint

danger to the people of Georgia. The FCC refused to issue **prior restraint** in that
case, saying the speeches did not literally endanger the nation. More recently,
plaintiffs wanted to shut down an antiabortion website because it contained what
they considered dangerous threats, but the courts again disagreed.

News reporters often argue that freedom of the press is being violated if they
are forced to disclose sources from which they obtained news or testify in court
regarding information they promised would be kept confidential. Reporters were

shield laws

required to testify for years, but during the 1970s, most states passed **shield laws**
that gave some protection to news reporters. In most states, they must testify only
if the evidence is crucial to the case and cannot be obtained in any other way.
Some reporters have gone to jail rather than reveal sources.[5]

11.6 Profanity, Indecency, and Obscenity

definitions

Profanity, indecency, and **obscenity** are all major areas where the First Amendment comes into play. They are outlawed by the U.S. Criminal Code, but this
criminal code and the First Amendment often clash. It can be difficult to determine when people's freedom of speech should be abridged because what they are
saying is profane, indecent, or obscene. Part of the problem is the changing definitions of these words. What was considered indecent in one decade may be perfectly acceptable in the next. In 1937, NBC aired a program in which Mae West
(see Chapter 6) performed a sketch about Adam and Eve. Nothing in the script

was considered indecent, but her sultry voice sounded suggestive, and the FCC reprimanded NBC. Today this wouldn't cause the slightest stir.

The three words, even when defined in their most concrete way, often overlap. Profanity is defined as irreverent use of the name of God. The FCC defines indecency as language that, in context, depicts or describes, in terms patently offensive as measured by contemporary community standards for the broadcast medium, sexual or excretory activities or organs. Obscenity is more extreme than indecency, and its definition was determined in a 1973 court case, *Miller v. California*. To be obscene, a program must contain the depiction of sexual acts in an offensive manner, must appeal to prurient interests of the average person, and must lack serious artistic, literary, political, or scientific value. Obscenity is the most serious of the three; however, the line between indecency and obscenity is fuzzy, and some profanity, combined with sexual references, can be part of indecent or obscene material. In practice, the FCC rarely chastises stations for profanity, but it has handled numerous cases related to indecency.[6]

One of the most famous indecency cases arose in 1973 when a Pacifica Foundation public radio station in New York, WBAI, aired a program on attitudes toward language. It was aired at 2:00 P.M. and included a comical monologue segment performed by George Carlin that spoofed seven dirty words that could not be said on the public airwaves. A father driving in the car with his son heard this monologue and complained to the FCC. The FCC placed a note in WBAI's license renewal file, which led the station to appeal through the courts all the way to the Supreme Court. WBAI claimed the FCC was censoring. The high court determined that censorship was not involved because the FCC did not stop WBAI ahead of time from airing the program. The Supreme Court did say, however, that the program should not have been aired during a time period when children were likely to be in the audience. One of the results of this ruling was that the FCC established a **safe harbor,** a period of time from 10 P.M. to 6 A.M. when indecent material can be aired because children are not expected to be in the audience.

WBAI

Stern

In the 1990s, the FCC put more teeth in its indecency actions. It fined Infinity Broadcasting $1.7 million because of numerous sexual and excretory remarks made by its on-air personality Howard Stern (see Exhibit 11.5). At first Infinity fought the fine, but eventually it decided to pay so it could get on with other business. In 2003, the FCC levied a $357,000 fine against Infinity because two of its syndicated shock jocks, Greg "Opie" Hughes and Anthony Cumia, had encouraged a couple to have sex at St. Patrick's Cathedral as part of an on-air contest (see Chapter 12).

Then along came Janet Jackson. During the 2004 Super Bowl halftime show, her singing partner, Justin Timberlake, tugged at her black bustier and exposed her right breast. FCC commissioners, politicians, and the public became incensed, especially because the Super Bowl is considered a family event. Janet and Justin apologized and said the idea was theirs and not that of the CBS network or the National Football League. They claimed there had been a "costume malfunction" and that

Exhibit 11.5
Howard Stern, the ultimate shock jock, cost Infinity, the company that employed him, $1.7 million in FCC obscenity fines.

(Rick Mackler/Globe Photos)

what was supposed to be exposed was her red bra, not her breast. But the heat that was being generated by indecency on the airwaves in general reached a boil, and a series of actions were taken (see "Zoom In" box).[7]

At the time, the maximum fine for indecency was $27,500 per incident, but the FCC fined CBS $550,000, which was $27,500 for each of its 20 owned-and-operated stations. CBS appealed, and while the appeal was in progress, Congress upped the maximum fine to $325,000 per incident. Other indecency complaints surfaced, many of them involving **"fleeting expletives"**—swear words uttered by celebrities or talk-show guests, especially those on awards shows. These were labeled indecency, not profanity, because the words, in their original intent, dealt with sexual or excretory functions. In 2008, the Supreme Court agreed to hear an indecency case involving "fleeting expletives," and an appeals court ruled in

 ZOOM IN: **Janet's Fallout**

A tremendous number of actions took place within a very short time after Janet Jackson's breast-baring incident at the 2004 Super Bowl halftime show produced by MTV. Among them were the following:

- TiVo reports that the incident was by far the most rewatched part of the Super Bowl.
- The FCC receives hundreds of thousands of emails complaining about the incident.
- CBS lengthens the delay time on the Grammy Awards to five minutes to allow video and audio to be stopped before it is aired, but the telecast has no problems.
- *ER* producers decide to blur the breasts of an elderly woman.
- PBS stations review *Antiques Roadshow* programs to make sure no nude etchings are shown.
- House and Senate hold hearings and fire tough questions at network and NFL executives.
- Clear Channel adopts a "zero tolerance" indecency policy, fires shock jock Bubba the Love Sponge, and takes Howard Stern off its owned stations.

Do you think politicians, the media, and the public overreacted, underreacted, or reacted appropriately to this incident? This all happened during a presidential and congressional election year. Do you think that affected people's actions in any way?

Janet Jackson and Justin Timberlake shortly before the "wardrobe malfunction."

(Alec Michael/Globe Photos)

favor of CBS, saying that the FCC had capriciously departed from its 30-year policy of regulating the airwaves with practiced restraint when it imposed the $550,000 fine.[8]

Obscenity is less of a problem for TV broadcasters than indecency. In their attempts to avoid indecent programming, they usually manage to stay far away from what might be considered obscene. Radio, however, has potential for obscenity cases because of the lyrics of some modern songs. For example, a Florida court ruled that 2 Live Crew's album *As Nasty As They Wanna Be* was obscene, but the decision was later overturned. Not wanting to tread into the area of obscenity, a number of stations decided not to air Eminem's 2001 release "The Real Slim Shady" (see Exhibit 11.6).[9]

Cable TV is more likely to see legal action regarding obscenity than is commercial broadcasting. Rules are not as strict for cable as for broadcast because, historically, people had to specifically subscribe to cable whereas broadcast signals were delivered free. The 1984 cable law did not prevent cable from showing indecent material, but it did prohibit obscene material. This gives cable more latitude than broadcasting, so it is more apt to cross the line between indecency

Exhibit 11.6

Eminem was known for his lyrics that push the envelope. To avoid problems, some stations policed themselves and did not play some of his music.

(Colin Broley/Reuters/Time Pix)

and obscenity. The 1996 Telecommunications Act stated that cable operators **lyrics** must scramble adult programming services, such as Playboy and Spice, or else only offer them between the hours of 10:00 P.M. and 6:00 A.M. But the Supreme **cable TV** Court threw out that provision, so adult channels can be offered 24 hours a day. Another problem that plagues cable TV is that systems are not allowed to censor **public access** shows, regardless of their content. The Telecommunications Act of 1996 contained a provision that allowed cable systems to refuse to transmit access programs they considered to be obscene, but again the Supreme Court struck this down as a First Amendment violation.[10]

The home video business is also vulnerable to obscenity problems. As with cable TV, however, someone must make a conscious effort to rent or buy a DVD. **DVDs, games,** Video stores have been sued for stocking sexually oriented videos, but store own- **movies** ers have won these cases by using the First Amendment. Video games and movies have codes that help keep them out of trouble. The problem is enforcing the codes—not selling an M video game to a 12-year-old or letting that 12-year-old into an R-rated movie. Both games and movies have been highly criticized for marketing their products to young people by advertising on TV programs that attract children.[11]

The internet is rife with indecency and obscenity problems. In the Telecom- **internet** munications Act of 1996, Congress tried to make it a crime for online computer services to transmit indecent material, but once again the Supreme Court declared this unconstitutional. In 1998, when members of Congress decided to put the Starr Report detailing the Bill Clinton–Monica Lewinsky affair on the internet, *Newsweek* quipped, "1996: House votes to ban smut on the Web. 1998: House votes to publish smut on the Web."[12]

11.7 Libel, Slander, and Invasion of Privacy

definitions

Most **libel** laws have their roots in the written press. In the early days of radio and television, there was debate as to whether libel applied at all. Libel is defined as **defamation** of character by published word, whereas **slander** is defamation by spoken word. Slander carries less penalty than libel, ostensibly because it is not in a permanent form to be widely disseminated. Some people thought radio and television should be governed by slander laws because words were spoken rather than printed. But because these spoken words are heard by millions, broadcast defamation comes under the libel category.

public figures

cases

Libel was not a big issue for the broadcast media until the late 1970s. Radio and television engaged in little investigative reporting before then, so libel suits were less likely to occur. Radio and television newscasts and documentaries were also limited in terms of time, so programs generally addressed only well-known issues and people. People who are **public figures** have great difficulty winning a libel suit because the rules and precedents applied to them are stricter than for ordinary citizens. To win a libel suit, a public figure must prove that a journalist acted with **actual malice.** As a result, few public figures bothered to bring libel suits against the broadcast media.

Exhibit 11.7

General William Westmoreland, whose 1985 libel case was supposed to be the "libel case of the century" because it involved the credibility of both CBS and the general, was settled quietly out of court with neither side declaring victory or defeat.

(UPI/Bettmann)

By the late 1970s, however, TV was a major source of information and a dominant force in society that some believed had become overly arrogant. Statements about people on TV definitely affected their reputations and livelihoods. As a result, libel cases were on the upswing. During the early 1980s, almost 90 percent of the people who filed libel suits against broadcasters won. This number was soon reduced to 54 percent as broadcasters began fighting these cases more seriously.

The courts heard many libel cases, including the following: Texas cattlemen who said talk-show host Oprah Winfrey's remarks about beef hurt their business; an eye clinic owner who said ABC falsely reported that the clinic rigged equipment to reveal cataracts that didn't exist; singer Wayne Newton, who claimed that an NBC newscast made it appear that he had strong ties with the Mafia; and General William Westmoreland, who claimed CBS wrongly accused him of purposely deceiving his military superiors about estimates of Vietnam enemy troop strength (see Exhibit 11.7).[13]

Invasion of privacy involves how information is gathered. Invasion of privacy laws vary from state to state; however, in general, privacy laws allow a person to be left alone. Often the laws deal with giving a person physical solitude. For example, reporters cannot trespass on a person's property and take a photo through the bedroom window. Aggressive news photographers are sued for invasion of privacy, and journalistic organizations have been successfully sued for using hidden cameras and microphones. The laws also consider the publication of private facts. In some states, facts are not private if they can be obtained from public records or if they are part of an important news story. For example, the names of rape victims are public record, so they can be broadcast. Victims' names are often withheld, however, because broadcasters know the victims do not want their names revealed.

invasion of privacy

Privacy on the internet is a thorny topic. Many sites ask people to give information about themselves. Sometimes people can indicate that this information not be used for other purposes, but enforcing such a provision is difficult. The 1998 Children's Online Privacy Protection Act requires that websites aimed at children younger than 13 obtain parental permission before collecting information from the kids. But children can easily lie about their age, and some say having to get parental permission is a violation of the child's privacy.[14]

internet

11.8 Copyright

The issue of **copyright** is one that affects most aspects of electronic media. The copyright principles stem from the U.S. Constitution, which authorizes Congress to promote science and art by giving authors and inventors exclusive rights to their works for a limited time. The Copyright Act of 1909 set that time at 14 years, with the possibility of 14 more years if the copyright holder was still alive at the end of the first 14 years. The Copyright Law passed in 1976 set the length of the copyright at 50 years after the death of a work's creator and at 75 years total for collaborative **works for hire,** such as movies. In 1998, the law was amended to extend the copyright another 20 years, to 70 years after death for copyrights held by individuals and 95 years for works for hire.[15]

provisions

In the early 2000s, the copyright extension was challenged by someone who wanted to put old, out-of-print books on the internet for people to download. The basis for the suit was that 70 years and 95 years are not "limited time." However, in 2003, the Supreme Court ruled that Congress has the right to create copyright times of this length. This was a big victory, especially for Disney because Mickey Mouse and Donald Duck were about to become 75 years old and would have no longer been copyrighted.[16]

After copyrights run out, works are placed in the **public domain** and can be used without obtaining permission. The copyright law also provides for **fair use,** which allows for some use of the work without permission from or payment to the copyright holder. Copyright principles try to balance the need for protection for the creator with the need to make those works available to the public. If artists cannot profit from their works, they are less likely to create them. However, totally protecting a work would mean it could not be quoted, even in everyday conversation.[17]

public domain and fair use

Television stations, cable networks, broadcast and internet radio stations, corporate producers, and others involved with audio or video production must obtain permission and often pay a fee before they can include copyrighted music, photographs, sketches, film clips, or other similar material. Because finding the copyright holders and making arrangements with each individually is a chore, intermediary nonprofit organizations were formed to collect and distribute copyright fees for music. Three such organizations currently exist—the American Society of Composers, Authors, and Publishers (ASCAP); Broadcast Music, Inc. (BMI); and the Society of European Stage Artists and Composers (SESAC).

music licenses

These music licensing organizations collect money from stations and other programming providers in two ways. One is called **blanket licensing**—for one yearly fee, a station can play whatever music it wants from the license organization without having to negotiate for each piece of music. The other way is called a **per-program fee**—the station pays a set amount for each program that utilizes music from the licensing organization. Radio stations (both over-the-air and internet), because they air so much music, generally opt for the blanket license. Television organizations are more likely to use per-program fees. ASCAP, BMI, and SESAC distribute the money they collect to composers and publishers in accordance with the amount the music is aired. Each organization uses a different method to determine what is aired, but in general they regularly survey media organizations to find out what they are playing.[18]

Organizations that want to use music but do not want to work with one of the music licensing organizations can pay a **needle drop fee** directly to the copyright holder. In other words, they negotiate a set amount to pay for the amount of a particular piece of music they want to use. This is a common method for movies that wish to include current hits. There is no equivalent of ASCAP for video material. Each clip that is used must be cleared individually. The copyright law also provides for **compulsory licenses,** such as those that cable TV systems and satellite programming providers (see Chapter 3) must pay to cover material provided by others, such as movie producers.[19]

piracy

One main problem facing various facets of the electronic media industry is violation of copyright, often called **piracy.** It can range from someone making copies of a song for friends to an international ring that distributes illegally dubbed movies to foreign countries even before the movies are released. It has always been a crime to make illegal copies of electronic media material, but now that just about everything is in digital form, it is easier to break the law than in analog days. The quality of the illegally duplicated material is far better because, unlike analog, digital quality does not degrade as it is copied from one medium to another.[20]

DRM

As a result, companies and the government have devised various methods of **digital rights management (DRM)**—ways to prevent the copyright laws from being violated. DRM includes **scrambling** video signals so they cannot be viewed without equipment that descrambles the signal. For example, someone who has not paid for a pay-cable channel, such as HBO, will see and hear a jumble when they tune to that channel. Of course, people devise illegal **decoder** boxes to unscramble cable and satellite signals, but scrambling methodology has improved, and most decoder boxes do not work for long. Similarly, there are copyright protection technologies that, when incorporated into a DVD, prevent it from being copied. DRM also includes placing digital **fingerprints** on recorded material so that if it is duplicated illegally, it will retain the fingerprint, and the person who duplicated it can be punished. This works well for copyright violations

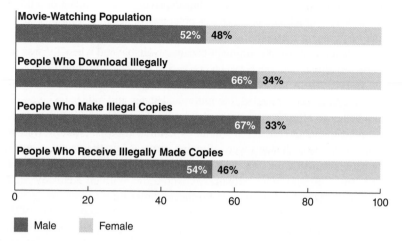

Piracy by Gender

Movie-Watching Population

52% 48%

People Who Download Illegally

66% 34%

People Who Make Illegal Copies

67% 33%

People Who Receive Illegally Made Copies

54% 46%

0 20 40 60 80 100

■ Male   ■ Female

Exhibit 11.8

This chart shows piracy by gender, according to a study by the Motion Picture Association of America. Although the movie-watching population is fairly evenly split by gender, males are more likely to be involved in piracy.

related to uploading and downloading material on the internet. If someone uploads fingerprinted material for which they do not own the rights, the hidden fingerprint will allow the rights owner to prosecute. Laws are also part of digital rights management. One way people were copying material was by taking a video camcorder into a movie theater and using it to make a grainy (but saleable) copy of the movie as it was projected onto the screen. Congress passed a law making it a federal crime to record a movie in a theater.[21]

The Federal Bureau of Investigation works to stamp out piracy, both in the United States and elsewhere, but it continues nonetheless. In some foreign countries, particularly in the developing world, people don't understand copyright principles. But most international piracy is not innocent. It is conducted by organizations with large duplicating facilities and elaborate distribution networks. Another major problem group is college students. According to a 2006 study by the Motion Picture Association of America, 71 percent of illegal downloading from the internet is undertaken by people attending college; the average film copyright violator is male and between the ages of 16 and 24 (see Exhibit 11.8).[22]

FBI

11.9 Access to the Courts

Issues related to cameras in the courtroom gained national attention because of the O. J. Simpson trial, but the controversy has a long history that involves **competing rights** between the First and the Fifth and Sixth Amendments. Although the First Amendment guarantees freedom of the press, the Fifth Amendment guarantees a fair trial, and the Sixth guarantees a public trial.

In 1937, the American Bar Association adopted a policy, known as **Canon 35,** which barred still cameras and radio from courtrooms. Later, TV cameras were excluded by the same policy. This was not a law, but simply an ABA policy. Most judges abided by this policy, with the rationale being that cameras would cause

competing rights

Canon 35

disruption and lead defendants, lawyers, and jurors to act in ways that they would not act if no cameras were present, thus depriving the defendant of a **fair trial.** In the early days of television, there was potential for actual disruption because of the bulky equipment and lighting needed for TV coverage. During the 1970s, however, when unobtrusive equipment became available and when TV was no longer such a novelty, broadcasters began pressuring for access to courtrooms.

Canon 3A

In 1972, the American Bar Association liberalized Canon 35, redesignating it **Canon 3A.** It recommended that individual states and judges be given discretion on whether to allow cameras into their courtrooms. One by one, states began allowing cameras in courtrooms on an experimental basis. Nothing went awry, so by the mid-1980s all states were allowing cameras into their courtrooms. In 1991, the cable channel Court TV was established to cablecast nothing but real courtroom trials. It, and other networks, covered the William Kennedy Smith 1991 rape trial and the 1994 Menendez brothers murder trial.

O. J. case

The stage was thus set for cameras in the courtroom for the 1995 trial in which O. J. Simpson was accused of killing his wife, Nicole Brown, and her friend Ron Goldman. One difference between this trial and all previous ones was the extent to which the public watched. The coverage of the trial, not only by Court TV but also by CNN, E! Entertainment Television, and many stations, outdistanced the ratings of all other programming. The verdict was one of the most watched events in TV history (see Exhibit 11.9). Because the trial became a circus, many questions were raised about the efficacy of allowing cameras in the courts, especially when a cameraman accidentally showed one of the jurors. Most judges believed the cameras interfered with court proceedings and, as a result, began routinely barring cameras from their courts. However, as time passed and memories faded, the courts began once again opening their doors to television.[23]

Exhibit 11.9

In front of 150 million TV viewers, O. J. Simpson and his lawyers react to his "not guilty" verdict.

(© SYGMA)

11.10 Licensing

The FCC is charged with licensing radio and TV stations so that they operate in the **public convenience, interest, and necessity;**[24] without a license, a station cannot operate. The FCC has no direct control over the broadcast networks. It does not license them, but it can control them indirectly through their owned and affiliated stations. For example, the FCC ruling that limited the number of hours of network programming that a station could air during prime time effectively chopped a half hour off network programming. When the FCC wanted to punish CBS for the Janet Jackson mishap, it levied fines against the stations CBS owns.

networks

Local cable TV systems, cable networks, and satellite providers are not licensed, but the FCC has jurisdiction over some aspects of management, such as rate control and must-carry. The FCC has always had a great deal of regulatory power over the telephone business, not in the form of actual licenses, but by setting rates and overseeing the overall business. The home video, movie, internet, and corporate video areas are not licensed, but occasionally they are part of antitrust investigations or other ownership considerations. Licensing of TV and radio stations, however, has raised many issues that have permeated the electronic media industry.

nonlicensed groups

Most of the radio and TV station frequencies are allocated, but new services occasionally become available, such as **low-power TV** (see Chapter 4). Originally companies wishing a license would submit applications to the FCC, and the FCC would decide which one should be given permission to offer the service. This used to be very laborious because FCC staff members sorted through all the applications trying to determine which group or person was most qualified. To alleviate the resulting backup, the FCC changed to allocating stations by **lottery.** The names of all applicants were "put in a hat," and the winner was chosen by chance. That applicant was then carefully checked to make sure it met basic qualifications. Now when the FCC allocates frequencies, it often conducts an **auction** and gives the service to the highest bidder, providing that bidder is qualified. Using lotteries and auctions signals a major change in the philosophy of the FCC. Instead of acting as guardian and selecting the best applicant, it now acts more as traffic cop, stopping applicants (who win by chance or money) only if they are undesirable.[25]

process

After an applicant receives a license, it usually experiences little supervision from the FCC until license renewal time, which currently is every eight years. The renewal process has changed greatly through the years. In the early days, the FRC and FCC renewed licenses almost automatically (see "Zoom In" box).[26]

In 1946, the FCC issued an 80-page document detailing its ideas on license renewal. This document was almost instantly dubbed the **Blue Book,** in part because of its blue cover, but more sarcastically because blue penciling denotes censorship. One of the provisions in it became known as the **promise versus performance** doctrine. It stated that promises made when stations were licensed should be kept and that performance on these promises should be a basis for license renewal. Broadcasters probably would not have objected to that philosophy, but the document went on to detail proper broadcasting behavior. It was particularly adamant about avoiding overcommercialization, broadcasting public-affairs programs, and maintaining well-balanced programming. Broadcasters

Blue Book

ZOOM IN: Two Notable Nonrenewals

Two cases arose during the late 1920s and early 1930s in which licenses were not automatically renewed.

One concerned Dr. J. R. Brinkley, who broadcast medical "advice" over his Milford, Kansas, station. He specialized in goat gland treatments to improve sexual powers and in instant diagnosis over the air for medical problems sent in by listeners. These problems could always be cured by prescriptions obtainable from druggists who belonged to an association Dr. Brinkley operated.

The other culprit was the Reverend Robert Shuler, who used his Los Angeles radio station to berate Catholics, Jews, judges, pimps, and others in his personal gallery of sinners. He professed to have derogatory information regarding unnamed persons who could pay penance by sending him money for his church.

The Federal Radio Commission decided not to renew either of these licenses, and although both defendants cried censorship, the U.S. Supreme Court sided with the FRC on the grounds that with only a limited number of frequencies available, the commission should consider the quality of service rendered.

In this 1930s photo, Robert Shuler (right) is leaving jail on $500 bail after serving 16 days of a 20-day sentence for alleged radio attacks on judges. He is shaking hands with the Los Angeles County jailer while an assistant jailer looks on.

(Bettmann/Corbis)

looked at this document as a violation of the section of the Communications Act that stated the commission was not to have the power of censorship, and on this basis the Blue Book provisions were never implemented.[27]

1960 policy statement

In 1960, the FCC issued a much briefer policy statement that listed 14 elements usually necessary to meet public interest and warned broadcasters to avoid abuses related to overcommercialization. The FCC also changed the license renewal forms so that broadcasters performed **ascertainment,** a process in which they had to interview community leaders to obtain their opinions on the crucial local issues and then design programming to deal with those issues. The results of these interviews and the proposed program ideas were then submitted to the FCC. Along with this material, stations sent copious information concerning station operation for the previous three years, including a **composite week's** list of programming. These seven days from the previous three years were selected at random by the FCC. The FCC's main concern was whether the programs listed adhered to what the station had proposed in its previous application (promise versus performance).[28]

Although the amount of paperwork increased with these 1960 changes, licenses continued to be renewed rather perfunctorily for several more years. Over a 35-year period starting from the late 1920s, only 43 licenses were not renewed out of approximately 50,000 renewal applications.[29] This changed during the mid-1960s as community members gradually became involved in license renewal, and the FCC scrutinized license renewal more carefully than in the past.

The first major community participation occurred in 1964 when a group from Jackson, Mississippi, led by the United Church of Christ, asked the FCC if it could participate in the license renewal hearing of station WLBT-TV because the group believed the station was presenting racial issues unfairly. The FCC refused this hearing because, until this time, only people who would suffer technical or economic hardship from the granting of a license were permitted to testify at hearings. The UCC appealed to the courts, stating that ordinary citizens should be heard concerning license renewals. The courts agreed and ordered that a hearing be held that eventually led to WLBT's being given to a nonprofit group. This case established the precedent for citizen participation in license renewal.[30]

WLBT-TV

In 1969, a group of Boston businesspeople successfully challenged the ownership of TV station WHDH, which had been operated by the *Boston Herald Traveler* for 12 years. This was a complicated case and involved rumors of improprieties on the part of all three companies that had originally applied for the station license. The station was awarded to the *Herald Traveler* in 1957, but various allegations kept the ownership in question. In 1962, the *Herald Traveler* was given a four-month temporary license in the hope that the situation would be clarified within that time. Seven *years* later, the commission revoked the *Herald Traveler* license and awarded it to Boston Broadcasters, Inc., an organization that included some of the original unsuccessful applicants for the station license back in the 1950s. This January 1969 decision sent shivers through the entire broadcasting community. Never before had a station license been transferred involuntarily unless the station licensee was found guilty of excessive violations.[31]

WHDH

The net result of the WHDH and WLBT cases was the filing of numerous renewal challenges. In some instances, groups requested a **comparative license renewal** hearing because they wanted to operate the stations themselves. In other cases, citizens' groups filed **petitions to deny** license renewal mainly so they could bargain for what they wanted. One license renewal case that dragged on for years involved RKO and its 16 stations. In 1980, the FCC refused to renew the licenses of several RKO stations because RKO's parent corporation, General Tire and Rubber, admitted to the Securities and Exchange Commission that it had bribed foreign officials. The license refusals were appealed, and lengthy deliberations began. As the hearings dragged on, RKO gradually (and painfully) sold its stations to other companies. In 1990, the FCC stopped the hearing without deciding whether or not RKO was qualified to be a licensee. Thus, the proceedings ended, not with a bang but with a whimper. The FCC called these "the most burdensome proceedings in the FCC history."[32]

RKO

Overall, however, the threat of losing a license abated in the 1980s. The FCC issued statements explaining that the commission would compare the station licensee and any challengers, but if the licensee had provided favorable service in

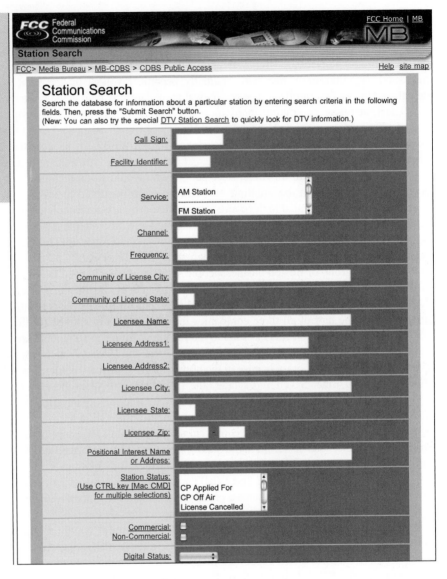

the past, it would be given **renewal expectancy.** In other words, radio and TV stations could now assume that unless they committed serious acts, their licenses would be renewed. The license renewal forms were also changed to decrease the work required of stations. Now stations only need to submit a small amount of information that they send to the FCC over the internet. The FCC makes information about stations available on the internet (see Exhibit 11.10) so that citizens can still become involved in license renewal.[33]

In addition to renewing licenses, if the FCC finds that a station is not fulfilling its obligations in serving the public convenience, interest, and necessity, it can revoke the license at any time. The FCC has done this for such causes as unauthorized transfer of control, technical violations, fraudulent contests,

renewal expectancy

revocation

overcommercialization, and indecent programs. One license it revoked in 1999 was that of religious broadcaster Trinity Broadcasting, charging that Trinity deliberately misled the FCC to evade ownership limits. The FCC can also issue lesser punishments such as fines and short-term renewals.[34]

The FCC becomes involved when a station is sold and the license is transferred from one party to another. It is the station management's prerogative to set the selling price and select the buyer, but the FCC checks the qualifications of the buyer.[35]

transfer

11.11 Ownership

Various regulatory bodies have a say in who owns media under what conditions. In recent years, the Justice Department has taken on a larger role because of numerous mergers and acquisitions. This department, which oversees the area of antitrust, looks at the mergers to see if monopolies are being created.

monopolies

Historically, however, the FCC has been most involved with ownership. In general, the trend has been to allow companies to own more and more stations and to own multiple stations in the same market. In the early days of radio and TV, one company could own no more than seven AM, seven FM, and seven TV stations. The numbers were increased over the years and then changed from number of stations to percentage of viewers or listeners reached. For example, in the early 1990s, a station group could only own enough stations to cover a maximum of 35 percent of U.S. TV homes. In 2003, the FCC decided to increase this number to 45 percent but met with great opposition from Congress and citizens, with the result that Congress compromised and raised the number to 39 percent.[36]

numbers

A law passed in 1992 required the FCC to set ownership rules for cable, and it decided that no one company should own cable or satellite systems reaching more than 30 percent of television subscribers. Time Warner took this rule to court and succeeded in having it rescinded. However, the FCC is still enforcing it under a clause that enables it to deem any merger not in the public interest.[37]

cable TV

A related area of ownership that the FCC deals with is **cross-ownership.** Many of the early owners of radio and TV stations also owned newspapers. This was considered against the public interest, especially if a town had only one TV station that was owned by the newspaper. The concern was that the public would learn only one version of the news. Similarly, controversies arose over whether one company should be allowed to own radio and also TV stations or TV stations and cable systems in the same market. The FCC has issued several guidelines on cross-ownership, some contradicting each other. Usually it grandfathers cross-ownership that already exists, so many different combinations of newspapers, radio, TV, and cable TV ownership exist. Mergers and acquisitions present particularly thorny problems concerning cross-ownership. When two companies merge, the new company often has multiple properties in one city because the two old companies had media interests in the same city. Usually the new companies petition the FCC to allow them to keep all properties, at least for a time. The FCC almost always grants the waiver because putting several stations on the market at the same time would artificially decrease their price. The general trend in recent years is for the FCC to allow more cross-ownership, but sometimes the courts disallow these decisions.[38]

cross-ownership

11.12 Equal Time

Section 315

Section 315 of the Communications Act is known as the **equal time** provision. In reality, it deals with **equal opportunity** rather than equal time. It reads as follows: If any licensee shall permit any person who is a legally qualified candidate for any public office to use a broadcasting station, he shall afford equal opportunities to all other such candidates for that office in the use of such broadcasting station.[39]

equal opportunity

This provision is in effect only during periods of election campaigns and deals only with people who have officially declared for political office. Because Section 315 actually mentions equal opportunity, it guarantees that all candidates for a particular office will be given the same number of minutes of airtime on a particular station. The airtime given must approximate the same time period so that one candidate is not seen during prime time and another in the wee hours of the morning. All candidates must be able to purchase time at the same rate. If one candidate purchases time and other candidates do not have the money to purchase equal time, however, the station does not need to give free time to the poorer candidates. Stations and networks are under pressure from Congress to give free time to all candidates, and some have agreed to do so. However, this is not required.

crisis speech

amendment

Through the years, interesting cases have arisen in regard to Section 315 (see "Zoom In" box). Shortly before the 1956 presidential election campaign between Dwight Eisenhower and Adlai Stevenson, Eisenhower, who was the incumbent president, was given free time on all networks to talk to the American people about an urgent crisis in the Middle East. Stevenson's forces immediately demanded equal free time. The FCC, however, decided that Eisenhower's speech was exempt from Section 315 because it did not deal with normal affairs, but with a crisis situation. This precedent has been followed in other election years with other crisis-oriented presidential speeches.

Exhibit 11.11

Lar Daly campaigning on Chicago TV in his Uncle Sam suit.

(Reprinted with the permission of BROADCASTING & CABLE, 1987 © 1994 by Cahners Publishing Company)

In 1959, Lar Daly (see Exhibit 11.11), a candidate for mayor of Chicago, protested that the incumbent mayor, who was running for reelection, was seen on a local news show, and Daly demanded equal time. Had he been an ordinary candidate, he might have been granted time and the issue dropped. He was an eccentric, however, who dressed in an Uncle Sam outfit—complete with a red, white, and blue top hat. Daly ran, unsuccessfully, for some office in every election. Daly was granted his equal time, but Congress amended Section 315 so that the equal time restrictions would not apply to candidates appearing on newscasts, news interviews, news documentaries, and on-the-spot coverage of news events. Eventually interviews on talk shows such as *Oprah* and *Tonight* were lumped in with bona fide news, and candidates have appeared on many of these shows.

In 1960, when Democrat John F. Kennedy and Republican Richard M. Nixon were running for president, the networks wanted the two to debate on television, and the two candidates agreed. But the wording of Section 315 would have necessitated equal debating opportunity for all the other splinter candidates on the presidential ballot. Because Congress believed this was a special situation, it suspended Section 315 temporarily for the 1960 presidential and vice presidential offices only (see Exhibit 11.12).

ZOOM IN: **Politics and Entertainment**

The entry of entertainers into politics created a dilemma for broadcasters in terms of Section 315. The issue became prominent when Ronald Reagan ran for president in 1980. Section 315 applied to the movies in which he appeared, so broadcasters did not air the films during the election period. Reagan's opponents made light of the situation, joking that showing the films might actually cause him to lose votes.

In 2003, Arnold Schwarzenegger ran for governor of California, and as with Reagan, the question of his movies arose. This time, however, it was cable TV channels that were considering airing his movies, and it is unclear whether Section 315 applies to cable TV networks. Not wanting to become test cases, networks such as TNT, FX, USA, and HBO, which regularly show Schwarzenegger movies, decided that the prudent thing to do was to cancel their showings until after the election.

In 2007, Fred Thompson, one of the stars of *Law and Order,* decided to run for president. NBC Universal carefully catalogued the shows and was prepared to keep those that included Thompson from airing. But his campaign fizzled, so, for the most part, the shows could air as scheduled.

Ronald Reagan, movie star and president.

Exhibit 11.12
Candidate John F. Kennedy speaks while Richard Nixon and moderator Howard K. Smith look on. In the foreground are the newspeople who asked the questions.

(UPI/Bettmann)

debates

**campaign
launch**

splinter parties

access

Broadcasters hoped that similar suspensions would be forthcoming, but the issue took a different turn. The FCC, under the 1959 Communications Act amendment that allowed on-the-spot coverage of bona fide news events, decided in 1975 that networks and stations could cover candidate debates if these were on-the-spot news events. The broadcasters could not arrange the debates, but if some other group, such as the League of Women Voters, organized the debates, the media could cover the event in much the same way as they cover awards ceremonies or state fairs. In 1976 and 1980, presidential debates were sponsored by the League of Women Voters and telecast as news events. Then in 1983, the FCC ruled that debates could be sponsored by broadcasters, on the grounds that broadcasters should have the same right to hold debates and exclude candidates as a civic group.

Exhibit 11.13

In the 1992 presidential campaign, Reform Party candidate Ross Perot joined Republican candidate George H. W. Bush and Democratic candidate Bill Clinton for the presidential debates.

(Corbis/Reuters)

For the 1980 campaign, Jimmy Carter's aides wanted to buy a half hour of network prime time early in December 1979 to launch Carter's campaign for reelection. The networks refused, saying that December was too early to start an election campaign. Carter forces appealed the decision, and the Supreme Court sided with them, indicating that it is the candidates, not the broadcasters, who decide when a campaign begins.

The issue of splinter-party candidates came to the fore again in 1992 and 1996. Ross Perot, a strong Reform Party candidate in 1992, was invited to join Democrat Bill Clinton and Republican George H. W. Bush in the debates (see Exhibit 11.13). In 1996, Perot was again a candidate and, of course, wanted to be in the debates. A nonpartisan commission formed to organize the debates decided that Perot did not have a realistic chance of being elected so should not participate. Perot fought the decision and was eventually allowed to participate in one debate.

Because of all the difficulties involved with Section 315, some stations decided they would not offer time to any candidate and thus avoid the question of giving time to opponents. But in 1971, the FCC nixed the idea, stating that stations could have their licenses revoked if they did not allow access to candidates running for federal office. However, for state and local elections, stations were given the option of not dealing with any candidates if they so desired. The FCC has issued thick tomes on political broadcasting, but in most election years new issues continue to arise regarding Section 315.[40]

11.13 The Fairness Doctrine

The **fairness doctrine** is no longer in effect, but the concepts surrounding it are still discussed. It was never a concrete doctrine, but rather a series of actions and rulings dealing with presentation of controversial issues. In 1949, as part of another ruling, the FCC issued a statement that said licensees must operate on a basis of overall fairness, making facilities available for the expression of contrasting views. When Congress amended Section 315 in 1959, it stated that nothing in the news exemptions should be construed as relieving broadcasters from the obligations imposed on them to give reasonable opportunity for the discussion of conflicting views on issues of public importance. Neither of these statements raised any ripples at the time, but in later years they were used in conjunction with fairness issues.[41]

origins

Fairness became an issue mainly in the 1960s. The political climate of unrest and distrust bred controversies that might have remained dormant in more settled times. The hallmark of the fairness doctrine is the Red Lion case. In 1964, station WGCB in Pennsylvania, operated by the Red Lion Broadcasting Company, broadcast a talk given by the Reverend Billy Hargis that charged author Fred J. Cook with Communist affiliations. Cook demanded that WGCB give him the opportunity to reply. The station said it would if Cook would pay for the time, but Cook took the case to the Supreme Court. The Court ruled that the station should grant Cook the time whether or not he was willing to pay for it, on the grounds that a broadcaster's right of free speech does not include the right to snuff out the free speech of others. The significance of this decision went beyond the dispute between Cook and the Red Lion Broadcasting Company because it upheld the constitutionality of the fairness concept.

Red Lion case

Banzhaf case

pendulum swing

One creative application of the fairness doctrine was a case brought by John F. Banzhaf III (see Exhibit 11.14). In 1964, the Surgeon General's Office determined there was a link between cigarette smoking and lung cancer, a point that was considered controversial. In 1967, Banzhaf petitioned WCBS-TV in New York, saying that because cigarette commercials showed cigarette smoking in a positive light, free time should be given to antismoking groups to present the other side. WCBS replied that it had presented the negative side of smoking within its news and public-affairs programs and did not think it needed to give equivalent commercial time to the issue. Banzhaf took the matter to the FCC, which sided with him—as did the appeals courts. Therefore, until 1972 when cigarette advertising was banned, any station that aired cigarette commercials had to provide time for anticigarette commercials.

Exhibit 11.14

John F. Banzhaf III used the fairness doctrine to enable antismoking messages to be aired.

(Courtesy of John F. Banzhaf III)

The rulings involving the fairness decisions of the 1960s led broadcasters and the public to believe that future decisions by the FCC and the courts would favor those who believed that fairness should be strictly interpreted. However, the pendulum swung, and the decisions of the 1970s began to turn around. When an ecology group tried to emulate Banzhaf's case by requesting that ads for gasoline be countered by antipollution announcements, the attempt was not successful. The FCC stated that the cigarette ruling was not to be used as a precedent because that case was unique.

During the 1980s, the number of complaints brought under the fairness doctrine waned, and the doctrine itself became the focus of attention. In a deregulatory mood in 1981, the FCC asked Congress to repeal the doctrine, but Congress did nothing. A court decision a few years later stated that the doctrine was not law but merely an FCC regulation that the FCC could abolish. In 1987, Congress passed a bill that would have codified the fairness doctrine and made it a law rather than an FCC policy, but President Reagan vetoed it. Soon after, the FCC said it was going to stop enforcing fairness and, in effect, abolish the whole fairness concept.

abolishment

This meant that stations could air programming about controversial issues without having to worry about complaints or contrasting points of view. One thing this enabled was the rise of talk radio, where issues discussed on programs are often one-sided. Although the fairness doctrine is dead, members of Congress periodically introduce bills to reinstate it. When this happens, talk-show hosts rail against the bills, with Rush Limbaugh calling them "crush Rush" bills.[42]

talk radio

11.14 Other Regulations

Many other laws and regulations affect telecommunications practitioners. For example, the U.S. Criminal Code outlaws **lotteries** sponsored by radio or TV. Stations are allowed to have contests, though, and sometimes cases arise as to whether a station is conducting a contest or a lottery. Generally, something is considered a lottery if a person pays to enter, if chance is involved, and if a prize is offered. Stations usually avoid lotteries by making sure people do not have to pay to enter contests.[43]

lotteries

The FCC also has rules against **hoaxes.** In 1990, after two Los Angeles disc jockeys faked a murder confession on the air that led to a nationwide police search for a nonexistent killer, the FCC established a mechanism for fining stations up to $250,000 for such hoaxes.[44]

hoaxes

From time to time, broadcasters and cablecasters must adhere to FCC regulations involving equal employment. These rules were plentiful and stringent during the 1960s and 1970s and more lenient during the deregulatory 1980s. At present, the main emphasis of the rules is on having an active outreach program so that minorities are adequately informed about job openings and recruited for them.[45]

equal employment

11.15 Issues and the Future

The internet has challenged and occasionally stumped the legal profession. Although some of the issues (e.g., copyright, privacy) are old, the twists given to them are new. Part of the challenge is that the internet is such a democratic tool; people anywhere can access and alter it. The same people who want their information kept private want private information about others so that they can market their widgets. Software providers want their material copyright protected

internet

but, at the same time, seek provisions that allow them to violate other companies' copyrights, such as those of the recording industry and the movie industry. Other new technologies pose problems, too. It is possible to send pornographic photos to cell phones. That is a new wrinkle that is likely to cause controversy. No doubt new applications in the wireless world will keep lawyers occupied.[46]

copyright

Is there a need to rethink the concept of copyright, given the ease with which digital information can be copied and the difficulty of enforcing current laws? One particularly thorny problem revolves around the fact that music licensing organizations distribute money to composers and publishers of music but not to the people who perform the music or the record companies that underwrite it. The original rationale was that airplay would lead to concerts and sales of records and that performers and record companies would profit from those. But musicians and record companies, who have always complained about this money distribution process, are complaining even louder now that distribution processes have changed and are appealing to Congress to pass a bill to include them as music licensing recipients. Of course, radio stations want the status quo and, in addition, want the amount they pay for airing music on the internet kept low.[47]

balance

Different telecommunications entities are regulated to differing degrees. Broadcasters particularly resent the many licensing restrictions that have tied their hands over the years while cablecasters, video stores, satellite systems, the internet, and others operate with much less regulation. The main reason for this is timing. Broadcasters were in business during the 1960s when heavy regulation was in vogue. Most of the other media systems arose during the 1980s and 1990s when deregulation was the watchword.

media giants

Although media companies like the idea of deregulation, many politicians and citizens' groups do not. They worry that without cross-ownership rules, a few media giants (Disney, Warner, Microsoft, Fox) will take over most of the media outlets, thus reducing news sources and competition.

auctions

Auctions and lotteries reduce the amount of time the FCC staff spends deciding who should get what frequencies, but they do not ensure that the best people receive access to the media. Particularly with auctions, money rather than morality decides who earns the right to program to the American people.

Section 315

There are occasional murmurs that Section 315 should be abolished. Of all the laws, that one is least likely to fall victim to deregulation because it affects the members of Congress who would have to vote to abolish it. Politicians want to make sure they have access to radio and TV. Although they would like to amend Section 315 so that they would have free time instead of having to pay for it, they are unlikely to destroy the equal time concept.

EAS

Some have questioned whether the Emergency Alert System is still needed. It was not put into effect during the terrorist attacks of September 11, 2001. The news media effectively undertook the role of keeping the populace informed. Maybe the system is an idea whose time has passed, or maybe it should ring people's cell phones to convey emergency information rather than being tied to broadcasting.

people problems

Regulation also suffers from people problems. Senators and representatives depend on the support of media to be elected and, hence, are often heavily influenced by broadcasting lobbies. FCC commissioners, because they know their decisions can be appealed to the courts, sometimes make conservative decisions that are unlikely to be overturned.

Regulation policies are in a constant state of flux—a situation having both advantages and disadvantages. The state of flux allows for changes to keep up with the times, but it also creates internal inconsistencies among the various regulatory bodies. It inhibits rigidity on the part of elected and appointed officials, but it also causes confusion and uncertainty for those who must comply with regulations.

11.16 Summary

The electronic media industry interrelates with all three branches of government—legislative, executive, and judicial—and with independent regulatory bodies. The main independent regulatory board is the FCC. Some of its chores are technical and have not changed much. Others, such as licensing, are more philosophical and are subject to changes as society's outlook alters. Licensing has caused broadcasters varying degrees of consternation through the years. Decisions regarding Brinkley, Shuler, WLBT, WHDH, and RKO affected the regulatory direction of the industry. Amount of ownership and cross-ownership is also an important topic. The FCC has issued fines against obscenity, as in the Howard Stern case, and decided when equal time provisions were violated, as in cases regarding presidential crisis speeches. On occasion, the FCC has issued guidelines or rulings such as the Blue Book.

The executive branch is most likely to claim clear and present danger if the president believes reporters are acting irresponsibly. The president can veto communication legislation, as in the case of the fairness bill. In addition, the president nominates commissioners and can activate the Emergency Alert System. The Justice Department handles matters related to monopoly.

The legislative branch passes and amends many laws that affect telecommunications. The amendment to Section 315 that removed equal time restrictions from bona fide newscasts had a great effect on election campaigns. Congress wrestles with copyright laws and all their ramifications. Other criminal and civil laws passed by Congress, such as those dealing with lotteries and hoaxes, affect broadcasting, as do state laws dealing with shielding reporters and invasion of privacy. Congressional committees hold hearings on communications-related issues, and the Senate approves the appointment of FCC commissioners.

The courts are the place of appeal for those who think they are treated unjustly by other branches of the government or by elements of society. Media and the courts have a double-edged relationship. On one hand, a battle was waged regarding the right of radio and TV to have access to the courts for news coverage—a battle that received a setback because of the O. J. Simpson case. On the other hand, the courts make many decisions that directly affect broadcasters, such as those involving the First Amendment. Indecency and obscenity rulings on subjects such as WBAI's "seven dirty words," radio lyrics, and the CBS fine for the Janet Jackson "wardrobe malfunction" may eventually be decided in the courts. Likewise, libel cases such as Westmoreland's, as well as privacy invasions, are fought in the courts. Some Section 315 decisions were appealed to the courts, such as the determination of when a campaign begins. The fairness doctrine kept the courts busy with cases that included Red Lion and cigarette advertisements and the abolishment of the doctrine. Although the various branches of government disagree with one another from time to time, equilibrium is maintained by the checks and balances system.

Suggested Websites

www.ascap.org (American Society of Composers, Authors and Publishers, a music licensing organization)

www.congress.org/congressorg/home (Congress)

www.fcc.gov (the Federal Communications Commission)

www.ftc.gov (the Federal Trade Commission)

www.whitehouse.gov (news about the president and others in the executive branch)

Notes

1. "About the FCC," www.fcc.gov/aboutus.html (accessed August 17, 2008); and "Fees on the Up-and-Up," *Broadcasting & Cable,* April 9, 2001, 39.
2. "EAS Advances," *TV Technology,* February 6, 2008, 1.
3. "FTC Bars Two Alcohol Ads," *Broadcasting & Cable,* August 10, 1998, 12; and "Ma Bell's Big Breakup," *Newsweek,* January 18, 1982, 58–59.
4. Steven Phipps, "'Out of Chaos': A Reexamination of the Historical Basis for the Scarcity of Channels Concept," *Journal of Broadcasting & Electronic Media,* Winter 2001, 57–74.
5. Lucas A. Powe, Jr., *American Broadcasting and the First Amendment* (Berkeley: University of California Press, 1987); "Press Shield Law Upheld," *Broadcasting & Cable,* November 8, 1999, 37; and "Jailing of Reporter Sends Jolt," *Hollywood Reporter,* July 7, 2005, 1.
6. Jan H. Samoriski, John L. Huffman, and Denise M. Trauth, "Indecency, the Federal Communications Commission, and the Post-Sikes Era: A Framework for Regulation," *Journal of Broadcasting & Electronic Media,* Winter 1995, 51–72; and Robert McKenzie, "Contradictions in U.S. Law on Obscenity and Indecency in Broadcasting: A Bleeping Critique," *Feedback,* August 2002, 28–34.
7. "Stern's Blue Streak Costs Infinity a Cool $1.7 Mil," *Daily Variety,* September 10, 1995, 6; "FCC Investigates Church Coupling; WNEW Axes DJs," *Hollywood Reporter,* August 23–25, 2002, 1; "Super Bowl Halftime Stunt Angers NFL, CBS, FCC," *Wall Street Journal,* February 3, 2004, B-1; "Super Bowl Episode Prompts CBS to Heighten Safeguards for Grammys," *Los Angeles Times,* February 4, 2004, C-1; and "Firing of 'Love Sponge' Signals Cleanup of Shock Radio," *Wall Street Journal,* February 25, 2004, B-1.
8. "Bared Breast Nets $550K Penalty," *Hollywood Reporter,* September 26–October 4, 2004, 71; "Lawmakers Poised to OK Indecency Bill," *Los Angeles Times,* June 3, 2006, C-1; "Court to

328   **Part 2** Electronic Media Functions

Decide What the @!*#%?," *Hollywood Reporter,* March 18, 2008, 1; and "CBS Wins Verdict on FCC Indecency Fine," *Wall Street Journal,* July 22, 2008, B-1.

9. "Crew Acquitted in Obscenity Case," *Daily Variety,* October 27, 1990, 1; and "Don't Mess Around with Slim," *Broadcasting & Cable,* June 11, 2001, 14.

10. "Court Clears Way for More Playboy," *Electronic Media,* May 29, 2000, 21; "How the Fight over Indecency Threatens to Turn the Cable Industry Upside Down," *Broadcasting & Cable,* December 5, 2005, 14; and "Split Ruling on Smut," *Daily Variety,* July 1, 1996, 6.

11. "Push to Allow DVDs to Be 'Sanitized' Alarms Studios," *Los Angeles Times,* June 23, 2004, C-1; "R Fare Takes Hill Hit," *Broadcasting & Cable,* June 25, 2001, 7; and "Games under the Gun," *Hollywood Reporter,* September 19–25, 2000, 14–15.

12. "High Court Strikes Down Internet Indecency Rules," *Electronic Media,* June 30, 1997, 37; "Special Too Much Information Edition," *Newsweek,* September 21, 1998, 8; and Ven-hwei Lo and Ran Wei, "Third-Person Effect, Gender, and Pornography on the Internet," *Journal of Broadcasting & Electronic Media,* March 2002, 13–33.

13. "Eye Clinic Sues ABC for $50 Million," *Broadcasting & Cable,* November 1, 1993, 25; "Newton's NBC Libel Damage Award Nixed," *Daily Variety,* August 31, 1990, 1; "The General's Retreat," *Newsweek,* March 4, 1985, 59–60; and "ABC Takes Double Hit in Court," *Broadcasting & Cable,* December 30, 1996, 10, 40.

14. "Providing Coverage, Protecting Victims: Rape Trials on Trial," *Broadcasting,* April 30, 1984, 134; "No One under 13 Admitted, but Who Told Them to Lie?" *Los Angeles Times,* February 22, 2001, T-1; and Miriam J. Metzger and Sharon Doctor, "Public Opinion and Policy Initiatives for Online Privacy Protection," *Journal of Broadcasting & Electronic Media,* September 2003, 350–74.

15. *The United States Constitution,* Section 8.

16. "Justices OK Copyright Extension," *Los Angeles Times,* January 16, 2003, 1.

17. Matt Jackson, "Commerce Versus Art: The Transformation of Fair Use," *Journal of Broadcasting & Electronic Media,* Spring 1995, 190–99.

18. "For Radio, a Web Royalty Check," *Broadcasting & Cable,* December 12, 2001, 17; and "ASCAP Singing with Radio Deal," *Hollywood Reporter,* October 19–25, 2004, 1.

19. "Copyright Office Proposes—Cable Opposes—Higher Fees," *Broadcasting & Cable,* August 11, 1997, 18; and "Panel Backs Up Boost in DBS Copyright Fees," *Hollywood Reporter,* February 11, 1999, 8.

20. "Windows Leave Open Too Many Opportunities," *Daily Variety,* March 3, 2003, A-4; and "Latest Plot Twist for 'Star Wars': Attack of the Cloners," *Los Angeles Times,* May 10, 2002, 1.

21. "Satellite Blows TV Pirates Right Off the Tube," *Los Angeles Times,* January 27, 2001, 1; "Safer Digital Information," *Broadcasting & Cable,* July 14, 2008, 16; and "Policing Web Video with 'Fingerprints,'" *Wall Street Journal,* April 23, 2007, B-1.

22. Motion Picture Association of America, "Worldwide Study of Losses to the Film Industry and International Economics Due to Piracy; Pirate Profiles," http://www.slyck.com/misc/mpaa_loss.doc (accessed August 17, 2008).

23. Susanna Barber, *News Cameras in the Courtroom: A Free Press-Fair Trial Debate* (Norwood, NJ: Ablex, 1987); "Courtroom Doors Begin to Open for TV, Radio," *Broadcasting,* September 3, 1990, 25; "Court TV's O. J. Duties Put on Trial," *Electronic Media,* January 30, 1995, 1; and "Will Bryant Trial Be an O. J.-Style Media Event?" *Broadcasting & Cable,* July 28, 2003, 1.

24. Section 307 of the Communications Act of 1934 states: "The Commission, if public convenience, interest, or necessity will be served thereby, subject to the limitation of this Act, shall grant to any applicant therefore a station license provided for by this Act." This section has become the keystone of regulation that the FCC has often used to justify regulations, especially those aimed at programming.

25. "Another Deluge of LPTV Filings Inundates FCC," *Broadcasting,* February 23, 1981, 29; and "Auction Worries Multiply," *Broadcasting & Cable,* July 9, 2001, 8.

26. "From Fighting Bob to the Fairness Doctrine," *Broadcasting,* January 5, 1976, 46. The Robert Shuler mentioned in the box should not be confused with the current Dr. Robert H. Schuller of the California Crystal Cathedral who uses TV and the internet for his messages.

27. Michael J. Socolow, "Questioning Advertising's Influence over American Radio: The Blue Book Controversy of 1945–1947," *Journal of Radio Studies,* December 2002, 282–302.

28. Erik Barnouw, *The Golden Web* (New York: Oxford University Press, 1968), 227–36.

29. Don R. Pember, *Mass Media in America* (Chicago: Science Research Associates, 1974), 283.

30. "Looking Back to WLBT(TV)," *Broadcasting,* April 16, 1984, 43.

31. "The Checkered History of License Renewal," *Broadcasting,* October 16, 1978, 30.

32. "FCC Gives RKO Green Light to Sell Stations," *Broadcasting,* July 25, 1988, 33; and "The Trials of RKO," *Channels,* January 1990, 70–73.

33. "The Next Best Thing to Renewal Legislation," *Broadcasting,* January 10, 1977, 20; "FCC Cuts Back on Paperwork," *Broadcasting & Cable,* April 6, 1998, 24; and "Station Search," http://fjallfoss.fcc.gov/prod/cdbs/pubacc/prod/sta_sear.htm (accessed August 18, 2008).

34. Charles E. Clift III, "Station License Revocations and Denials of Renewal, 1970–1978," *Journal of Broadcasting,* Fall 1980, 411–21; and "FCC Yanks Trinity License," *Broadcasting & Cable,* April 19, 1999, 14.

35. Wenmouth Williams, Jr., "The Impact of Ownership Rules and the Telecommunications Act of 1996 on a Small Radio Market," *Journal of Radio Studies,* Summer 1998, 8–18.

36. "New Ownership Cap Fits Fox, CBS Perfectly," *Broadcasting & Cable,* January 26, 2004, 5; and Jennifer M. Proffitt, "Juggling Justifications: Modifications to National Television Station Ownership Rules," *Journal of Broadcasting & Electronic Media*, December 2007, 575–95.

37. "FCC Fears Cable Deal Firestorm," http:/www.thedeal.com (accessed May 25, 2004).

38. "Cross-Ownership Gets FCC OK," *Hollywood Reporter,* June 15, 1998, 8; "It's Almost as If There's No Rule," *Broadcasting & Cable,* March 20, 2000, 8; and "Court Rejects FCC Limits on TV Ownership," *Los Angeles Times,* February 20, 2002, 1.

39. Public Law No. 416, Section 315.

40. "Appeals Court Agrees with FCC: Broadcasters May Sponsor Debates," *Broadcasting,* May 12, 1984, 69; "Supreme Court Rules Vid Nets Erred in Not Selling Carter Air Time," *Variety,* June 2, 1981, 1; "FCC Wearing 2-Party Hat, Claims Critic," *Daily Variety,* August 2, 1991, 1; "Debate Continues over Debates," *Broadcasting & Cable,* September 23, 1996, 14; and "NBCU Ready for Thompson White House Run," *Broadcasting & Cable,* July 16, 2007, 6.

41. *In the Matter of Editorializing by Broadcast Licensees,* 13 FCC 1246, June 1, 1949; and Public Law No. 416, Section 315.

42. Timothy J. Brennan, "The Fairness Doctrine as Public Policy," *Journal of Broadcasting & Electronic Media,* Fall 1989, 419–40; "Fairness Held Unfair," *Broadcasting,* August 10, 1987, 39-D; and "Who's Afraid of the Fairness Doctrine?" *Broadcasting & Cable,* July 28, 2008, 6.

43. *Lotteries and Contests: A Broadcasters Handbook* (Washington, DC: National Association of Broadcasters, 1985).

44. "FCC Adopts $25,000 Fine for Hoaxes," *Broadcasting,* May 18, 1992, 5.

45. "FCC Issues EEO Rules Once Again," *Broadcasting & Cable,* January 24, 2000, 10; and "EEO: The Key to FCC License Renewal," *TV Technology,* February 5, 2002, 26.

46. "Glitterati vs. Geeks," *Newsweek,* October 14, 2002, 40–41; and "Private Screening in Public Places," *Los Angeles Times,* July 19, 2001, T-1.

47. "Back-Door Man for Music Royalties?" *Hollywood Reporter,* May 22, 2008, 14.

Chapter 12

ETHICS AND EFFECTS

One by-product of deregulation is an increased emphasis on ethics. When electronic media operations are heavily controlled through laws and regulations, individuals encounter legal constraints that tell them what they can and cannot do. When there are few laws, individuals and "the corporate conscience" are in a position to make many fundamental decisions. Often, they must rely on a sense of ethics, and ethics can be influenced by knowing the effects of media on the public.

For example, when cable rates are regulated (see Chapter 4), cable companies know how much they can or cannot raise rates. When no government agency is policing how much they charge subscribers, managers must use their conscience and their sense of economic fairness when designing their costs. Similarly, if the public has little input concerning license renewal, can (should) station managers ignore the public?

> The speed of communication is wonderful to behold. It is also true that speed can multiply the distribution of information that we know to be untrue. The most sophisticated satellite has no conscience—and in the end the communicator will be confronted with the age-old problem of what to say and how to say it.
>
> **Edward R. Murrow, former CBS newscaster**

Ethics still comes into play when there are laws. For example, a **payola** law prohibits a disc jockey from accepting money from a record company in exchange for playing that company's music on the airwaves without reporting it (see Chapter 5). If a disc jockey gets caught engaging in payola, he or she can be fined or even imprisoned. But what if the disc jockey feels 99 percent sure of not getting caught? In that case, accepting the money is both a legal and an ethical issue. Although it is illegal for talk-show talent to say something obscene over the air, doing so can increase ratings. The extra money from advertisers based on the increased ratings can offset the fine from the FCC. Should the talk-show talent be obscene? Should management encourage this **obscenity**? Again, ethics and legality are intertwined.

12.1 Ethical Guidelines

To help managers and employees make proper ethical decisions, some electronic media organizations draw up guidelines of **self-regulation** dos and don'ts. For example, a professional association for newspeople, the Radio-Television News Directors Association (RTNDA), has a Code of Broadcast News Ethics (see Exhibit 12.1). It was created in 1946 and has been rewritten seven times to keep it up-to-date. This code has no legal force, and there are no penalties for violating it, but it is something the organization hopes that journalists follow.[1]

RTNDA code

Exhibit 12.1
A few provisions of the RTNDA code.

Sample RTNDA Code Provisions

Professional electronic journalists should:

- Fight to ensure that the public's business is conducted in public.
- Clearly disclose the origin of information and label all material provided by outsiders.
- Not present images or sounds that are reenacted without informing the public.
- Treat all subjects of news coverage with respect and dignity, showing particular compassion to victims of crime or tragedy.
- Disseminate the private transmissions of other news organizations only with permission.
- Not accept gifts, favors, or compensation from those who might seek to influence coverage.
- Gather and report news without fear or favor, and vigorously resist undue influence from any outside forces, including advertisers, sources, story subjects, powerful individuals, and special interest groups.
- Seek support for and provide opportunities to train employees in ethical decision-making.

Oscar code

In the early part of the 21st century, many of the practices related to judging the Oscars went a little over the top. Companies and individuals were blatantly promoting their movies to the industry people who vote and, in some cases, even starting smear campaigns against competing films. As a result, the Academy of Motion Picture Arts and Sciences adopted a code of behavior for the awards season. Among other things, it advised industry members not to write letters to the editor or be quoted in ads. Campaigners were to avoid advertising that bad-mouthed other films and were not to throw special parties for voters.[2]

broadcast standards

Much of ethical self-regulation comes from individual stations or networks, which have their own written or unwritten codes. At the commercial broadcast networks, the business of making sure all programs and commercials adhere to good taste falls to the Department of Standards and Practices (often called **broadcast standards**), a group operating independently of programming or sales and reporting directly to top network management. This department often reviews program and commercial ideas when they are in outline or **storyboard** form and then screens them several times again as they progress through scripting and production. If any of the proposed ideas run counter to what are determined to be the prevailing standards, the Department of Standards and Practices requests changes before the idea can proceed to the next step (see Exhibit 12.2).

conflict and cooperation

Naturally there is conflict between broadcast standards and program or commercial producers. But there is also often cooperation so that problems can be averted. For example, NBC decided to air Steven Spielberg's movie *Schindler's List* uncut, even though it had gruesome scenes and some nudity. Spielberg and a network executive spoke at the start of the movie, stating that the story was too important to edit, warning the audience about the graphic nature of the scenes, and suggesting that children not be allowed to watch. Having framed the broadcast this way, the network did not receive complaints but rather letters of congratulations.

local stations

At local radio and TV stations, the broadcast standards functions may not be as formal. The general manager frequently performs the function on an as-needed basis, and sometimes the function is delegated to whichever group or person handles the station's legal matters or public relations. In any organization, the person who is the "boss" plays an enormous role in setting the tone for ethical behavior.

cable TV

Within the cable TV and satellite broadcast industries, individual networks have their own standards. Arts & Entertainment, for example, has a policy against frontal nudity—a policy that it knowingly violates when it shows certain Woody Allen movies because Allen insists that none of his movies be edited.

individuals

Network, station, and internet self-regulation occurs constantly and informally at all levels through the actions of individuals who work in the electronic media. Writers, producers, directors, actors, editors, and website designers are constantly basing decisions on their own ideas of propriety and appropriateness, even when management provides no direction (see "Zoom In" box).[3]

To: Executive producer of the series

From: Broadcast Standards & Practices editor for the show

Subject: Broadcast Standards & Practices report

CC: Programming executive responsible for the show, the head of the programming department, the head of the Standards Department, and the in-house legal counsel

Page 2: Please ensure that there is plenty of coverage for the fight scene in the bar. Delete having Mark break a beer bottle on Sam's face.

Page 4: Delete "Christ" in Molly's dialogue, "Christ, what did I do to deserve this!"

Page 6: In the bedroom scene, do not have Molly "moving sensuously" on top of Mark. There can be no grinding action in this sequence, nor can there be any nudity. Molly must remain appropriately covered as she gets out of bed. Again, we suggest that you provide adequate coverage to avoid problems at the rough cut stage.

Page 22: Overly graphic images of the crime scene need to be avoided.

Page 24: Please make sure you have the proper clearances for the use of the song, "Stop in the Name of Love."

Page 38: The medical information about Alzheimer's disease must be correct. Please have your medical consultant provide documentation that confirms the scripted information.

Page 42: Substitute for scripted dialogue about illegal immigrants being the cause of the nation's ills. We do not want to include advocacy positions in this entertainment program.

Page 43: Do not have the teenagers smoking cigarettes, as scripted.

Page 47: Avoid a "how to" when Molly commits suicide. Do not include every step of the process she uses as we do not want viewers to be able to imitate her actions. Extreme care must be exercised to avoid making suicide appear to be a viable solution to her problems.

Page 55: At the airport, revise having the security guard portrayed as a buffoon ("You look like a nice guy who wouldn't want to harm all these nice people who are flying home for the holidays") who allows passengers to board without having their baggage inspected, as we do not want to ridicule necessary homeland security procedures.

Exhibit 12.2

This is not an actual Broadcast Standards and Practices report, but it is representative of the type of material that might appear in a report related to a one-hour drama.

(Courtesy of Philippe Perebinossoff)

In his book *Real-World Media Ethics,* Philippe Perebinossoff proposes an ethics rubric for media practitioners to use as an approach to a wide range of ethical dilemmas (see Exhibit 12.3). His Evaluate, Truth, Harm, Investigation, Codes of Ethics, and Situational Ethics rubric provides media practitioners (and students) with a comprehensive way to analyze what is at stake ethically so that they can make reasoned decisions.[4]

rubric

ZOOM IN: Ethical Decisions Gone Awry

Most ethical decisions never come to the fore. The people making them come to the right decision, and life goes on. When the decisions go awry, however, they receive publicity and scrutiny. Here are four such decisions.

The people at NBC news were faced with a tough ethical decision in the spring of 2007 when the network received a package of video, photographs, and writings from Cho Seung-Hui, the gunman who killed people on the campus of Virginia Tech. Staffers disagreed about what to do with the material, and news president Steve Capus said that the debate was about weighing the value of the material as vital information versus the perception of glorifying a madman. In the end, the network aired an edited video of Cho's rantings and some stills of him looking threatening. For the most part, the network received negative feedback from the public for its decision to air any of the material.

Don Imus made a personal decision in 2007 when he called the Rutgers women's basketball team "nappy-headed hos." Known for a series of sexist and racist comments, Imus said he was just trying to make a sarcastic point. But the description lost him his jobs on CBS radio and MSNBC. About a year later, however, he returned to work with a new show on WABC-AM.

Shock DJs Andy and Opie also lost their job at WNEW-FM in New York when, in 2002, they broadcast a live account of a couple allegedly having sex in St. Patrick's Cathedral. This stunt was part of a regular feature on the show in which couples could win a trip by having sex in risky places, but the public and station management thought a place of worship was ethically tasteless.

In a *60 Minutes* segment on CBS in September 2004, Dan Rather reported that President George W. Bush had received preferential treatment when he was in the National Guard during the Vietnam War. However, it turned out Rather and others at CBS had rushed to judgment, and the documents they were basing their information on were actually forgeries. The network, and Rather in particular, were slow to acknowledge the mistake and apologize. An independent panel investigating the incident strongly rebuked all involved for faulty reporting. Four CBS producers were fired, and when Rather's contract ended, it was not renewed.

St. Patrick's Cathedral.

12.2 Ethical Considerations

Making ethical decisions within the fast-paced communications industry is not easy. It involves heavy thinking regarding such subjects as fairness, taste, conflict of interest, trust, and accuracy. Individuals within all facets of life have a responsibility to act in an ethical manner, but the burden is often greater within the electronic media because they are so visible.[5]

An Ethics Rubric

E *Evaluate*
The ethical situation needs to be examined in a number of different ways, not just from a single perspective. Though it is important to consider a creator's intention and goals, do not simply buy into one person's point of view. Take the time to analyze.

T *Truth*
Make sure you have a clear definition of the facts as well as a definition that can be agreed upon by others. The truth is not always black and white. In a world where exaggeration seems to be the norm, it is often difficult to get to the essential truth.

H *Harm*
Consider what harm is likely to ensue from a given action. Real-world situations are involved, and media rumors can ruin lives. Stop to evaluate negative outcomes that can ensue, even from seemingly ethical decisions. Of course, intentional harm should similarly be evaluated.

I *Investigation*
Investigation involves going beyond a cursory review and digging deep to gather the necessary information. Do not fall into the trap of getting there first and getting it wrong. This holds especially true for news, but it is applicable elsewhere.

C *Codes of Ethics*
Codes provide valuable guidelines that can facilitate ethical decision making. Before you make a final determination about a course of action, it's wise to pause to check if there's a code of ethics that addresses the case in question.

S *Situational Ethics*
Approach issues on a case-by-case basis that allows you to adjust ethical beliefs to meet a particular situation. Doing so creates a sensitivity to a set of particular circumstances and avoids the tendency to rush to judgment based on other, possibly very different cases.

Exhibit 12.3
An easy-to-remember method for dealing with ethical situations.

Most of the ethical decisions those engaged in electronic media must make involve their responsibilities to their audience. For example, audiences believe **toward audience** that the news they are receiving is actually news, but if they hear that a network is offering to pay Michael Jackson $5 million for an interview in exchange for shelving a critical report of him, they can question the efficacy of the news.[6]

In addition to ethical actions toward the audience, media practitioners must **toward others** also consider ethical practices toward each other. Many "deals" are made in the entertainment business, and if the agreements arrived at are not honored, individual and corporate credibility suffers. People want to do business with other people who keep their word, but sometimes in the rush to make the deal, ethics becomes a low priority. Once people know each other and trust that they will act ethically, they don't need to check details as carefully as they would with new people. This makes it harder for new people to break into the business because people like to work with those they already trust, and it makes it extremely important that someone act ethically when hired for the first time. An easy way to destroy your career potential is to gossip about a show's star or divulge plot lines on your blog.

The temptation for unethical practices is great within the field. Hollywood has **temptations** always had a reputation for living outside the bounds of morality. The entertainment business is one in which a few people can get rich very fast, legitimately. Many other people who want to get rich fast are walking the thin line between right and wrong. This does not justify unethical behavior, but it makes it seem more acceptable because "everyone is doing it." An aspiring actor who knows that a competitor received a part in a movie because his uncle is a friend of the producer may feel justified in telling people he once played Hamlet, even though it was just a short scene in his high school English class. Movie studios sometimes "overestimate" their box office take to appear to beat the competition. One studio created a fake critic and several fake fans to hype its movies. Authors pay people to review their books on amazon.com.[7]

murkiness

quick decisions

pressure

thin line

Ethical decisions are not clear-cut. Deciding to do something may harm one segment of society while deciding not to may harm another. For example, condom advertisements on TV are opposed by people who say the ads promote promiscuity and favored by people who say stations' refusal to air the ads hinders progress in the control of teenage pregnancies and AIDS.

Many media-related ethical decisions, especially those concerning broadcast journalism, must be made quickly and without complete information. When a person holding hostages calls the news director of a station and threatens to kill the hostages unless allowed to broadcast demands immediately, the news director does not have time to read the ethics manual or hold a meeting with the station's top management. The news director does not know what the demands will be. Perhaps they will involve the threat to kill more people, thus making the situation worse. Providing airtime may lead others who are unbalanced to demand airtime; however, if the airtime is not given, the hostages may be killed. This is a no-win ethical situation.

Exhibit 12.4

Jessica Lynch, who was taken prisoner and then rescued in Iraq, was pursued by many to give interviews and tell her life story. She refused most of the offers, preferring to live a more private life.

(Mark Reinstein/POL/Globe Photos)

Sometimes important decisions must be made in situations where there is a great deal of pressure. A network pulled an episode of *Law & Order* after Procter & Gamble canceled its advertising on the show because it didn't like content that dealt with the accidental shooting of a child. Responsible programming or caving into pressure? Sometimes the pressure comes internally—for example, if a news department needs a scoop. After Army Private Jessica Lynch (see Exhibit 12.4) was rescued in Iraq, she was a big "get" for all news organizations. Although news organizations appeared to steer clear of offering her actual money, they were not above tempting her through movie or book deals with businesses that were within their corporate families.[8]

Sometimes there can be a thin line between an ethical slip and a joke or a promotion campaign. For example, in 2004, when filmmaker M. Night Shyamalan had a new film, *The Village,* coming out, he and the Sci Fi Channel tried to develop a marketing stunt by creating a documentary that supposedly contained information about Shyamalan's life that he did not want revealed. They issued press releases about a rift between the filmmaker and the documentarians when, in reality, there was no rift or revealing information. It was all a hoax.[9]

Ethical decisions can also vary depending on time and geography. The ethics involved in deciding whether to air a program on child pornography at 3:00 in the afternoon could be quite different from the factors that go into deciding whether to air that program at 11:00 at night. A news story about a man in St. Louis who has unusual ways of extracting sexual favors from women might be exploitatively titillating when aired in Denver. It might even cause copycat crimes. If aired in St. Louis, however, the story can help protect women by letting them know the man's procedures.

time and place

Ethical values change over the years. When *The Mary Tyler Moore Show* was proposed in 1970, some people wanted to make Mary a divorced woman. Others thought this would be unethical because it would glamorize divorce, so Mary remained single. In 1991, Murphy Brown, a divorced character, announced she was going to have a baby out of wedlock. Obviously, ethical standards had changed during these two decades.[10]

Technological advances make for ever-changing ethical dilemmas because they make possible things no one had thought of previously. For example, computer technology now allows for parts of a picture to be removed and something else inserted, a procedure that couldn't be done in times past. When CBS was covering the millennium celebrations, it placed a CBS "virtual" billboard into Times Square—replacing a real NBC sign. Was this an ethical breach, a prank, or a creative use of technology? Technologies that have been around for a long time, such as simple editing, also have ethical ramifications. Showing a cutaway of someone chewing nervously on a pencil can give a news story a different slant than showing the same person smiling.[11]

technology

Something may appear unethical to a certain group of people that genuinely appears ethical to the person engaged in it. For example, a salesperson may sell a package of station ads to a particular diet company. This diet is attacked by a group of health food store owners as being useless or even harmful. Both the person selling the ad and the station have a great deal to lose monetarily if the diet commercial is removed. In addition, the salesperson may have used the diet and believes it works well. The health food store owners may be against the diet because it takes business away from them. Continuing to air the ads, although this may appear unethical to many, may at worst be confusing.

mixed opinions

Entertainment companies are businesses and have an obligation to shareholders to maximize profits. Given that violence sells, movie producers sometimes highlight a violent scene in a movie trailer even though that may be the only violence in the movie. Is this a deception or just clever marketing that will sell tickets and thus enhance profits?

shareholders

Sometimes conflicting loyalties affect ethical decisions. A reporter who learns of infidelity on the part of a political candidate can find that the decision about whether or not to report the information cuts across loyalties to him/herself, the employer, the audience, a political party, the nation, and an ideological stand. If the reporter releases the information, his or her stature as an investigative journalist will increase, the station or network involved can boast of a scoop, and the audience will be informed about a character trait of the candidate. The reporter, however, may be a member of the same political party as the candidate and may believe the nation will suffer if this candidate is not elected. The reporter may also think that private sex life has nothing to do with the ability to govern.

conflicting loyalties

Ethical decisions are rarely easy, but being unable to resolve an ethical issue doesn't mean you shouldn't grapple with it. See how you and your classmates would handle the ethical dilemmas posed in Exhibit 12.5. Sometimes hindsight

ethical decisions

Exhibit 12.5
Here are a few examples of situations that involve ethical decision making. Read them, decide what you would do, and discuss them with others. The situations cover a variety of aspects of the communications business because no facet of the business is exempt from ethical problems.

Ethical Problems

1 You are the producer for a potential new TV series call *The New Hit*. You have talked to the network about this show and indicated several leading ladies that you think you could obtain. The network executives responded most favorably to Betty Bigname, so you make a concerted effort to hire her. But her agent tells you there is no way; she is overcommitted. You pursue your second choice, Mary Mediumname. She is available and eager for the part. The network gives you the go-ahead to make a pilot with Mary Mediumname, but you can tell the executives are not as excited about the series as they were when they thought Betty Bigname would star. But you sign Mary to a contract and begin planning the production. One day Betty Bigname's agent calls and says the major movie Betty was to star in fell through, so she is now available for your series. You know that legally you can buy out Mary—pay her what she would have earned but don't use her. Since she really wants the part, she will probably drive a hard bargain. You would then be paying for two leading ladies. In addition, Mary's public relations firm issued a great deal of publicity saying she will be starring in this pilot. What do you do?

2 You are the salesperson for radio station KICK. Last month you sold a large package of ads to the Goody Food Restaurant. This earned you your biggest commission ever. You want Goody Food to buy an even more lucrative package of ads, but the manager is stalling because she doesn't think the ads she bought did much good. You hit on an idea. You could call some of your friends and have them make reservations at Goody Food, making sure to mention that they heard an ad on KICK. You could then call the manager and ask again if she wants the ads. Of course, your friends could later cancel their reservations, but not until you had the ad deal signed. Should you do this?

3 You are the assistant to the president of an internet company. At a company party the president, to whom you are somewhat attracted, becomes a little drunk and asks you to go out with him/her. What should you say? A little later in the evening, after he/she has had a few more drinks, he/she asks you for a ride home. What should you do?

4 As a news reporter, you are covering the death of a famous rock star who died under mysterious circumstances. The other members of the rock group, who probably know how the rock star actually died, will not talk to reporters. The anchor at your station suggests you pose as someone from the coroner's office and see if you can get the rock group members to talk to you. How would you react to this suggestion?

5 You are producing a commercial for a vitamin company. One suggestion for the production is to juxtapose a shot of the vitamin bottle next to a bodybuilding champion. You have no proof that taking the vitamins will make anyone look like this champion, so you will not say anything on the audio track to imply or disclaim a relationship between the two. Is it all right to produce the commercial this way?

6 You and a friend made a video in which you are horsing around, cracking off-color jokes, and pretending you are co-hosts of a late-night talk show. When you watch your video, it is actually funny, so you put it on YouTube. It is starting to gain a following, and you have even had an inquiry from the booking agent of a stand-up comedy club where you have been wanting to appear. But the next day your friend pleads with you to take the video off YouTube. She has applied for a job as an elementary school teacher and fears the principal who interviewed her will search the internet, find the video, and decide not to hire her. Should you take down the video, try to convince her that the principal probably won't find it and that it won't hurt her if it is found, tell her your career is more important than hers, or find some other solution?

makes ethical decisions look easy. But because of lack of facts, peer pressure, the desire to succeed financially, and genuine differences of opinion, deciding what is right in a particular circumstance can be difficult. Ethics is basically doing the right thing from a moral point of view when decision making is required. In fact, sometimes good ethics is nothing more than good manners.

12.3 Effects of Media

The influence that media exert on our society is extensive. The mere ability to communicate instantaneously affects the process of communication. Beyond this, the permeation of opinions, emotions, and even fads can often be attributed to various elements of the media.

influence

Sometimes communications practitioners try to downplay the effect they have on society, especially when they are under attack for their violent programming. They say that no one takes what they report in the news, dramatize in movies, or show in a video game seriously enough for it to affect anyone's actions. However, they weaken their own argument when, in the next breath, they tell advertisers how effective commercials are at getting people to buy products.

A number of people have blamed media for deaths. The families of victims of the Columbine High School shootings filed a lawsuit against several companies that create video games and sex-oriented websites. They cited a home videotape of the teenage shooters with one holding a shotgun he called "Arlene," after a character in the *Doom* video game. In several court cases, indicted murderers used "television intoxication" as a contributing factor in their defense. They claimed they committed brutal acts because of things they saw on television.[12]

accusations

Not all media effects are considered bad, however. Radio and television programs cause people to evaluate social problems. Whether the evaluations lead to the proper conclusions depends on one's point of view, but electronic media expose the issues. Portions of documentaries and dramas have been played before Congress to obtain funding for such causes as AIDS and the homeless. Showing footage of starving children in Africa helped develop a climate for approving humanitarian aid.

good effects

One problem with determining the effects electronic media have on individuals is that it is essentially impossible to isolate media from other aspects of society. Gone are the days when researchers could find a group of people who never watched TV to compare with a group that did. Many other aspects of society—family, church, school, friends—reinforce, influence, or reflect effects that can also be attributed to media. Despite this, many people and organizations investigate the effects of electronic media and have deep-seated opinions about these effects.

ubiquity

12.4 Organizations That Consider Effects

As already indicated, the communications industry considers the effects of its programming through self-imposed codes and standards. Government agencies also consider effects—Congress through its hearings on such subjects as violence and news, and the FCC through its regulations such as those dealing with obscenity. Critics who write reviews of TV programs, movies, and video games for newspapers and magazines sometimes address both positive and negative effects that they think the materials will create. Two other entities that are involved with effects are citizen groups and academic institutions.

12.4a Citizen Groups

TV-focused

Many national citizens' organizations are critical of the role of media in society. Some of these are organized specifically to affect television programming. For example, Action for Children's Television (ACT), founded by a group of Boston parents, had as its mission the improvement of children's programming. Media Research Center's main objective is to neutralize the impact of what it sees as the media's liberal bias on the American political scene.

other focuses

Other organizations have their own larger agendas but see the media as having a major role in helping them reach their goals. For example, the National Organization for Women (NOW) is vocal about the portrayal of women on TV and about the sexist lyrics of songs heard on the radio. The National Association for the Advancement of Colored People (NAACP) speaks out about the portrayals of minorities on TV. The Parent-Teacher Association (PTA) and the American Medical Association (AMA) both have objected to violent programming. The left-leaning organization MoveOn.org applauded Michael Moore's film *Fahrenheit 9/11* (see Exhibit 12.6) and its stated purpose of helping to get President Bush out of the White House, while right-leaning organizations condemned it.[13]

12.4b Academic Institutions

purpose of studies

Much of the research on the effects of electronic media comes from professors at colleges and universities (see Exhibit 12.7). These studies are scientific and supposedly unbiased and are usually undertaken with the intent of adding to the overall basic knowledge of the effects of media on society. Citizens' groups, broadcasters, and government agencies often use the studies to support their points of view or guide their actions.

Academicians use a wide variety of techniques when they undertake research. Some studies are conducted in a **laboratory** setting (people might be brought to a university classroom where they are given a pretest about their feelings toward Native Americans, shown a movie about Native Americans, and then given a posttest on their feelings); others are conducted in the **field** (researchers might interview family members and observe them watching TV in their own homes to determine whether fathers, mothers, or children are most likely to decide what TV programs are watched). Sometimes they involve **surveys** (radio station programmers

Exhibit 12.6

Michael Moore's film *Fahrenheit 9/11* galvanized political sides.

(ES/Globe Photos, Inc.)

might be sent questionnaires about their websites that are then analyzed to determine, among other things, whether music stations or news stations are more likely to have chatrooms). Research can also involve **content analysis** (researchers might view video games and count the number of male versus female characters).

As with audience measurement (see Chapter 10), academic research can be divided into **quantitative** and **qualitative.** In a quantitative research study, a researcher might count the number of times each family member pushes the remote control button to change the channel. The number of button pushings could then be studied statistically to compare it with other factors, such as family income. The researcher's **hypothesis** that children are more likely to control what is seen on TV in high-income families than in low-income families could then be mathematically substantiated or rejected. Qualitative research would take place, however, if the researcher interviewed the families and obtained anecdotal information that led to the conclusion that children are more likely to decide what the family watches on TV when they are tired and grumpy than when they are in good spirits.

Research is often biographical, historical, or aesthetic in nature. A biographical study of Senator Joseph McCarthy's relationship with media before he started his **blacklisting** inquiries might give insight into actions that could prevent this type of activity from recurring. A historical study of detergent commercials through the years can reveal a great deal about changing social mores. A study of the films of John Ford can uncover particular characteristics of his directing techniques. Some studies try to determine **cause and effect:** does listening to antipolice lyrics cause people to lower their opinions of the police? Others look at **uses and gratifications:** what are the primary reasons people use cell phones—convenience, mobility, relaxation?[14]

Academicians also develop models to explain media functions and effects. Exhibit 12.8 shows one such model. It illustrates communication as an ongoing process that includes a source, a message, a channel, a receiver, barriers, and feedback. The **source** includes the people who decide what is sent out to others. In the case of the internet, anyone in the world with access to a computer can act as a source. For many media forms, such as movies, the source is a group of people—actors, directors, writers, cinematographers, producers. Some people are more important than others in terms of their influence on what is communicated. For example, the people who decide which news will be seen on CNN have more influence than the camera operator for a corporate video on fixing a carburetor. People, such as news directors, who decide what a large number of other people will see or hear are often referred to as **gatekeepers** (see Chapters 1 and 8).

Volume 52, Number 3, September 2008
ISSN: 0883-8151

Journal of Broadcasting & Electronic Media

Personality Traits, Television Viewing, and the Cultivation Effect

A TV in the Bedroom: Implications for Viewing Habits and Risk Behaviors During Early Adolescence

The Relationship Between Exposure to Sexual Music Videos and Young Adults' Sexual Attitudes

What is the Appropriate Regulatory Response to Wardrobe Malfunctions? Fining Stations for Television Sex and Violence

The Role of Motivation and Media Involvement in Explaining Internet Dependency

Understanding the Interactive Effects of Emotional Appeal and Claim Strength in Health Messages

Visual Framing of the Early Weeks of the U.S.-Led Invasion of Iraq: Applying the Master War Narrative to Electronic and Print Images

How Gender and Age Affect Newscasters' Credibility— An Investigation in Switzerland

Cultural Proximity and Audience Behavior: The Role of Language in Patterns of Polarization and Multicultural Fluency

Book Reviews

R Routledge
Taylor & Francis Group   **BEA**

Exhibit 12.7

Journal of Broadcasting & Electronic Media is one of the main academic journals that publishes the research of college professors in this field.

techniques

quantitative and qualitative

other types

source

Exhibit 12.8

Communication
model.

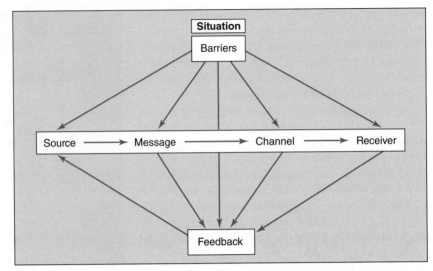

The people who act as the source are concerned with sending a **message** that has a purpose. Some messages are intended to inform, some are designed to entertain, and some are planned to persuade. Messages are transmitted through **channels.** For electronic media, some of the most common channels are radio, broadcast and cable TV, the telephone, and the internet. Various devices and processes can influence the effectiveness of a channel. For example, instant replay, which is possible on the TV medium, enables sportscasting to communicate to the fans in ways that other channels, such as radio, cannot.

The **receiver** of the message is the audience. For some forms of telecommunications, primarily broadcast TV, the audience is a composite of individuals that is generally large, heterogeneous, and fairly anonymous. Cable systems that **narrowcast** are seeking a more homogeneous audience. For corporate video, the receiver might be even narrower—for example, the people who sell a new line of women's clothes. For email, the receiver is often only one person.

Many **barriers,** sometimes referred to as **noise,** can obstruct the communication process or reduce its effectiveness. The source can have built-in physical or psychological barriers. For example, if all the people making a movie live and work in California, they may not have a proper frame of reference to design a product that will communicate to people in Vermont. Messages also have barriers. A poorly written joke may not entertain. An informational piece delivered in Spanish will not communicate to someone who does not understand that language. A campaign speech intended to persuade might only inform. A message can be distorted simply because information presented on radio or TV takes on added importance.

The receiver, too, is affected by many barriers. Some of these are in the receiver's background or environment. For example, a person who has lived in France might perceive a program about French politics in an entirely different way from a person who has never even visited France. A person playing a video

message

channels

receiver

barriers

game in a hot room might react differently from someone playing in a cold room. Interruptions or distractions, such as a child crying, can also be barriers for the receiver.

The message, channel, and receiver all serve as foundations of **feedback** to the source. For example, if no one laughs at a joke during rehearsal, this can be a signal to the creator that it is not funny. Managers of affiliated radio and TV stations often express their opinions to the people who provide the network programs. Email involves frequent feedback, with the source and the receiver switching roles as each exchange occurs. Through ratings, fan letters, and online chat groups, receivers can alter the decisions made by the source, as is evidenced by the cancellation of shows with poor ratings. There are also barriers to feedback. People can laugh out of politeness, not bother to complain, or fill out a rating form incorrectly.[15]

feedback

12.5 High-Profile Effects

Individuals, organizations, and academicians consider many topics related to the effects of communications—how and why people use the internet to acquire political information, the puritanical impulses of David Lynch's films, the role of women in radio during World War II, gay and lesbian visibility on TV, partisan balance in local TV election coverage, whether the slim women shown in commercials cause feelings of inadequacy that lead to anorexia, why reality shows have an appeal, how sports fans differ from fans of other forms of programming.[16]

Some of the subjects related to possible effects of communications are discussed and studied more than others. Primary among these are violence, children and media, news, women and minorities, and sex.

subjects

12.5a Violence

Violence in media is highly criticized, especially when the violence seems gratuitous. Movie previews on DVDs are deemed too violent. The internet comes under attack for websites that teach how to build bombs and execute violent acts. The violence depicted in music videos is decried, especially since their target audience is the age group that instigates much of the violent behavior in society. The same goes for violent lyrics played on radio and violent video games. There is debate as to how much violence news programs need to show. Sports programming is accused of inciting fired-up men to beat up on their significant others; battered women's crisis centers report that for victims of domestic violence, Super Bowl day is one of the worst days of the year.[17]

examples

Congress has investigated violence many times. As far back as 1950, Senator Estes Kefauver suggested in the U.S. Senate that there was too much violence on TV. A major outcry arose in 1963 after the assassination of President Kennedy with claims being made that violence on TV had led to the possibility of assassination. In 1972, the Surgeon General issued a report stating that the causal relationship between violent TV and antisocial behavior was sufficient to merit immediate attention. Hardly anyone paid immediate attention, but a growing protest against violence reached a crescendo in 1977, leading to network program changes that led to diminished protest. During the 1980s, the subject of violence

history

was resurrected by Reverend Jerry Falwell, head of the Moral Majority, and Reverend Donald Wildmon, who organized the Coalition for Better Television. They took credit for the fact that several advertisers canceled sponsorship of violent programs. In 1992, the Senate and the networks reached an agreement under which the networks would receive a three-year exemption from antitrust laws in return for curbing violence on the tube. The networks did very little, so in 1993 Capitol Hill was once again summoning TV network leaders asking them to void their programs of violence. The networks (both broadcast and cable) responded by agreeing to put violence warnings on the air before showing programs that they deemed to be heavily violent. But very few shows were so deemed. In 1996, Congress passed legislation that mandated V-chips in new television sets so that parents could block out violence. Very few people use these chips, and violence complaints continue. In 2007, the FCC issued a report suggesting powers that Congress could give the FCC, so it could curb violence, but Congress has done little to implement the recommendations.[18]

measurement problems

Innumerable studies have been conducted concerning violence, but they have built-in problems. Measuring violence is not like measuring cups of sugar. Is pushing someone in front of a runaway cactus the same violent act as pushing someone in front of a car? Is it violent for one cartoon character to push another off a cliff when the one pushed soars through the air and arrives at the bottom with nothing injured but pride? Should a heated argument be treated the same as a murder? Is it worse to sock a poor old woman than a young virile man? Should a gunfight be considered one act of violence, or should each shot of the gun be counted?

indexes

Despite all these measurement pitfalls, indexes abound in an attempt to tell whether media violence is increasing or decreasing. One of the oldest violence-measuring systems was developed by Professor George Gerbner of the University of Pennsylvania. He trained observers watching one week of TV fare a year to count acts of violence according to his complicated formula. The networks took offense at some of the acts he called violent, such as comedic violence, and commissioned a study through UCLA. Neither of the academic institutions gave the networks much credit for curbing violence. The National Citizens' Committee for Broadcasting, somewhat with tongue in cheek, developed a "violence index" that calculated how many years each network would spend in jail if convicted of all the crimes it portrayed in one week—the range was from 1,063 to 1,485 years.[19]

theories

Measurement is not the only pitfall connected with violence. The effect of violence on society is also debated and hard to determine. Some people who subscribe to the **catharsis theory** say it can be good for people to watch violence. Watching it can provide a vicarious thrill that gives people the excitement they need in their lives and stifles their urges to actually act in a violent manner. Other people hold to the **observational theory**—violent programming incites people to commit crimes and shows them how to do it. They point to numerous copycat crimes in which someone has committed a crime very much like one shown in a movie or on TV. Some researchers have investigated the **mean world syndrome**—the fear that people have about undertaking their daily tasks because they think, from watching TV and movies, that the world is a much more violent place than it actually is. Still others look at the **desensitizing effect**—watching violence makes people think that actual violence isn't such a bad thing (see "Zoom In" box).[20]

The bottom line is that violent programming sells. People watch it, and advertisers buy time on it. Networks say they are just giving the people what they want and that stifling violent programming is against the First Amendment. The government and many individuals and organizations believe that media program suppliers should invoke their collective ethical conscience and keep violent material off the channels.

points of view

12.5b Children and Media

Much of the discussion and research related to children and media involves violence. The people and organizations that speak out against violence in general

ZOOM IN: Violence Results and Counter-Results

Many research projects and surveys about violence have been conducted, the findings of which are generally severely (perhaps even violently) challenged by both friends and foes of TV fare. Some of the results are enlightening.

1. People who watch violence on a large screen show greater tendencies toward aggressive behavior than do those who watch on a small screen.
2. The inclusion of humor in a program dampens the tendency toward aggressive behavior on the part of the viewer.
3. Cartoons have more acts of violence in them than do dramas.
4. Only one in six acts of violence on TV is punished while one in three is actually rewarded.
5. Violence is overrepresented on reality-based police shows; about 87 percent of the shows' crimes are violent whereas the FBI classifies only 13 percent of the nation's crimes as violent.
6. The more frequently children watch TV at age 8, the more serious are the crimes they are convicted of by age 30.
7. The average person watches 500 murders a year on TV.

8. There is little difference between news programs that contain only nonviolent news and ones that contain both violent and nonviolent items in terms of increased inclination toward aggression on the part of viewers.
9. The more children identify with violent characters in a program, the greater is their inclination toward aggression.
10. There is little correlation between what children consider to be violent acts and what mothers consider to be violent acts.
11. Four out of 10 people believe that watching violence is harmful to the general public and to children in particular.
12. Hardly anyone believes that watching violence hurts him or her personally.

usually are especially vocal about the negative effects violence has on children. As already indicated, the scientific research yields conflicting results. The preponderance of research, however, indicates that media violence is bad for children, especially those already predisposed toward aggressive behavior.

cartoon study

In one study, sixth graders were divided into three groups. One group acted as a control and was not shown any television. The other two groups were shown a *Woody Woodpecker* cartoon in which Woody knocks a man unconscious. One group was simply shown the cartoon; the other was told to think about the feelings of the man while they were watching. The children who were told to think about feelings liked the man better and thought the violence was less funny than either of the other groups.[21]

violent movies

Another study found that violent movies attract on average 12.5 percent of the country's 10- to 14-year-olds. Those most likely to watch these gory movies are boys, minorities, and children whose parents don't restrict their viewing. As movies become more accessible because they are distributed over multiple platforms, young people find it easier to view them. For example, a Federal Trade

 ZOOM IN: **Kids and TV**

Here are some other results of research involving children and media.

1. Children do learn reading and vocabulary from TV, but the children who watch TV the most are the ones who do poorly in reading in school.
2. Watching TV generally decreases book reading, but certain TV programs that refer to books actually increase book reading.
3. Children three and younger understand very little of what they watch on TV, yet they sometimes sit mesmerized before the set.
4. Children with TVs in their rooms but no computer at home score low on math and language tests; those with no bedroom TV but with home computers score the highest in those areas.
5. White (but not black) adolescents who have a TV in the bedroom are at greater odds of initiating heath risk behaviors, such as cigarette smoking and sexual intercourse, than are white adolescents who do not have a TV in the bedroom.
6. When asked to name their favorite TV character, almost all boys select a male, but only half the girls select a female.
7. It is difficult for children to understand time leaps in television programming, but children who watch the most TV understand them the best.
8. More TV ads are aimed at boys than at girls.
9. Children develop attitudes about alcohol based on what they see on TV, but these attitudes are also formed by what their parents do.

Exhibit 12.9
The interactions of the cast members of *Sesame Street* were well researched, even before the show went on the air.

(Courtesy of © Children's Television Workshop)

Commission report indicated that 40 percent of children 13 to 16 were able to purchase R-rated movie tickets at theaters, but 80 percent were able to buy DVDs.[22]

A great deal besides violence is studied in relation to children and media, given that an estimated 4,000+ studies have dealt with the subject (see "Zoom In" box). For example, much research was conducted to assess the learning outcomes of *Sesame Street* (see Exhibit 12.9) when it first aired and even before it went on the air. The pattern of testing to see if a program achieves its goals has been followed with many other children's TV series.[23]

learning outcomes

parents

Another area that is frequently discussed concerning children and media is the role of parents. Network executives hold that parents are the ones who should control and influence their children's use of media, but parents point out that they are not always with their children when they are using media and that radio and TV networks and stations and internet sites should take responsibility for programming material that is healthy for children's minds and emotions. The government gets in and out of the middle of this debate in fairly regular cycles, regulating and deregulating aspects of children's programming and commercials. Research studies that attempt to assess the extent to which parents control what their children listen to and watch show that such involvement is not universal. However, they also reveal that if parents talk to their children about TV, they can improve their children's ability to learn and shape their attitudes toward violence and other aspects of TV programming.[24]

Electronic media are very much a part of the home. As such, they touch the lives of children at least as much as those of adults. The effects of media on children have been subjected to a great deal of research and dialogue, but thought and discussion regarding these effects are bound to continue because of the importance of the values and behavior of future generations.

12.5c News

Much of the research and discussion concerning news is tied to the types of stories shown on the news. People decry sensationalism, such as the showing of dead bodies and flaming automobiles. Other forms of sensationalism also come under attack, such as barging in on grieving relatives or reporting details of rapes. Individual news organizations have policies regarding the depiction of violent and sensational events, but many show a great deal of this material, often at the expense of weightier public-interest subjects such as international relations.[25]

sensationalism

bias

Bias is another concern. Conservatives believe that news is a product of the eastern liberal establishment, and liberals decry the conservative commentators, especially those on radio. Studies show that the nature of radio and TV news presentations can bias audience members, but the studies of evidence of bias are murkier, with many of them concluding that newscasts, as a whole, are not biased toward any particular point of view. As with violence, research techniques used for bias studies have their shortcomings. What one researcher considers a biased story, another might give a clean bill of health.[26]

uses and gratifications

A fair amount of news research is in the uses-and-gratifications category, attempting to learn how and why certain groups of people (senior citizens, working women) use various media forms (cable TV, radio, the internet) to obtain news. This type of research can also explore the impact of different types of stories (serious versus light, long versus short).[27]

presentation

Somewhat related is research that explores the effects of various forms of news presentation. Do people remember more about a story if it contains graphics than if it is just read by an anchor? Does the gender or age of a newscaster affect perceived credibility? Do people pay more attention to stories that are preceded by a teaser than stories that are not?[28]

Many research reports hit on a number of topics. For example, one that involved violence, children, and news was a study of how children viewed news during the

1991 Persian Gulf conflict. This study surveyed third to sixth graders about their interest in war-related news, paying particular attention to whether they had watched neutral background news or the more violent casualty news. The study had elements of both cause and effect and uses and gratification. It found that children who were more upset (cause) during the war were more interested in neutral background news and avoided exposure to news coverage (effect) more than children who were not as upset. The study also found that the main reason children watched (used) casualty news was to be informed about the war (gratification) as opposed to a need for reassurance or a desire to be entertained. The study also found that girls were more interested in casualty news than were boys.[29]

violence, children, and news

12.5d Women and Minorities

Also receiving extensive consideration regarding the effects of media is the area of women and minorities. Some scholars study the ways that women and minorities use media. One study, for example, found that when women are in a bad mood, they prefer media (specifically, music) that are not involving or absorbing so that they can think about their stress, whereas men tend to distract themselves with high-absorbing messages.[30]

use

Most of the research, and the controversy, about women and minorities involves how these groups are portrayed on television and in films. Highly criticized is the portrayal of minority groups that conveys negative **stereotypes**—the doddering old man, the lazy Mexican American, the servile African American, the brawny lesbian, the subservient wife. Largely as a result of criticism and research, radio, TV, and movies have corrected many of their blatant stereotypes. For example, studies of the 1970s found that very few African Americans were portrayed as professionals or managers; studies of the 1990s found that African Americans and European Americans were equally likely to be portrayed in those categories.[31]

portrayal

The issue of stereotypes still surfaces, however, especially in regard to situation comedies. In this form, just about everyone is made fun of—for the sake of a laugh—but sharp barbs at the expense of minorities are not appreciated. Another problem is that old movies from the 1940s and 1950s often appear on television or can be rented at the video store, and they readily exhibit old-fashioned stereotypes.

Another issue related to women and minorities is that they are underrepresented. This criticism often emanates from Chicano, Asian American, Native American, African American, female, and physically disabled actors who have trouble finding work. Criticism of underrepresentation, however, also affects the self-esteem and inclusion feelings of viewers. Children, for example, who do not see people who look like them on TV feel left out and are not able to build positive TV role models.[32]

representation

12.5e Sex

An issue closely related to women is the portrayal of sex on TV and in movies. This subject, like violence, is a perennial thorn. Sexual permissiveness, both in society and on the TV screen, has increased over the years. The V-neck dresses that raised eyebrows during the 1950s are considered modest by today's standards. Television drama has broached sexually sensitive subjects, such as homosexuality and incest, usually to the almost instant outcry of critics. But the hue and cry dies down, and the networks go on to challenge another taboo.

perennial issue

type and amount

Both the type and amount of sex depicted draw fire. The barrage of sexual innuendos on sitcoms, the bedroom scenes in movies, the titillating "news stories" about prostitutes, the sexual setups of reality shows, the abundance of sex sites on the internet, the use of sexy women as targets in video games—all are decried. Radio talk shows, which periodically rediscover sex, receive a great deal of criticism and some government regulatory restrictions. After quieting down, the subject once again emerges. The lyrics of rock music aired on radio and the semiclad, sensuous women seen in music videos are criticized by public-interest groups. When sex is combined with violence against women, it receives some of its strongest criticism because of the fear of copycat crimes and a belief that such material desensitizes men in general to act against women.[33]

promiscuity

One of the main criticism of sex centers on the idea that people, seeing so much sex, become promiscuous. Research has born this out. For example, one study showed that exposure to sexually explicit music videos was associated with more permissive attitudes toward premarital sex and stronger endorsement of the sexual double standard, regardless of gender, overall TV viewing, or previous sexual experience.[34]

safe sex

In addition, researchers theorize that the sex shown on TV and in movies is idealized and makes people feel cheated or inadequate in relation to their own sex lives. These media enhance sexual expectations that reality does not meet. Showing (or even talking about) sex can be negatively instructive, teaching young children things they shouldn't know. Studies also deal with the depiction of safe sex. For example, in one research study, one group of college students was shown a program with sexual intercourse that depicted the use of condoms, another group was shown a program with sexual intercourse but no condoms, and the third group was shown a program with no sexual content. The women who saw the show with the condoms then had a more positive attitude toward condom use than the women in either of the other groups, but there was no significant difference among men in the three groups.[35]

decline

Despite the outcries against sex on TV, it, like violence, continues because sex sells. There are positive indications, however, that the media are dealing with sex somewhat more responsibly. A content analysis of programs most heavily viewed by teens during the 2001–2002 season and the 2004–2005 season showed a decline in sexual content (see Exhibit 12.10).[36]

Exhibit 12.10

This table shows some of the results of one study related to differences in sexual content of network programs often watched by teenagers.

Sexual Content		
	2001–2002	2004–2005
Programs with any sexual content	83%	70%
Programs with talk about sex	80%	68%
Programs with any sexual behavior	49%	45%
Programs with precursory sexual behaviors (e.g., flirting) only	29%	37%
Programs with sexual intercourse	20%	8%

12.6 Issues and the Future

Most issues related to ethics and effects will be around for a long time. Naysayers predict that ethics are on the way out—or that they have already disappeared. Those among us who have faith in human nature and in future generations believe that people will always be faced with ethical dilemmas but that most people will make the right decisions for the right reason most of the time.

One issue related to ethics that periodically fluctuates is the relationship between regulation and ethics. Sometimes the government is in the mood to regulate actions that at other periods are considered individual ethical issues. Children's TV is a good example. On occasion the government has laid down a strict list of rules for stations; at other times it has left the decision of how to program for children up to the conscience of program producers and executives (see Chapter 8). Ethical issues that are likely to undergo regulation and deregulation in the future include music lyrics on radio, the amount of titillating content in all media, and, of course, violence.

regulation

Of all the issues affecting communications, violence generates the most heat—and no doubt will continue to do so. As in the past, violence will figure in debates about how to program for children, whether to cut down on news sensationalism, and how to depict women in movies and video games. Sexual content, in terms of both talking about sex and showing sexual activity, is also a major issue.

violence

The influence of citizen groups on the media has been a roller-coaster phenomenon. During the 1960s, the groups had a great deal of power, but their influence waned in the 1980s, in part because they lost regulatory battles and in part because people seemed to tire of their watchdog role. However, some current grassroots groups are concerned enough about media effects to try to force reform.

citizen groups

The government, citizen groups, and academicians are turning increased attention to the ethics and effects of computer services. The number of pornographic sites on the internet has caused watchdogs to become alarmed, as has the fact that people using chat groups sometimes lure children into sexual encounters.

computer services

Academic researchers are always looking for new areas to explore, and the newer media lend themselves to unexplored topics. Satellite radio, high-definition TV, personal video recorders, internet phones, and iPods are all worthwhile topics for future research. In addition, researchers will continue to study further aspects of the array of subjects that they presently study, including violence, news, children, women and minorities, and sex.

future studies

12.7 Summary

Ethics and effects are intertwined. Someone who programs a TV network and knows that violence appears to affect children negatively must consider that fact when making programming decisions. The people who develop self-regulation codes make provisions for avoiding news bias and negative stereotypes mainly because their sense of ethics leads them in that direction. People who enforce broadcast standards watch for gratuitous sex and violence in part because it is their job and in part because they want to further ethical standards.

Unfortunately, ethical decisions are not always easy. They are hampered by temptations, lack of clear delineation, pressures, technological changes, and variations over time and space. Citizen groups often have one opinion of right and wrong while industry groups have another.

Academic research, both quantitative and qualitative, helps delineate negative and positive effects and can help with ethical decisions. Despite measurement shortfalls, a great deal of research is conducted on the issue of violence, with the catharsis, observational, mean world, and desensitizing theories receiving study. Research on children and TV also encompasses the effects of violence, while other studies and comments focus on the responsibility of parents. News research centers on sensationalism, bias, and presentation techniques, but it, too, has violent elements. Research on minorities and women has led to changes in TV and movies that favor these groups. Research on sex shows negative effects and some positive progress.

Models, most of them formed by academicians, also aid in pinpointing effects. They can, for example, show the role of gatekeepers. Although many resources are available to help people understand effects and make appropriate ethical decisions, in the end it is the innate character of each individual that really counts.

Suggested Websites

www.kff.org/entmedia (the Kaiser Family Foundation, which sponsors research related to media)

www.mediaresearch.org (Media Research Center, a conservative watchdog organization)

www.naacp.org/home/index.htm (National Association for the Advancement of Colored People [NAACP])

www.now.org (National Organization for Women [NOW])

www.rtnda.org (Radio-Television News Directors Association)

Notes

1. Radio-Television News Directors Association, "Code of Ethics and Professional Conduct, http://www.rtnda.org/pages/media_items/code-of-ethics-and-professional-conduct48.php?id=48 (accessed September 25, 2008).
2. "Acad Voters Catch a Code," *Daily Variety,* September 4, 2003, 1.
3. "NBC Bashed for Unleashing a Killer's Rants on Its Airwaves," *Los Angeles Times,* April 20, 2007, 1; "Finding a Replacement for Imus Won't Be Easy," *Wall Street Journal,* April 16, 2007, B-1; "FCC Investigates Church Coupling; WNEW Axes DJs," *Hollywood Reporter,* August 23–25, 2002, 1; and "'Myopic' Eye Wipes Four," *Hollywood Reporter,* January 11–17, 2005, 1.
4. Philippe Perebinossoff, *Real-World Media Ethics* (Burlington, MA: Focal Press, 2008), 19–21.

5. Some publications that can help with communications ethics are Philip Patterson and Lee C. Wilkins, *Media Ethics: Issues and Cases* (New York: McGraw-Hill, 2001); Larry Z. Leslie, *Mass Communication Ethics: Decision Making in Postmodern Culture* (Boston: Houghton Mifflin, 2000); and Louis A. Day, *Ethics in Media Communications* (Belmont, CA: Wadsworth, 2002).

6. "News for Sale," *Broadcasting and Cable,* January 26, 2004, 1.

7. "Long Soft on Hollywood, Times Seen as Improving," *Los Angeles Times,* February 15, 2001, A-20; and "Untruths and Consequences," *Los Angeles Times Calendar,* June 24, 2001, 5.

8. "Wells Scolds CBS for Pulling 'Law' Episode," *Hollywood Reporter,* August 20, 2001, 2; and "In the Big Interview Hunt, What Exactly Is Fair Game?" *Los Angeles Times,* July 30, 2003, E-2.

9. "Shyamalan Documentary a Hoax, Sci Fi Admits," *Los Angeles Times,* July 19, 2004, E-1.

10. "Few Now Quail at TV's Unwed Moms," *Los Angeles Times,* October 26, 2001, F-1.

11. "CBS' Virtual Logos a Real Pain," *Broadcasting & Cable,* January 17, 2000, 20; and "Do New Technologies Create New Ethical Issues?" *Beyond Computing,* June 2000, 10.

12. "Columbine Families Sue Game Makers, Web Sites," *Hollywood Reporter,* April 24–30, 2001, 101; and "'Sopranos' Scenario in Slaying?" *Los Angeles Times,* January 28, 2003, B-1.

13. "Kidvid: A National Disgrace," *Newsweek,* October 17, 1983, 81–83; "Media Research Center," http://www.mediaresearch.org (accessed September 26, 2008); "President of TV Indecency Watchdog Group to Step Down," *Los Angeles Times,* September 2, 2006, C-1; and "Michael Moore, MoveOn, and Fahrenheit 9/11," http://www.freerepublic.com/focus/f-news/1162291/posts (accessed August 27, 2004).

14. Material dealing with research methodology includes Joy Keiko Asamen and Gordon L. Berry, *Research Paradigms, Television and Social Behavior* (Thousand Oaks, CA: Sage, 1998); and Arthur Asa Berger, *Media and Communication Research Methods* (Thousand Oaks, CA: Sage, 2001).

15. This model is based on Bert E. Bradley, *Fundamentals of Speech Communication* (Dubuque, IA: William C. Brown, 1991), 5.

16. Barbara K. Kaye and Thomas J. Johnson, "Online and in the Know: Uses and Gratifications of the Web for Political Information," *Journal of Broadcasting & Electronic Media,* March 2002, 54–71; Jeff Johnson: Pervert in the Pulpit: The Puritanical Impulse of the Films of David Lynch," *Journal of Film and Video,* Winter 2003, 3–14; Jim Grubbs, "Women Broadcasters of World War II," *Journal of Radio Studies,* June 2004, 40–54; B. J. Dow, "'Ellen,' Television, and the Politics of Gay and Lesbian Visibility," *Critical Studies in Media Communication,* June 2001, 123–40; Sue Carter, Frederick Fico, and Jocelyn S. McCabe, "Partisan and Structural Balance in Local Television Election Coverage," *Journalism and Mass Communication Quarterly,* Winter 2002, 41–53; Amanda J. Holmstrom, "The Effects of Media on Body Image," *Journal of Broadcasting & Electronic Media,* June 2004, 196–217; Zizi Papacharissi and Andrew L. Mendelson, "An Exploratory Study of Reality Appeal: Uses and Gratifications of Reality TV Shows," *Journal of Broadcasting & Electronic Media,* June 2007, 355–70; and Walter Gantz et al., "Sports Versus All Comers: Comparing TV Sports Fans with Fans of Other Programming Genres," *Journal of Broadcasting & Electronic Media,* March 2006, 95–118.

17. Mary Beth Oliver and Sriram Kalyanaraman, "Appropriate for All Viewing Audiences: An Examination of Violent and Sexual Portrayals in Movie Previews Featured on Video Rentals," *Journal of Broadcasting & Electronic Media,* June 2002, 283–99; R. E. Caplan, "Violent Program Content in Music Videos," *Journalism Quarterly,* Spring 1995, 144–47; "Videogame Makers Testify on Violence," *Daily Variety,* December 9, 1993, 8; "Super Violence Comes Home," *Mother Jones,* January 1987, 15; and "Plugged In, but Tuned Out," *Wall Street Journal,* October 6, 2005, D-1.

18. "Simon Introduces TV Violence Bill, *Broadcasting,* March 30, 1987, 148; "The TV Violence Proposal: Let's Get Cynical," *Los Angeles Times,* July 1, 1993, F-1; "TV Violence No Longer a Front-Burner Issue in Washington," *Broadcasting & Cable,* January 23, 1995, 9; "TV Violence in the Cross-Hairs," *Hollywood Reporter,* April 28, 1999, 4; "Lawmakers, Parent Groups Laud Media Violence Findings," *Los Angeles Times,* January 18, 2001, C-1; "Remember the V-chip? TV Guide Cuts Icons," *Daily Variety,* September 16, 2003, 5; and "Violence: The New Indecency?" *Broadcasting & Cable,* January 22, 2007, 18–19.

19. "More Violence Than Ever Says Gerbner's Latest," *Broadcasting,* February 28, 1977, 20; "All That TV Violence: Why Do We Love/Hate It?" *TV Guide,* November 6, 1976, 6–10; George Gerbner, "Reclaiming Our Cultural Mythology," *The Ecology of Justice,* Spring 1994, 40; and Jack Glascock, "Direct and Indirect Aggression on Prime-Time Network Television," *Journal of Broadcasting & Electronic Media,* June 2008, 268–81.

20. Material about results of violence studies includes *National Television Violence Study* (Thousand Oaks, CA: Sage, 1997); W. James Potter and Stacy Smith, "The Context of Graphic Portrayals of Television Violence," *Journal of Broadcasting & Electronic Media,* Spring 2000, 301–23; Nancy Signorielli, "Prime-Time Violence 1993–2001: Has the Picture Really Changed?" *Journal of Broadcasting & Electronic Media,* March 2003, 36–57; and "A New View on TV," *Wall Street Journal,* September 6–7, 2008, 1.

21. Amy I. Nathanson and Joanne Cantor, "Reducing the Aggression-Promoting Effect of Violent Cartoons by Increasing Children's Fictional Involvement with the Victim: A Study of Active Mediation," *Journal of Broadcasting & Electronic Media,* Winter 2000, 125–42.

22. "Study Tracks Kids' Viewing of Violence," *Los Angeles Times,* August 5, 2008, C3.

23. Gerald S. Lesser, *Children and Television: Lessons from "Sesame Street"* (New York: Random House, 1974); and "A Lot to Learn," *Broadcasting & Cable,* November 5, 2007, 16. Articles related to material in the box include Erica Weintraub and Heidi Kay Meili, "Effects of Interpretations of Televised Alcohol Portrayals on Children's Alcohol Beliefs," *Journal of Broadcasting & Electronic Media,* Fall 1994, 417–35; Robert Abelman, "You Can't Get There from Here: Children's Understanding of Time-Leaps on Television," *Journal of Broadcasting & Electronic Media,* Fall 1990, 469–76; Cynthia Hoffer, "Children's Wishful Identification and Para-Social Interaction with Favorite Television Characters," *Journal of Broadcasting & Electronic Media,* Summer 1996, 389–402; Marina Krcmar and Kelly Fudge Albada, "The Effect of an Educational/ Informational Rating in Children's Attraction to and Learning from an Educational Program," *Journal of Broadcasting & Electronic Media,* Fall 2000, 674–89; Christine Jackson, Hane D. Brown, and Carol J. Pardun, "The TV in the Bedroom: Implications for Viewing Habits and Risk Behaviors during Early Adolescence," *Journal of Broadcasting & Electronic Media,* September 2008, 349–67; and "More Studies Say TV Bad for Kids," *Los Angeles Times,* July 6, 2008, B3.

24. Jan Van den Bulck and Bea Van den Bergh, "The Influence of Perceived Parental Guidance Patterns on Children's Media Use: Gender Differences and Media Displacement," *Journal of Broadcasting & Electronic Media,* Summer 2000, 329–48; Lynne Schafer Gross, Susan Plumb Salas, and Cindy Miller-Perin, "Active Guidance: A Study of the Degree to Which Parents Will Engage in Activities in Order to Enhance Their Involvement in Their Children's TV Viewing," *Journal of Research,* Fall 2003, 20–29; and Moniek Buijzen and Patti M. Valkenburg, "Parental Mediation and Undesired Advertising Effects, *Journal of Broadcasting & Electronic Media,* June 2005, 153–65.

25. Travis L. Dixon, Cristina L. Azocar, and Michael Casas, "The Portrayal of Race and Crime on Television Network News," *Journal of Broadcasting & Electronic Media,* December 2003, 498–523; and Davis Buzz Merritt, "Missing the Point," *American Journalism Review,* July/ August 1996, 29–31.

26. D. Charles Whitney et al., "Geographic and Source Biases in Network Television News, 1982–1984," *Journal of Broadcasting & Electronic Media,* Spring 1989, 159–74; and Renita Coleman, "The Intellectual Antecedent of Public Journalism," *Journal of Communication Inquiry,* Spring 1997, 60–76.

27. Thomas F. Baldwin, Marianne Barrett, and Benjamin Bates, "Uses and Values for News on Cable Television," *Journal of Broadcasting & Electronic Media,* Spring 1992, 225–33; Carolyn Johnson and Lynne Gross, "Mass-Media Use by Women in Decision-Making Positions," *Journalism Quarterly,* August/September 1985, 850–53; Henry Grunwald, "Opening Up 'Valleys of the Uninformed,'" *Media Studies Journal,* Fall 1993, 29–32; and David Tewksbury, "The Seeds of Audience Fragmentation: Specialization in the Use of Online News Sites," *Journal of Broadcasting & Electronic Media,* September 2005, 332–48.

28. Hao-chieh Chang, "The Effect of News Teasers in Processing TV News," *Journal of Broadcasting & Electronic Media,* Summer 1998, 327–39; Annie Lang, Deborah Potter, and Maria Elizabeth Grabe, "Making News Memorable: Applying Theory to the Production of Local

Television News," *Journal of Broadcasting & Electronic Media,* March 2003, 111–23; and David Weibel, Bartholomaus Wissmath, and Tudolf Groner, "How Gender and Age Affect Newscasters' Credibility," *Journal of Broadcasting & Electronic Media,* September 2008, 466–84.

29. Cynthia Hoffner and Margaret J. Haefner, "Children's News Interest During the Gulf War: The Role of Negative Effect," *Journal of Broadcasting & Electronic Media,* Spring 1994, 193–204.

30. Silvia Knobloch-Westerwick, "Gender Differences in Selective Media Use for Mood Management and Mood Adjustment, *Journal of Broadcasting & Electronic Media,* March 2007, 73–92; Osei Appiah, "Americans Online: Differences in Surfing and Evaluating Race-Targeted Web Sites by Black and White Users, *Journal of Broadcasting & Electronic Media,* December 2003, 537–55; and Marie-Louise Mares and Emory H. Woodard IV, "In Search of the Older Audience: Adult Age Differences in Television Viewing," *Journal of Broadcasting & Electronic Media,* December 2006, 595–614.

31. Christina Pieraccini and Douglas Alligood, *Color Television: Fifty Years of African American and Latino Images on Prime-Time Television.* (Dubuque, IA: Kendall/Hunt, 2005); Martha M. Lauzen, David M. Dozier, and Nora Horan, "Constructing Gender Stereotypes through Social Roles in Prime-Time Television," *Journal of Broadcasting & Electronic Media,* June 2008, 200–214; Tom Robinson and Caitlin Anderson, "Older Characters in Children's Animated Television Programs: A Content Analysis of Their Portrayal," *Journal of Broadcasting & Electronic Media,* June 2006, 287–304; Joan McGettigan, "Interpreting a Man's World: Female Voices in *Badlands* and *Days of Heaven,*" *Journal of Film and Video,* Winter 2001, 33–43; C. Lee Harrington, "Homosexuality on *All My Children:* Transforming the Daytime Landscape," *Journal of Broadcasting & Electronic Media,* June 2003, 216–35; and Rick Busselle and Heather Crandall, "Television Viewing and Perceptions about Race Differences in Socioeconomic Success," *Journal of Broadcasting & Electronic Media,* June 2002, 265–82.

32. W. H. Anderson, Jr., and B. M. Williams, "TV and the Black Child: What Black Children Say about the Shows They Watch," *Journal of Black Psychology,* Spring 1983, 27–42; "Fewer Women at TV Series Helm," *Hollywood Reporter,* July 19, 2004, 5; and "Gender Equity in Televised Sports: A Comparative Analysis of Men's and Women's NCAA Division I Basketball Championship Broadcasts, 1991–1995," *Journal of Broadcasting & Electronic Media,* Spring 1999, 222–35.

33. Barry S. Sapolsky and Joseph O. Tabarlet, "Sex in Primetime Television: 1979 Versus 1989," *Journal of Broadcasting & Electronic Media,* Fall 1991, 505–16; and Mary Beth Oliver et al., "Sexual and Violent Imagery in Movie Previews: Effects on Viewers' Perceptions and Anticipated Enjoyment," *Journal of Broadcasting & Electronic Media,* December 2007, 596–614.

34. "TV's Found Wanton," *Daily Variety,* September 7, 2004, 6; and Yuanyuan Zhang, Laura E. Miller, and Kristen Harrison, "The Relationship between Exposure to Sexual Music Videos and Young Adults' Sexual Attitudes," *Journal of Broadcasting & Electronic Media,* September 2008, 368–86.

35. Kirstie M. Farrar, "Sexual Intercourse on Television: Do Safe Sex Messages Matter?" *Journal of Broadcasting & Electronic Media,* December 2006, 635–50.

36. Keren Eyal et al., "Sexual Socialization Messages in Television Programs Most Popular Among Teens," *Journal of Broadcasting & Electronic Media,* June 2007, 316–36.

Chapter 13

TECHNICAL UNDERPINNINGS

The electronic media are grounded in technology. The word "electronic" signifies that they depend on the movement of electrons. Anyone who deals with electronic media must have some knowledge of technology, if only to accomplish simple tasks such as rebooting a computer or adjusting the sound quality of a car radio. Those who work in the industry should have a more thorough knowledge of the technical underpinnings of the industry so that they can take advantage of technological innovation and converse with those responsible for it. Engineers, of course, need to be thoroughly versed in electronics. They are the ones who invent and improve the equipment and processes that make telecommunications possible.

Television is the triumph of equipment over people.

Fred Allen, comedian

One way to look at technical underpinnings is to divide them into production, distribution, and exhibition. Regardless of the form of electronic media used, its messages must be produced utilizing some sort of equipment—microphones, cameras, computers. The production must then be distributed by wired or wireless means—cable TV, satellite TV, the internet. However, the program material is of little use if it simply remains in the distribution pipe; it must be seen and heard through an exhibition device—a radio, a TV set, a computer monitor, a motion picture screen. This chapter is not intended to make anyone an expert in technology, but it should serve as an introduction to the equipment and processes needed to deliver electronic media content.

13.1 The Digital Process

Today most electronic media programming is produced, distributed, and exhibited through **digital** means. But prior to the 1990s, the underlying process used by electronic media was **analog.** Some media forms, such as radio and movies, are still in the process of converting from analog to digital; others, such as television, have converted recently; and others, such as the internet and DVRs, came to the fore during the digital era and have been digital since their inception.

The technical difference between digital and analog is that digital utilizes discrete on and off impulses, whereas analog involves a continuous method of signal recognition (see Exhibit 13.1). Analog is similar to creating a line graph for

technical differences

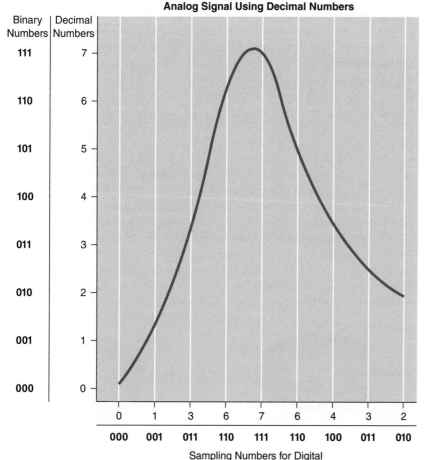

Exhibit 13.1

Analog is a continuous wave, whereas digital is a series of discrete samples from the wave that are turned into off-on (0-1) binary numbers. The off-on numbers can be faithfully reproduced, whereas the analog wave tends to shift position, distorting the signal.

a statistical analysis. All measurements are on a continuum, and the line curves up and down. Digital sound is similar to a person looking at individual numbers and writing them down in a set order. It is, in essence, many samples of the analog signal.

quality

When the analog method is used to take an electronic signal from one place to another (e.g., one recording device to another), the signal changes shape and degrades slightly in quality. This is similar to someone trying to trace a curve on a graph. The reproduction will not be exactly like the original. When the digital method is used to transport the signal, it does not waver. As it travels from one source to another, separate bits of information transmit with little loss of quality. This would compare to someone copying the numbers of the statistical study. The numbers would be copied accurately, so the results would not be distorted.

compression

This gives digital an advantage over analog because material can be dubbed or duplicated with little or no loss of quality. With analog dubbing, colors run together, the picture becomes washed out, and the audio develops hisses and pops. In general, both audio and video signals are sharper and crisper with digital than with analog, even in first generation. Another advantage of digital signals is that they can be compressed. **Compression** allows more information to be packed into less space. Because digital is similar to copying numbers, when the same number appears a second time in the same part of a signal, it does not need to be recopied. In other words, only changes in a signal need to be noted. If a recording consists of a person talking in front of a wall, the part of the frame containing the wall needs to be considered only once. For the rest of the recording, only the person's changing facial movements need to be noted.[1]

13.2 Audio Production Equipment

microphones

The main pieces of equipment used to produce stand-alone audio material (such as radio) or audio that accompanies video are microphones, audio players/ recorders, audio boards, and speakers. **Microphones** convert sound into electronic impulses that can be transmitted through the rest of the equipment. Microphones come in many forms designed for different purposes. For example, some microphones, called **omnidirectional,** pick up sound from all directions, while others, referred to as **cardioid,** pick up mainly from one direction (see Exhibit 13.2). The omnidirectional microphone would be good to use for crowd noises, while the cardioid would be more suited to situations in which one person is talking. Some microphones are designed to pick up the wide range of sounds of musical instruments, whereas others optimize the sound of the human voice.

Exhibit 13.2

Two of the most common microphone pickup patterns.

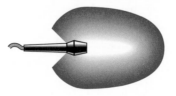

Cardioid   Omnidirectional

Audio players and recorders include **compact disc (CD)** players, **flash memory**, and computer **hard drives.** Sometimes sound goes directly from the microphone to a recorder. The microphone can be built into the recorder, the signal can be sent through wireless means, or, most commonly, the sound is sent through a cable that is connected to the microphone on one end and the recorder on the other. Recorders/players are also used to play back recorded material, such as music and sound effects, so that it can be incorporated into a production.

Audio boards (see Exhibit 13.3) are used to mix various sounds together. For example, an audio production may consist of two people talking (each into a separate microphone) and music coming from a CD. All three signals would be sent into an audio board, and an operator could then raise and lower the volume on each so that the music did not overpower the voices and the voices meshed well together. Sound from an audio board can be sent to a recorder, where it can be stored and played back whenever it is needed. Audio boards also have other functions. They can be used to change the quality of a sound (e.g., give a voice an echo so it sounds like it is coming from a cave) or determine which sounds go to the right and left channels of a **stereo** recording.

Exhibit 13.3
An audio board.

(Courtesy of Sony Enterprises, Inc.)

recorders

boards

For modern digital audio work, the functions of the audio board and the audio recorder are often combined in a computer program. The technician gathers the sounds needed by plugging a microphone directly into a computer, by transferring sound from a CD or memory stick into the computer program, or by downloading sound from an internet site. Then, using the computer program, the technician mixes and edits the sounds into a finished product. The computer programs have innumerable effects, often referred to as **signal processing,** which can be used to enhance and alter the various sounds.

Speakers are used to hear the audio as it is being recorded, transferred, edited, and mixed, as well as played back in its final form. The output of the speaker is analog because our ears cannot hear the off and on impulses of digital.

Audio equipment has changed greatly during a short period of time. Early radio microphones were huge and did not replicate the sound with the clarity known today. Analog recorders, also huge, mainly used audiotape as a recording medium. Boards had far fewer functions, and speakers only output **monaural** sound. Audio engineers of the past several decades are responsible for designing new and better hardware and software that gives us the sound quality we enjoy today.[2]

computers

speakers

changes

13.3 Video Production Equipment

Video production utilizes all the audio equipment but also requires additional equipment such as cameras, video recorders, switchers, monitors, lights, editors, and visual effects equipment. **Cameras** are at the heart of any video system because they capture the image through a lens and send it on to other equipment—in much

cameras

Exhibit 13.4

A charge-coupled device (CCD), which is a chip that changes the light energy coming into the camera into an electrical form.

(Courtesy of Sony Electronics, Inc.)

film

recorders

the same way that a microphone captures sound. For several decades, cameras were analog and contained **cathode-ray tubes** that electronically scanned an image. Starting in the 1980s, manufacturers began replacing the tubes with **chips** (see Exhibit 13.4) that provided the same image sensor functions while allowing cameras to be much smaller.

Sometimes picture information is shot with film cameras, which operate in an analog chemical manner. They have a shutter behind the lens that opens and closes to allow the image to reach photosensitive frames of film stock that can later be developed into individual pictures at a laboratory. When video cameras were analog, film cameras provided a far superior picture to video. But video cameras that are high-definition digital rival the quality of film. Some cinematographers who are used to film are switching to video cameras, but they want the cameras to act as much like film cameras as possible (see "Zoom In" box).[3]

Video cameras often come in a package, known as a **camcorder,** which also includes a recorder. Originally video recorders, like audio recorders, used tape as the recording medium, but today they are more likely to use discs or hard drives. Cameras can also be separate from their recorders, a configuration that is common in studio situations where several cameras that are wheeled around on **pedestals** capture different angles of the same scene and all the images are sent to a switcher.

 ZOOM IN: **The Red One**

Jim Jannard made a fortune on sunglasses as the founder and CEO of Oakley, Inc. For his second act, in 2005, he founded a camera company, Red Digital Cinema, with the purpose of producing a high-resolution video camera for use in productions to be shown in movie theaters. While the camera was in development, it created a great deal of buzz.

Because Jannard's camera company is new, it does not have to protect existing products by making its camera backward compatible, as Sony, Panasonic, and other camera manufacturers must do. It can develop new approaches. The Red One has its own proprietary image sensor and its own method of compression. It operates more like a film camera than a video camera in that it shoots 24 frames per second, with each frame being the same quality as a super 35 mm film frame. It also allows for easy use of the same lenses used for film cameras. One of its most attractive features is its price of $17,500—far below that of other professional high-definition cameras.

Directors and cinematographers paid for the privilege to be on the waiting list to buy one of these cameras, and when the first ones shipped in 2007, the customers seemed quite pleased with the product.

The Red One.

(Courtesy of Red Digital Cinema)

A **switcher** for video is similar to an audio board for sound in that it mixes various pictures together (for example, credits that roll over the end of a drama, a fade from black to the beginning of a show, or a scene from a fire that appears in a box behind a news anchor) and also creates effects, such as a picture that swirls on or off the screen. The output of a switcher usually goes to a recording device—a videotape recorder or a computer **server**—where the picture and its sound are stored.

switcher

Monitors are similar to TV sets, except they don't show regular TV channels. Instead, they show the pictures that are being created and manipulated in the production facility. There are usually a large number of them in a studio setting because all the inputs (each camera, graphics, material being played back from a recorder, satellite feeds, etc.) are shown on separate monitors so that a director can select the image that he or she wants to go through the switcher and be recorded.

monitors

Although it is possible for cameras to capture an image using only the light that is available, professionals usually augment a scene with additional light either to boost the available light or to create a particular mood. The main type of lamp that lighting technicians have used for many years is a **quartz lamp.** It is made of quartz glass that contains halogen gas and uses a tungsten filament. It provides excellent light that is easy to control for television and film production, but it is a hot lamp that uses a great deal of energy. As a result, companies have developed other forms of lights (**high-frequency fluorescent, light-emitting diode**) that are cooler, smaller, and more efficient. In a studio situation, lights hang from a ceiling **grid;** in other locations, they mount on light stands.

lights

Once the video images and their sound are captured, they often need to be edited. For the most part, this is undertaken on computers with **nonlinear editing** software (see Exhibit 13.5). This type of editing allows picture and sound to be

editing

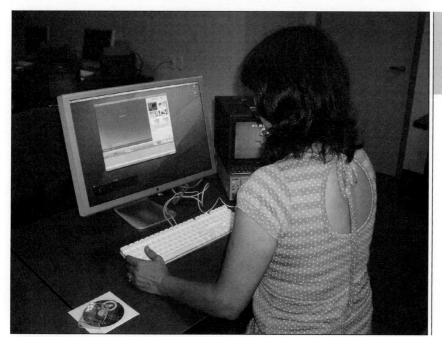

Exhibit 13.5
This editor is booting up editing software for use with a computer.

manipulated easily. Any shot can be placed anywhere on the final product and then moved about easily—for example, the third and fifth shots could be interchanged with little or no difficulty. If the director decides to lengthen a shot by two seconds, the computer automatically moves all the following shots ahead by two seconds. As with many other technological improvements, nonlinear editing was developed by people with engineering backgrounds who found ways to improve on older methods such as **film editing,** which involved the rather tedious cutting and taping of film frames, and **linear editing,** in which all material had to be laid down in order and was very difficult to change. Nonlinear editing has become so popular that movies that are shot on film are transferred to video, so they can be edited by computer.

visual effects

Computers are also used to create the elaborate **visual effects** seen in movies, such as fiery crashes, flying people, and lumbering dinosaurs. Sometimes entire animated films are created on a computer. Generally, the computers for movie projects are configured to handle large volumes of complicated footage, and the software programs for visual effects are sophisticated. However, simpler effects operate within nonlinear editing programs and can be utilized by people producing wedding videos, corporate productions, and material to distribute over the internet.[4]

Once a production has been completed, using whatever equipment was needed, it is ready to be distributed to an audience.

13.4 The Electromagnetic Spectrum

One way to divide distribution technologies is into **wired** and **wireless.** Wired technologies are obviously ones that carry information over wires. Wireless technologies are those that send signals through the airwaves using the **electromagnetic**

frequencies

spectrum, a continuing series of energies at different **frequencies.** The frequencies can be compared to the different frequencies involved with sound. The human ear is capable of hearing sounds between about 16 cycles per second (low bass) and upward of 16,000 cycles per second (high treble). This means that a very low bass sound makes a vibration that cycles at a rate of 16 times per second.

hertz

These rates are usually measured in **hertz** in honor of the early radio pioneer Heinrich Hertz. One hertz (Hz) is one cycle per second. A low bass note would be at the frequency of 16 Hz, a higher note would be at 400 Hz, and a very high note would be at 16,000 Hz. As the numbers become larger, prefixes are added to the term *hertz* so the zeros do not become unmanageable. One thousand hertz is referred to as one kilohertz (kHz), one million hertz is one megahertz (MHz), and a billion hertz is one gigahertz (GHz). Therefore, the 16,000-hertz high note could also be said to have a frequency of 16 kHz.

radio waves

Radio waves, which are part of the electromagnetic spectrum, also have frequencies and are measured in hertz. Their frequencies are higher, however, ranging from about 30 kHz to 300 GHz. They are capable of carrying sound and pictures, and it is through these radio waves that most telecommunications distribution oc-

other waves

curs. Above radio waves on the electromagnetic spectrum are infrared rays and then light waves, with each color occupying a different frequency range. After visible light come ultraviolet rays, X rays, gamma rays, and cosmic rays (see Exhibit 13.6).

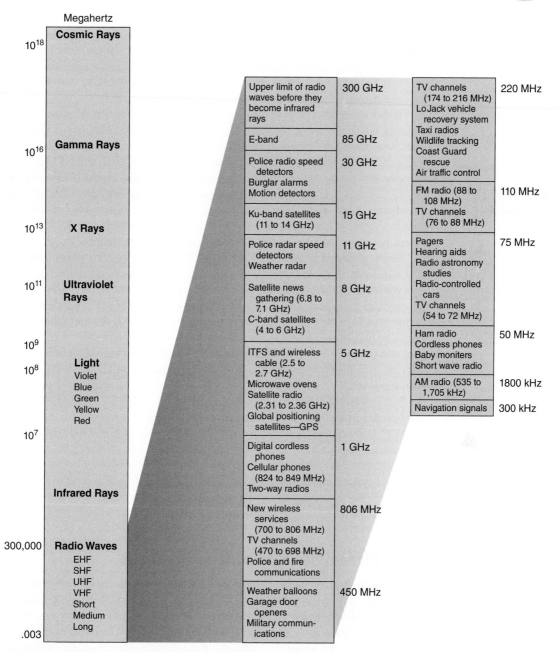

Megahertz

10^{18}	**Cosmic Rays**		
10^{16}	**Gamma Rays**		
10^{13}	**X Rays**		
10^{11}	**Ultraviolet Rays**		
10^{9}			
10^{8}	**Light** Violet Blue Green Yellow Red		
10^{7}			
	Infrared Rays		
300,000	**Radio Waves** EHF SHF UHF VHF Short Medium Long		
.003			

Upper limit of radio waves before they become infrared rays	300 GHz
E-band	85 GHz
Police radio speed detectors Burglar alarms Motion detectors	30 GHz
Ku-band satellites (11 to 14 GHz)	15 GHz
Police radar speed detectors Weather radar	11 GHz
Satellite news gathering (6.8 to 7.1 GHz) C-band satellites (4 to 6 GHz)	8 GHz
ITFS and wireless cable (2.5 to 2.7 GHz) Microwave ovens Satellite radio (2.31 to 2.36 GHz) Global positioning satellites—GPS	5 GHz
Digital cordless phones Cellular phones (824 to 849 MHz) Two-way radios	1 GHz
New wireless services (700 to 806 MHz) TV channels (470 to 698 MHz) Police and fire communications	806 MHz
Weather balloons Garage door openers Military communications	450 MHz

TV channels (174 to 216 MHz) LoJack vehicle recovery system Taxi radios Wildlife tracking Coast Guard rescue Air traffic control	220 MHz
FM radio (88 to 108 MHz) TV channels (76 to 88 MHz)	110 MHz
Pagers Hearing aids Radio astronomy studies Radio-controlled cars TV channels (54 to 72 MHz)	75 MHz
Ham radio Cordless phones Baby moniters Short wave radio	50 MHz
AM radio (535 to 1,705 kHz)	1800 kHz
Navigation signals	300 kHz

Exhibit 13.6

The electromagnetic spectrum and the main services of the radio wave portion.

**characteristics
of radio waves**

Although all radio frequencies in the electromagnetic spectrum are capable of carrying sound and pictures, they are not all the same. Frequencies toward the lower end of the spectrum behave more like sound than do frequencies at the higher end, which behave more like light. For example, the lower frequencies can go around corners better than the higher frequencies, in the same way that you can hear people talking around a corner but cannot see them.

**uses of radio
waves**

Radio waves also have many uses other than the distribution of entertainment and information programming usually associated with the electronic media industry. These uses range from the opening of garage doors to secret reconnaissance functions. Some of the most common media and nonmedia uses are shown in Exhibit 13.6.

**placement
of services**

The spot on the radio portion of the spectrum at which a particular service is placed depends somewhat on the needs of the service. The lower frequencies have longer ranges in the earth's atmosphere. For example, **shortwave** radio, which travels long distances, uses very low frequencies; motion detectors that need to detect movement only in one or two rooms operate on high frequencies.

Many placements, however, are an accident of history. The lower frequencies were understood and developed earlier than the higher frequencies. Not long ago, people were not even aware the higher frequencies existed, and today the extremely high frequencies are still not totally understood. AM radio was developed earlier than FM (see Chapter 5), so the former was placed on the part of the spectrum that people then knew and understood. Ultra-high frequencies were discovered during World War II and led to the FCC's reallocating television frequencies after the war into two categories, **very high frequency (VHF)** and **ultra-high frequency (UHF)** (see Chapter 3).

These VHF and UHF frequencies were used for analog TV, but when it came time to convert to digital, some of the frequencies were not suitable for digital, so the FCC gave stations new frequencies. This allocation also enabled the stations to continue broadcasting analog signals while they were working to establish their digital transmissions. Once the nation converted to digital TV, the stations had to return their old frequencies to the FCC. The FCC auctioned off the old television frequencies, most of which were secured by companies that want to expand or develop new high-speed wireless services (see Chapter 4).[5]

13.5 Terrestrial Radio Broadcasting

definitions

Terrestrial broadcasting occurs close to the earth. It encompasses radio and TV signals that travel only a short distance in the earth's atmosphere, as opposed to satellite broadcasting that goes into outer space. AM and FM radio are terrestrial broadcasting forms, as is a new form of digital radio called **in-band on-channel (IBOC)** radio (see Chapter 5). The parts of the electromagnetic spectrum that are most important to these services are 88 to 108 MHz, where FM resides, and 535 to 1705 kHz, the home of AM. IBOC can be squeezed into those same frequencies. The position in the spectrum, however, has nothing to do with **AM (amplitude modulation), FM (frequency modulation),** or **PCM (pulse code modulation),** the type of modulation that is used for IBOC.

Sound, when it leaves the radio station studio in the form of radio energy, travels to the station's **transmitter** and then to the **antenna.** At the transmitter it is

Consider this to be an electrical wave representing the original sound.

Exhibit 13.7

Diagram of an AM wave.

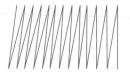

Consider this to be the carrier wave of a particular radio station. Notice it is of much higher frequency than the electrical wave.

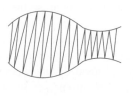

This would be the modulated carrier wave taking the sound signal. Note that the sound signal makes an image of itself and that the amplitude, or height, of the carrier wave is changed—hence, amplitude modulation.

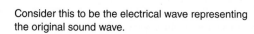

Consider this to be the electrical wave representing the original sound wave.

Exhibit 13.8

Diagram of an FM wave.

Consider this to be the carrier wave.

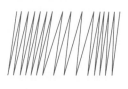

This would be the modulated carrier wave. The frequency is increased where the sound wave is highest (positive), and the frequency is decreased where the sound wave is lowest (negative). The amplitude does not change.

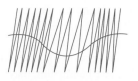

Perhaps this can be better seen by superimposing the sound wave over the carrier wave.

Exhibit 13.9

PCM modulation.

0101110101010
1111000011010
1110001010

Pulse code modulation is a digital form so consists of a series of on (1) and off (0) pulses.

modulated, which means the electrical energy is superimposed onto the **carrier wave** that represents that particular radio station's frequency (e.g., 710 AM, 97.1 FM). The transmitter generates this carrier wave and places the sound wave on it,

modulation

using the process of modulation. Amplitude, frequency, or pulse code modulation can occur regardless of where the carrier wave is located on the spectrum. In amplitude modulation, the amplitude (or height) of the carrier wave is varied to fit the characteristics of the sound wave (see Exhibit 13.7). In frequency modulation, the frequency of the carrier wave is changed instead (see Exhibit 13.8). Pulse code modulation, because it is a digital form, encodes information as 0s and 1s (see Exhibit 13.9).

modulation differences

AM, FM, and PCM have different characteristics caused by the modulation methods. For example, AM is much more subject to static because static appears at the top and bottom of the wave cycle. Because FM depends on varying the frequency of the wave and PCM is just a code, the top and bottom can be eliminated without distorting the signal. AM, however, depends on height, so the static regions must remain with the wave.

spectrum differences

AM and FM also have differences because of their placement on the spectrum, differences that will also affect PCM depending on whether it is in the frequency band of AM or FM. AM signals can travel great distances around the earth, while FM signals sometimes cannot be heard if an obstruction, such as a large building, comes between the transmitter and the radio that is attempting to receive the station. FM is higher on the spectrum and closer to light than is AM, and just as light waves do not travel through buildings or hills, FM signals are similarly affected, making them **line of sight.** AM, at the lower end of the spectrum, is not so affected, but these lower frequencies are affected by a nighttime condition of the ionosphere that, when hit by radio waves, bounces the wave back to earth. AM waves can be bounced great distances around the earth's surface—from New York to London, for example. Because of this phenomenon, some AM stations are authorized to broadcast only during daylight hours so they will not interfere with other radio station signals that are traveling long distances because they are bouncing. FM is not affected by this because of its position in the spectrum, not because of its manner of modulation. Theoretically, if FM waves were transmitted on the lower frequencies, such as 535 kHz, they could bounce in a manner similar to AM.

bandwidth

Another difference between AM and FM is that the **bandwidth** is greater for FM stations. Although each station is given a specific frequency, such as 550 kHz or 88.5 MHz, the spectrum actually used covers a wider area. For an AM radio station, this width (bandwidth) is 10 kHz; for FM, the bandwidth per station is 200 kHz. An AM station at 550 kHz actually operates at from 545 to 555 so that it can have room to modulate the necessary signals and prevent interference from adjacent channels. Because FM stations have a broader bandwidth, they can produce higher fidelity than AM stations. This makes FM more adaptable to stereo

stereo

broadcasting because there is room for two channels of sound. AM stereo does exist, but not all stations use it because it is not as effective as FM stereo.

IBOC

FM's broader bandwidth also makes it easier to add PCM digital radio so that it comes from the same frequency area as the FM station. Engineers involved with IBOC had to work much harder to superimpose digital radio in the AM band. The plan, however, is for both AM and FM stations to broadcast the same programming in both analog and digital and, eventually, after listeners have bought radios that can receive the PCM signals, to abandon analog and only broadcast in digital. The FCC, however, is not mandating that radio switch to digital; it is purely voluntary on the part of the stations.

In addition, the broader bandwidth of FM enables independent signals other than those needed for digital and stereo sound to be **multiplexed** on an FM radio station signal. This information is carried on part of the FM signal, but it can be received only by those with special receivers. Some of the services include piped music to doctors' offices and reading for the blind.

multiplexing

After the sound waves are modulated (either AM or FM) onto the proper frequency carrier wave at the transmitter, the carrier wave is radiated into the air. It is sent into the airwaves at the assigned frequency and power from the radio station antenna. Because FM frequencies travel line of sight, FM antennas are usually positioned on high places overlooking a large area. AM antennas can operate effectively at relatively low heights.

antennas

The FCC has established a complicated chart of frequency and power allocations to allow a maximum number of stations around the country to broadcast without interference. There is room for 117 stations of 10 kHz each between 535 and 1,705 kHz, but there are more than 5,000 AM stations nationwide. This is possible because many stations throughout the country broadcast on the same frequency, but in a controlled way that considers geographic location and power.[6]

number of stations

13.6 Terrestrial Television Broadcasting

As with radio, the terrestrial television transmitter is the device for superimposing or modulating information onto a carrier wave. With the analog television signal that was used from the early days of television until recently, audio and video were sent to the transmitter separately. The video signal was amplitude modulated, and the audio signal was frequency modulated in either monaural or stereo. The two were then joined and broadcast from the station antenna. In the early days, this signal went from the station antenna through the local airwaves to receiving antennas on people's rooftops and from there to a TV set with "rabbit ears" (see Exhibit 13.10). Most households do not receive their TV stations this way anymore. They are more likely to get them through cable TV or satellite TV. But stations still send out their terrestrial signal, and often cable and satellite systems pick up that signal and transpose it into their own systems. People who do receive their signal through terrestrial broadcast have had to obtain converter boxes so that they can view the new digital TV on analog sets.

modulation

digital TV

Exhibit 13.10

This dipole antenna for UHF and VHF TV shows why such antennas were referred to as "rabbit ears."

Digital TV is modulated in a manner similar to digital radio because it consists of a series of on and off pulses. The **digital TV** situation is much more complicated than digital radio, however. For one thing, unlike radio, the FCC mandated that all TV stations switch from analog to digital whether they wanted to or not. For another, the digital stations are not all on the same frequency that their analog stations were. For example, a station that used to be channel 4 may now be channel 38. Also, there is the confusion between digital TV and **high-definition TV** (see Chapter 4). Digital is

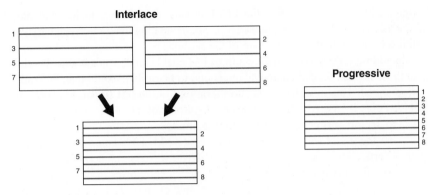

Interlace

Progressive

Exhibit 13.11

Interlace scanning lays down the odd-numbered lines then the even-numbered lines to build a complete frame. Progressive scanning scans all lines sequentially. Obviously there are many more lines (720 or 1,080) than shown here, but eight lines are shown for simplicity's sake.

(Courtesy of Brian Gross)

a new transmission form based on pulses rather than analog waves; HDTV incorporates digital but also provides more lines of resolution so that the picture looks sharper than the old analog picture, which only had 525 lines.[7]

scanning

To further complicate things, several different formats of HDTV have different numbers of lines and different methods of **scanning** (the process for creating the picture). A video picture is never an entire frame. It is a series of very rapidly illuminated dots (**pixels**) that look like full-frame pictures to us because our eyes retain the image produced by the pixels. Analog TV employed **interlace scanning,** in which all the odd lines of information created by the dots were laid down first, followed by the even lines in order to form a frame. This interlace process is used by some forms of high-definition television, but others use **progressive scanning** that lays information on the screen one line at a time from top to bottom (see Exhibit 13.11). When TV was developed in the 1930s and 1940s, progressive scanning was not a viable technical option, but by the time HDTV was in development, it was possible for HDTV to utilize either form of scanning. Many combinations of scanning and number of lines have been developed, but the two that are used the most are 720p and 1080i. The numbers stand for the number of lines while "p" stands for progressive scan and "i" stands for interlace (see "Zoom In" box).[8]

13.7 Satellites

Terrestrial signals travel only a short distance, usually staying within the confines of one city. **Satellites** can be used to distribute to a wider area. Signals are sent from satellite dishes on earth and are received on satellites in outer space that

transponders

have **transponders**—essentially reception channels that can hold video, audio,

ZOOM IN: **So Which Is Better, 720p or 1080i?**

It depends on which engineer you ask. NBC and CBS broadcast in 1080i, and ABC and Fox use 720p. If you go to their websites, they each claim their system is best. Fortunately, TV sets know the difference and switch seamlessly, so the consumer isn't saddled with dealing with the different formats. 1080i obviously has more lines of resolution so should have a sharper picture. But many engineers feel that the human eye retains the image of progressive more effectively so that 720p appears to be as good as or better than 1080i. To complicate things, there is another format that is rising fast—1080p. It is used by high-definition DVDs. Theoretically, it is superior to 720p and 1080i, but some people claim that the eye really can't tell the difference. And, of course, there are engineers who like the softer analog look (and sound) and don't think much of digital at all.

It is easy to go to an electronics store and see 720p, 1080i, and even 1080p next to each other, so you can make your own decision. If you do ask an engineer, he or she might tell you it depends on the Kell factor, the black level, the seating distance, or the deinterlacing, so be careful what you ask for.

and data information. The information goes from the satellite back down to earth where it can be received by any equipped facility. Being equipped to receive the signal means owning a satellite dish positioned in such a way that it lines up with the signal being sent from the satellite.

The satellites used for telecommunications transmit in two main frequency bands, **C-band** and **Ku-band.** Those in C-band operate between 4 and 6 GHz, and those in Ku-band operate between 11 and 14 GHz. The C-band satellites were put up first and carry much of the cable TV programming. Additional cable programming is carried on Ku-band satellites, and **satellite TV** services such as DirecTV and Dish Network transmit from this frequency range. **Satellite radio** operates from slightly lower frequencies, in the 2.3 GHz range.

bands

Most of the satellites are positioned 22,300 miles above the equator and are powered, at least in part, by energy they gather from the sun. These **synchronous satellites** travel in an orbit that is synchronized with the speed of the earth's rotation, thus appearing to hang motionless in space. This way they can continually receive and send signals to the same points on earth. It is not difficult for signals to travel 22,300 miles because once they leave the earth's atmosphere, they do not encounter the interference and inhospitable weather conditions that waves traveling through the earth's atmosphere encounter.

orbit

The satellites that transmit to the United States are positioned along the equator between 55 and 140 degrees west longitude. From there they can cast a signal, called a **footprint** (see Exhibit 13.12), over the entire United States. Satellites positioned at other longitudes have footprints over other sections of the world. Most of the earth's surface can be covered by three strategically placed satellites. In this way, instant worldwide communication is possible through a satellite network, and pictures can be beamed from anywhere in the world to the United States.[9]

position

Exhibit 13.12

A satellite and
its footprint.

*(Courtesy of Hughes
Aircraft Company)*

early uses

The first geosynchronous communication satellite, Telstar I, was launched in 1962 by AT&T and was followed by satellites launched by other companies. The early communication uses of satellites included the transmission of international events, such as splashdowns of U.S. space missions and Olympic games. HBO's placement of its pay service on RCA's Satcom I in 1976 opened the floodgates (see Chapter 3). Once the cable industry took to satellite, the demand for transponders outstripped the supply. When Satcom III, launched by RCA in December 1979, was lost, the cable companies that were planning to place their services on that satellite complained so bitterly that RCA leased time on other companies' satellites and re-leased it to the cable companies.

other uses

In addition to cable TV, many other media entities found uses for satellite transmission. PBS and NPR were among the first to send programs to affiliates using satellites. Syndicators that previously had mailed or carried tapes to stations began distributing by satellite. New radio networks sprang up and delivered their programming to AM and FM stations by satellite, and established networks, both radio and TV, converted from older distribution means to satellite to deliver to their affiliates. Some individuals bought satellite receiving dishes for their backyards (see Chapter 4) and were then able to receive most of the signals being transmitted by cable services, syndicators, networks, and others.

low-orbit

Telephone companies started using satellites to distribute phone calls, and corporations used them for data distribution, such as credit card transactions. These groups were more likely to use **low-orbit satellites,** ones that were closer to the earth but were not synchronous. Instead, a series of low-orbit satellites hand signals off to each other. One reason for using them is that they do not have as much latency as synchronous satellites have. When having back-and-forth telephone conversations, it is annoying to have the degree of delay that occurs when signals travel 22,300 miles.[10]

13.8 Other Wireless Distribution

The electromagnetic spectrum is used for many forms of telecommunications other than radio, broadcast TV, and satellite distribution. One of the oldest uses is **microwave** transmission. Located between 1 and 30 GHz, microwaves are higher in the electromagnetic spectrum than radio and TV stations, but they are terrestrial in that they do not go into outer space. They are line of sight and travel only about 30 miles. Microwave transmitters were once used to send network signals across the country. Towers were built about every 30 miles, each one picking up a signal from the previous tower and sending it on to the next. When satellites were developed, it was much easier to send signals to a satellite so they could be received anywhere in the country. Microwaves are still used to transmit signals from TV studios to transmitters and from news trucks to stations (see Exhibit 13.13). Cable systems sometimes import distant signals or bring in the signals of local stations using microwave.[11]

microwave

Shortwave radio, sometimes used to distribute information from one country to another (see Chapter 14), uses low frequencies on the spectrum that travel long distances. Radio and TV stations use higher frequencies for **translators.** These are towers that pick up a signal and send it to a particular area, such as a spot within a station's coverage area that has poor reception. **Cellular phones** are another example of wireless terrestrial transmission in that the signals are sent from one low-power cell to another (see Exhibit 13.14). Cordless phones, PDAs, pods, and other wireless equipment also take advantage of low-power terrestrial signals. Wireless mics that enable performers to move around unrestricted by wires are basically tiny radio stations that send sound from the mic to a receiver using the electromagnetic spectrum. Similarly, cameras used in the field can have antennas on them to send their signals through the airwaves to a switcher or recorder.

international

other media uses

Wireless is also used to send internet signals to laptop computers. Wireless fidelity, or **wi-fi,** utilizes a low-power radio transmitter that is connected to the net and modulates the data onto radio waves at various frequencies in the electromagnetic spectrum that are assigned to wi-fi. The wireless radio waves emanate for some distance from their transmitter; the area covered by the wi-fi signal is called a **hotspot.** Anyone within a hotspot who has the certified wi-fi hardware in his or her computer can connect to the internet without a

internet

Exhibit 13.13
The top of this truck holds microwave apparatus for sending news from its location back to the TV station.

Exhibit 13.14

This drawing shows how the wireless cells of a cellular telephone system overlap so that signals can be handed from one wireless cell phone tower to another.

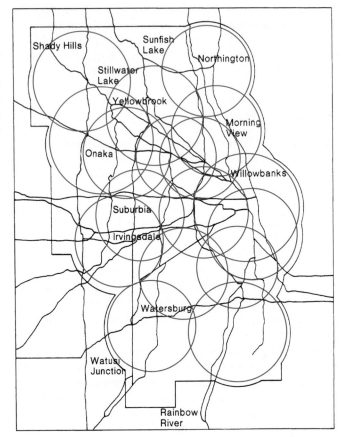

wire. When businesses, city governments, or private individuals invest in wi-fi hotspots, they allow people to search the net, listen to web radio, and engage in many other activities without being cabled. In addition to office, research, and university complexes, wi-fi is popular with coffee shops and other businesses that realize a hotspot can attract customers.

In short, media companies use the electromagnetic spectrum in many ways. However, they are not its only occupants (refer back to Exhibit 13.6). It is also very important for health and safety; police and fire departments use it for communications, as do air traffic controllers and the Coast Guard. A fair amount of the spectrum is used to track weather conditions and wildlife. It is also the home of many of life's conveniences, such as baby monitors, microwave ovens, LoJacks, and global positioning systems.[12]

other uses

13.9 Wire Transmission

Wires are used for sending electronic signals short and long distances. Short-distance use includes running a wire from a consumer DVD to a TV set, from a microphone to an audio board, from a studio camera to a switcher, or from a projector to speakers in a movie theater. Much longer hauls are made when wires are

Exhibit 13.15
A fiber optic.

(Courtesy of Corning Glass Works)

used to transmit phone calls and email messages around the world or to deliver cable TV channels throughout a community.

Not all wires are alike. Voice and data can use simpler wires than video. For many years, the copper **twisted-pair** was the standard for telephone operations. Television signals required a new type of wire called **coaxial cable,** which was the type that cable TV systems used when they first built their systems. Both twisted-pairs and coax are still used, but most applications that use wire have switched or are switching to **fiber optics** (see Exhibit 13.15).

types of wires

With this technology, audio, video, or other data can be sent through an optic strand that is less than a hundredth of an inch in diameter. Because this strand is made of glass, it is relatively inexpensive, lightweight, strong, and flexible. A fiber optic cable that is less than an inch in diameter can carry 400,000 phone calls simultaneously—10 times the amount that can be carried on conventional copper wire. Fiber optics carry digital information, so their capacity is measured in number of bits sent per second. If the information the fiber optic is to carry is analog, it must first be converted to digital form. Then it is carried through the glass fiber on light produced by laser diodes. When the information gets to its destination, it can be translated back into analog form. Using light makes transmission less susceptible to electrical interference than with coaxial cable or twisted-pair lines.[13]

fiber optics

Wires are the workhorses of cable TV systems, which consist of a **headend** where all the inputs, such as local TV stations, pay cable networks, internet phone services, and basic cable networks, are received. At this headend, all the services are placed on a wire that is either buried underground or hung on telephone poles. This wire branches and goes to various subscribers' homes, where it is connected to TV sets. Because the wire is physically in the home, signals can be sent back up the wire to the headend, enabling cable to be interactive. Although wires must

cable TV

be shielded in some way, they are not as subject to interference as signals traveling through the airwaves. Therefore, it is possible to put a large number of signals in a fairly small wire, including HDTV signals. Most cable systems have converted from coaxial cable to fiber optics except for the wires at the end of the line that go into consumer homes.[14]

internet

Wires are also the primary carriers of the internet, although, as already mentioned, it can also be distributed in a wireless manner. One of the problems of internet transmission, wired or wireless, is the possibility of **gridlock.** When material streams over the internet, each computer is delivered a different stream. Unlike over-the-air broadcasts or satellites where one signal can go to many receivers, each person accessing the internet receives a separate stream of information, even though they may all be seeing or hearing the same content. If too many people try to access one website at the same time, most of them cannot get through because there are not enough streams available, and that is how gridlock sets in. Organizations that host websites often charge by the number of streams a particular website owner wants to have available.

The element of bandwidth is a factor for the internet also and relates to how quickly information can move from one point to another. Originally, internet data were sent over dial-up phone lines at what, by today's standards, were very slow speeds—1,200 bits per second (bps). This meant that only words, not pictures, sound, or video, could get through, and even then it took a long time for information to download. Speeds increased to more than 28.8 kilobits per second (28,800 bps) using phone wires, but it wasn't until both phone companies and cable companies developed technologies (primarily **cable modems** and **DSL** phone lines) that speed increased to 6 megabits per second (6,000,000 bps) and then upwards of 100 Mbps. These faster technologies, which usually use fiber optics, are referred to as **broadband** and have facilitated the transmission of audio and video.[15]

competition

Wireless and wires often compete for the same business. Some phone calls are sent over satellite, some over wires, and some over a little of each. Using wires is still the most common way to receive the internet, but wireless is increasingly popular. Wireless is more flexible in that it has the potential to be picked up anywhere, and it can be changed relatively easily. Once wires are installed, it is difficult and expensive to replace them with a newer technology. But wires are more secure. It is harder to pirate a signal that is riding securely in a wire than one that is floating through the air.[16]

13.10 Pick Up and Carry

From the ethereal to the mundane, many forms of program distribution involve the physical transport of a program or movie from one place to another. Usually this means that one or more human beings on foot or in a vehicle actually pick up tapes, films, or discs and carry them. One term used to describe this type of distri-

bicycling

bution is **bicycling,** because at one time messengers on bicycles carried material from one point to another. The old educational television network that mailed copies of tapes from one station to another used bicycling (see Chapter 3). Some cable TV public access programming involves bicycling; the community people who produce the programs carry the tapes or discs to several different cable TV systems, where they are placed on playback machines and cablecast on access channels.

One of the best examples of pick up and carry is the distribution pattern by which DVDs get to the video store and then to the consumer. The program material is duplicated onto many discs, which are then shipped to wholesale houses or retail stores. Consumers, on foot, in a car, or perhaps even on a bicycle, go to the store, buy or rent the program, take it home, and put it in their machine. This method of distribution is being challenged by companies that are distributing the same material through the internet, so it can be downloaded to a hard drive through either wired or wireless means.

home video

Another good example of pick up and carry is currently found in the movie business. When a new release comes out, thousands of prints of the film are made. These are shipped to movie theaters throughout the country, so they can all profit from the film's promotion and start showing it on the same day. Making prints is an expensive process, estimated to total more than $1 billion a year, and distributing them the present way is cumbersome.[17]

movies

Although the pick-up-and-carry method of distribution is not as glamorous as satellites, fiber optics, or broadcast transmission, it is often both effective and inexpensive.

13.11 Motion Picture Exhibition

A broadcast antenna, a satellite, a wire, or even a Federal Express van does not succeed in distributing program material unless someone actually watches or listens. For the movie theater business, this means that people come to the theater, purchase a ticket, and sit in the dark where they see and hear the movie. The most common method currently used for this process is to place the film on a projector where frames are pulled down in front of a light that shines through a lens. The picture is projected onto a beaded screen that is particularly responsive to picking up the light images. The sound for the film usually rides beside the picture frames and is sent to multiple speakers that give the audience members the feeling that they are surrounded by the sound (see Exhibit 13.16).[18]

projector

digital cinema

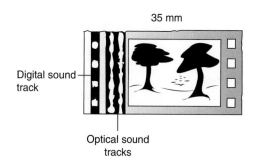

35 mm

Digital sound track

Optical sound tracks

Exhibit 13.16

A frame of film that contains the picture and two types of sound. Optical sound was developed first and is still used in many theaters. This film frame has two channels of optical for stereo sound. Digital sound rides between the sprocket holes and often contains six tracks for surround sound.

However, many motion picture theaters are converting to digital projection. Referred to as **digital cinema** (see Chapter 6), this methodology involves showing movies using digital media such as hard drives and servers rather than film. Obviously, the projector needed is different from that used for film. One basic design for a digital cinema projector involves more than a million tiny mirrors, each mounted on a hinge that can tilt thousands of times per second. Some of these mirrors respond to digital data representing reds, some to blues, and some to greens. The

0s and 1s to which the image has been reduced instruct each mirror to pivot so that it either reflects or deflects the colored light needed for that image. The light from the mirrors is magnified and focused on the cinema screen to reproduce the image.

The standards that were set for digital cinema in 2005 by seven movie studios (see Chapter 6) covered many technological aspects. They included specifications for such things as the distribution master, protection of content from piracy, and parameters for two digital projection formats—2K (2,048 horizontal pixels by 1,080 vertical pixels) and 4K (4,096 × 2,160). It is generally conceded that 4K is about the quality of 35mm film. It has also been found that 4K projection lends itself well to **3-D** movies.

studio support

Because digital cinema financially benefits the movie studios more than it benefits the movie theaters, the studios are bearing most of the cost for the conversion. The studios will save because they no longer need to make expensive film prints and are passing those savings on to theater owners who need to convert to digital projection. Theater owners are also hopeful they will bring in more customers if they make their theaters 3-D capable.[19]

13.12 Radio Receivers

For radio broadcasts, radio receivers pick up and amplify radio waves, separate the information from the carrier wave, and reproduce this information in the form of sound. Some radios also have digital displays that list facts about songs and other information. The major change in radios over the years is that they have shrunk; some are now small enough to fit in the ear. A radio has been developed that can handle both digital and analog AM and FM. At present, a different receiver is needed to hear satellite radio. Some of the satellite radio receivers are movable, in that they can plug into the car or into a home electric socket. Engineers are continuously improving radios, so there may soon be some that can handle all the forms of audio broadcast.[20]

radios

13.13 TV Sets and Computer Monitors

TV sets

TV sets come in a variety of forms. For many years, they were all cathode-ray tube based. An **electron gun** located at the back of the picture tube sends a stream of electrons to a **phosphor screen** at the front of the tube. When the electrons hit the phosphor screen, they cause it to glow, creating the television picture. Tube-based sets are deep because there needs to be room between the gun and the screen. In recent years, thin TV sets, often referred to as **flat-screen TV,** have been developed. The two main forms are **liquid crystal display (LCD)** and **plasma.** LCD screens are made up of two transparent pads with crystal molecules between them. These molecules light up according to the voltage applied to them, and the voltage relates to the characteristics of the electrons making up the picture. Plasma screens are made up of two glass panels with gas between them. The gas activates colored dots according to the pattern of voltages applied.[21]

computer screens

Computers have screens that look very much like TV sets and are constructed similarly, with tube-based, LCD, or plasma technology. For the most

ZOOM IN: **TV on the Road**

It was baseball that motivated Blake Krikorian to develop Slingbox, a small box that attaches to your TV set and allows you to transfer all the channels you receive on your TV to your computer, no matter where you are. Krikorian was on the road and wanted to watch the San Francisco Giants, but he couldn't do so from where he was located. So he came up with the idea of a device that enables people to see what they are used to watching on TV whether they are one room away or a continent away. He co-founded Sling Media and developed Slingbox.

In order for Slingbox to work, you must connect it to your TV set and your router and install software on your computer. If your router and TV set aren't in the same area, you can connect through an electrical circuit, a wireless gaming adapter, or equipment offered by Sling Media. What Slingbox does is send your TV set's signals through the internet, so you can watch them on your computer monitor. You have complete remote control of your TV set, even if it's not turned on, from wherever you operate your laptop computer—a hotspot, a hotel room. You can watch your hometown stations, the cable or satellite channels you subscribe to, and anything you have recorded on your DVR. You can also send all the signals to your cell phone, PDA, or other portable device and watch them there.

Krikorian notes that just as TiVo established "time shifting," Slingbox establishes "place shifting."

---

part, however, they cannot receive over-the-air, cable TV, or satellite TV signals. They are used primarily for displaying the internet, video games, movies on DVDs, and documents created with computer programs such as Word and Excel (see Chapter 2). Technically, it is possible to make a computer monitor capable of displaying both television and computer applications, and many such systems have been proposed. Some, such as Slingbox (see "Zoom In" box), have achieved modest success, but none has caught the overall fancy of the public as yet.[22]

Video material, whether from a TV set or a computer, can also be sent from a projector onto a screen. The oldest technology for **projection TV** consisted of a unit that took the signals from a conventional TV tube, magnified them, and projected them onto a large screen. This form of projector was heavy, however, and when LCD technologies were developed, they were integrated into smaller projectors that could be placed on a table or hung from a ceiling (see Exhibit 13.17). Other projection technologies have since been developed, two of which are DLP and LCOS. TV sets can also be small and portable, as evidenced by cell phone screens that can be used to view video material.

projection

Television sets have improved in many ways. For example, TV sets used to contain tiny, three-inch speakers that delivered low-quality monaural sound. When stereo TV was authorized in the 1980s, this changed rapidly. Now TV sets have excellent speakers that produce high-quality sound. Some people build themselves "home theaters" that include speakers placed around the room to give a surround-sound feeling. Another big improvement of the 1980s was the **remote control** that allowed viewers to change channels from the comfort of their armchairs. Through the years, the remote control has added many features, such as muting, freeze-frame, and the ability to view several TV programs at once. The advent of HDTV has brought a change in TV set **aspect ratio**—the relationship

improvements

High Definition

Exhibit 13.17
An LCD video projector.

(Courtesy of Sharp Electronics Corporation)

Exhibit 13.18
The top monitor is the newer 16:9 aspect ratio; the bottom is the older 4:3 aspect ratio.

Standard Definition

of the width to the height of the screen. Previously sets were rather square with an aspect ratio of 4 units wide for every 3 units high. One of the parameters for HDTV involved changing that to 16 by 9, a much wider screen similar to that used for widescreen movies (see Exhibit 13.18).[23]

13.14 Issues and the Future

screens

It will be interesting to watch the battle of the screens. Will high-definition TV pictures and sound prove to be such an improvement that people will stay home for their media enjoyment and stop going to movie theaters? Will digital cinema and the possibility of 3-D movies draw more people to the theaters? Will TV sets become computer screens, or will computer screens become TV sets? Will people discard their TV sets and the broadcast, cable, and satellite programming they provide and instead watch the video material that is programmed for the internet? Will small-screen portable TVs become the predominant video-viewing device?

new ideas

Technological improvements that are in the nascent stage or not yet thought of are bound to affect the future. For example, PBS is setting up a system whereby its affiliates can access PBS programs from a server and have them delivered by satellite. In other words, PBS will not initiate sending the programs; the affiliates decide what programs they want and, through computer and satellite technology, pull them from the server to the station. The FCC has opened up frequencies above 70 GHz referred to as **E-band.** The range of these signals at such a high frequency is very limited, but the speed at which they can transmit data is about 1,000 times faster than today's cable modem or DSL line. This will probably trigger new technologies unheard of today. Perhaps you will receive movies by email.[24]

One asset, and also a liability, of digital technology is that it can be manipulated easily. This is an advantage to those in the production business because it is easy to edit and create special effects. However, digital technologies can also make things appear to be what they are not. One person's head can be placed on another person's body. People can be made to look as if they are in a location where they have never been. The ethical implications of this are obvious. At present, there is nothing but a sense of ethics to keep electronic media practitioners from damaging people by manipulating them digitally.

**digital
manipulation**

The radio waves portion of the electromagnetic spectrum is finite. It does not go above 300 GHz because at that point it turns into infrared light. Many media entities want to use radio waves, and although the potential demands have not yet exceeded the supply, various organizations are arguing for more space on the spectrum. The Department of Homeland Security established after 9/11 definitely feels the need for spectrum space, as do other safety and security agencies, and they believe their needs should take precedence over providing entertainment. Add to this the government's desire to make money by auctioning off the frequencies to the highest bidder, and the future management of the electromagnetic spectrum could be in for some rocky controversies.

spectrum issues

Some question whether over-the-air TV broadcasting is still needed in the digital age. Satellite TV is already digital and takes little spectrum space because one set of signals is going to many homes. Cable can accommodate digital without taking up any frequency space. Broadcasters have convinced the FCC that the current conventional stations should be given precious frequency space to convert so that they would not be left out, but why do they need to broadcast through the airwaves when most people receive their signals over cable or satellite?

broadcasting

There are bound to be problems with each new technology. Closed-captioning for the hearing impaired poses problems for HDTV because it cannot be accomplished technically using the methods that were used for standard-definition TV. Dealing with the new aspect ratio may cause production and exhibition challenges. However, these types of problems are not new to the electronic media world. Color TV, UHF, and satellite TV went through similar throes. The technologies for production, distribution, and exhibition will no doubt continue to change at a fairly rapid rate.

**dealing with
problems**

13.15 Summary

Various media have different production, distribution, and exhibition requirements, but all of them that were not digital to start with are converting from analog to digital. Digital with its on-off underpinnings has advantages in terms of quality and compression.

Radio is produced with audio equipment such as microphones, recorders, audio boards, computers, and speakers. It is sent to transmitters, where it is modulated as AM, FM, or PCM and distributed over airwaves by antennas. The various forms of radio have different characteristics caused by their positions on the electromagnetic spectrum (designated by hertz), their bandwidth, and accidents of history. Some radio is sent from satellite. Radio receivers demodulate and amplify signals and play them through devices that have greatly shrunk in size over the years.

Broadcast, cable, and satellite TV programming is produced with both audio and video equipment. Cameras capture the image onto chips; recorders, such as those in camcorders, store the image; switchers mesh pictures together; and monitors display them. Lights, which are used for illumination and mood, are trending toward cooler, more energy-efficient bulbs; editors are nonlinear; and visual effects are computer-based. Broadcast signals utilize the electromagnetic spectrum; their digital transmissions encompass HDTV with either interlace or progressive scanning. Cable TV systems depend on wires for distribution from the headend, using both coaxial cable and fiber optic cable. Satellite TV broadcasts use synchronous satellites with transponders and frequencies that are higher up on the electronic spectrum than broadcast—either C-band or Ku-band. All are displayed on TV sets or projectors that have changed from tube based to LCD or plasma based. Sound accompanying TV has improved greatly, and sets have recently adopted a new aspect ratio.

Movies are shot on film, which is a chemical medium, or video and edited electronically. They have been distributed through a pick-up-and-carry method, in which many prints are made and delivered to theaters that use film projectors to exhibit the movies. This is changing, however, as theaters convert to digital cinema, which is being underwritten by the movie studios who stand to gain the most financially from the conversion.

The internet is most commonly sent over wires but also utilizes wi-fi technology and hotspots so that it can be displayed on portable devices. Individual streams are sent to each computer, which can result in gridlock. Broadband connections are best for reception of the internet. It is exhibited on computer monitors, which look very much like TV sets and are technically capable of being combined with TV sets.

Many other media forms use wired or wireless technologies. Telephones use wires and low-orbit satellites. Microwave, shortwave radio, cell phones, and new services on E-band are all part of the technological underpinnings of electronic media.

Suggested Websites

www.audio-technica.com (Audio-Technica, an audio company specializing in microphones and headsets)

www.panasonic.com (Panasonic, a manufacturer of video and audio equipment such as cameras and TV sets)

www.smpte.org (Society of Motion Picture and Television Engineers, a trade organization that, among other things, sets technical standards)

www.sony.com (Sony, a company known for audio and video equipment, particularly cameras)

www.yamaha.com (Yamaha, a company that provides audio gear such as audio boards and musical equipment)

Notes

1. "Digital Compression," *CTI,* October 2000, 46–51; and "Crunch Time for MPEG-4 Standard," *Electronic Media,* October 29, 2001, 9.
2. For more on audio equipment, see David E. Reese, Lynne S. Gross, and Brian Gross, *Audio Production Worktext* (Boston: Focal Press, 2009); "Directional Mic Behavior and Usage Examined," *TV Technology,* June 13, 2007, 34; "Audio Consoles Get Heard," *TV Technology,* May 26, 2003, 60–62; and "Location Surround Sound Recording Gets Small and Flash[y]," *P3,* March 2007, 18.
3. "Red Delivers on Its Promise," *TV Technology,* November 21, 2007, 10; "Format War Hits New Ground," *Broadcasting & Cable,* October 1, 2007, 18; and "Tubeless Camera Unveiled by RCA," *Broadcasting,* April 16, 1984.
4. For more on video equipment, see Lynne S. Gross and James C. Foust, *Video Production: Disciplines and Techniques* (Scottsdale, AZ: Holcomb Hathaway, 2009); Lynne S. Gross, *Digital Moviemaking* (Boston: Cengage, 2009); "Illuminating HD," *TV Technology,* February 20, 2008, 33; "Pushing the HD Editing Envelope," *TV Technology,* January 19, 2005, 40; and "Indies Embrace New High-Def and DV Formats," *Daily Variety,* September 8, 2004, 32.
5. "The Crowded Airwaves," *Fortune,* March 31, 1997, 133; "The Battle for Bandwidth," *Mix,* October 2008, 56–59; "Spectrum for Sale," *Wall Street Journal,* August 4, 2006, A-11; and "FCC to Decide in Battle for TV Spectrum," *Wall Street Journal,* August 18, 2008.
6. "Radio Tunes in Warily to Its Digital Future," *Los Angeles Times,* April 7, 2003, C-1; "AM Stereo Rears Its Divided Head," *Broadcasting & Cable,* April 12, 1993, 61; and "System Sends High-Speed Data with FM Subcarriers," *Microwaves,* February 1995, 6–7.
7. "The Digital Pace," *Broadcasting & Cable,* September 29, 2008, 24; "Broadcasters Agree to Go All Digital," *Los Angeles Times,* July 13, 2005, C-1; "Stages and Transitions," *Diffusion,* May 2001, 1–12; "FCC Awards New DTV Channels," *TV Technology,* May 8, 1997, 20; and "Converter Boxes Are Coming," *TV Technology,* October 17, 2007, 1.
8. Article Alley, "Interlace vs. Progressive Scan: What's the Difference?" http://www .articlealley.com/article_12985_45.html (accessed October 2, 2008).
9. "Space Odyssey," *Aviation Week & Space Technology,* March 19/26, 2007, 86–88; and "Satellite Soaring," *Mix,* April 2006, 34–35.
10. "After 10 Years of Satellites, the Sky's No Limit," *Broadcasting,* April 9, 1984, 43–68; "Above It All for 20 Years," *Broadcasting & Cable,* September 27, 2004, 45–48; and "Get Ready for a New Breed of Equipment," *Los Angeles Times,* March 3, 1999, C-1.
11. "Microwave Community Faces Major Facilities Relocation," *Enterprise Communications,* March 1995, 62.
12. "Free, Wireless Television Makes a Resurgence," *TV Technology,* June 13, 2007, 42; "ENG Cameras Cut the Cord," *TV Technology,* June 23, 2004, 12; "Wireless Mics Search for RF Haven," *TV Technology,* November 23, 2005, 16; and "'WiMax' Wireless Web Access the Next Big Thing?" *Los Angeles Times,* May 9, 2004, C-1.
13. "Coaxial Cable: The Name of the Game," *Computer Technology,* October 2000, 74; "How, When, Where: Wire," *Sound and Video Contractor,* July 2002, 36; and "Connecting Fiber Optics, *TV Technology,* October 18, 2006, 22.
14. "Will Cable Be Ready for HDTV?" *Broadcasting & Cable,* March 9, 1998, 43; and "Time Warner's Phone Service Shows Cable's Growing Clout," *Wall Street Journal,* February 23, 2005, B1.

15. "Broadband Raises the Bar," *Broadcasting & Cable,* June 23, 2008, 19; and "DSL's Big Push," *Los Angeles Times,* December 21, 2000, T-1.
16. "The Benefits of Optical Fibre," *Communication Technology,* December 2001, 38; and "Going the Distance," *Sound and Video Contractor,* May 2001, 38.
17. "Coming Soon: Movies You Rent on the Web—and Then Download," *Wall Street Journal,* August 30, 2004, B-1; "Successor to the Compact Disc?" *TV Technology,* August 8, 2007, 34–35.
18. "Digital Sound in the Cinema," *Mix,* October 1995, 116–29; and "Digital Sound and Picture," *Mix,* January 1996, 165–70.
19. "Major Studios Agree to Back Switch to Digital Projection," *Los Angeles Times,* October 2, 2008, C-1; "Sony Tries to Build Buzz on 4K Projector," *Hollywood Reporter,* June 26, 2008, 6; "Digital Cinema: Devil Is in the Details," *Daily Variety,* March 24, 2004, A-6; and "3-D Cinematography," *American Cinematographer,* April 2008, 52–63.
20. "New Chip for HD Radio," *Broadcasting & Cable,* September 8, 2003, 17.
21. "LCDs Versus Plasmas," *Sound and Video Contractor,* June 2003, 26–30; and "HDTV Is Getting Clearer," *Newsweek,* September 26, 2005, 65.
22. "Intel Moving into Digital Living Room," *Los Angeles Times,* October 5, 2004, C-1; "Internet TV: Coming to a Screen Near You," *TV Technology,* April 11, 2007, 74; and "Blake Krikorian," *Broadcasting & Cable,* April 14, 2008, 40.
23. "Revolution by Remote," *Los Angeles Times,* April 13, 2003, E-29; "Dolby's Transition to Greatness?" *Hollywood Reporter,* February 19, 2008, 9; and "The Ultimate in TV," *Consumer Reports,* March 2004, 18–20.
24. "PBS Pushes File-Based Future," *Broadcasting & Cable,* June 18, 2007, 14; and "Wiring the World," *Aviation Week,* September 15, 2008, 37.

Chapter 14

THE INTERNATIONAL SCENE

Electronic media are very important to the world. The ability to communicate almost anything to the most remote part of the earth instantly has changed the way people and governments act. Conversely, the entire world is very important to electronic media. United States media organizations depend on worldwide distribution for profits, and technologies such as satellites and the internet have created a worldwide market for advertised products.

Media systems in most countries have developed very differently from those in the United States. Knowledge of the history and customs of various systems around the world is crucial to anyone who intends to deal with global media. Organizations vary as boundary lines are crossed, yet common elements glue the world's media systems into a unified whole.

The orbiting satellites herald a new day in world communications. For telephone, message, data, and television, new pathways in the sky are being developed. They are sky trails to progress in commerce, business, trade, and in relationships and understanding among peoples. Understanding among peoples is a precondition for a better and more peaceful world. The objectives of the United States are to provide orbital messengers, not only of words, speech, and pictures, but of thought and hope.

President Lyndon B. Johnson

14.1 Early Film

Producing theatrical movies is expensive, so many countries engaged in little or no early film production. In the early 1900s, Hollywood came to dominate the production and distribution of popular movies, but other countries historically have contributed to film's repertoire. As mentioned previously (see Chapter 6), the Lumière brothers of France were instrumental in creating turn-of-the-century short movies and developing a film projection system.

France

During the 1920s, France was the gathering place for artistic experimentation. Directors such as Luis Buñuel, Salvador Dali, Jean Renoir, and René Clair produced movies with such techniques as dream effects, slow motion, distortion, multiple exposures, and visual symbolism. In René Clair's 1923 *The Crazy Ray,* a scientist with a magical ray freezes activity in Paris so that only a few people can move about. Clair uses the frozen motion for social commentary by showing such scenes as an unfaithful wife frozen in the arms of her lover and a pickpocket with the wallet of a frozen victim (see Exhibit 14.1).[1]

Germany

Germans also developed film techniques during the 1920s, many of them starting in the 1919 film *The Cabinet of Doctor Caligari.* Some of their films used cameras in a subjective manner to show external events as seen from an individual's viewpoint. Almost all German shooting of this period was confined to studios, where the filmmakers could carefully control lighting, sets, and other elements. The main studio, Ufa (Universum Film AG), had repertory players, not movie stars, and directors who favored expressionistic lighting that emphasized highlights and shadows. Many of the directors and technicians who worked in Germany came to the United States in the 1930s and were instrumental in developing Hollywood films.[2]

Russia

Russia was another country whose practitioners made major contributions to early film techniques. The Russians, particularly Sergei Eisenstein, are best known for developing editing techniques. One of Eisenstein's best-known films, *Potemkin* (1925), was about a 1905 rebellion on the battleship *Potemkin* and the reprisals by the czarist government. One scene involves a sailor smashing a dish, an action that takes but a few seconds, but Eisenstein uses an edited sequence of 11 shots to emphasize the sailor's rage. In another scene, in which innocent people are killed as they run down steps, Eisenstein uses a rhythmic variety of shots—long shots from the bottom and top of the steps, close-ups of faces expressing horror, moving shots that reinforce the chaotic movement of the people (see Exhibit 14.2).[3]

Exhibit 14.1

In René Clair's *The Crazy Ray,* a man is frozen on the side of a building.

(©Photofest)

14.2 Early Radio

Most countries of the world experimented with radio broadcasting during the 1920s. In some countries, private radio clubs were the initial broadcasters; in other countries, the government took the lead. Radio was slow to develop in the Middle East because the conservative Muslim leaders, who could not understand how wireless worked, suspected radio to be the work of the devil. Ibn Saud, Saudi Arabia's leader, who wanted radio, devised an experiment to satisfy the religious leaders. He asked a group of them to travel to Mecca, where they were to await a wireless transmission that he would send from the capital city, Riyadh. At the appointed time, he read passages of the Koran to them. Because the devil cannot pronounce the word of the Koran, he convinced them that radio must be the work of human beings, not the devil.[4]

Three basic structures for broadcasting eventually developed throughout the world—private, public, and authoritarian. **Private** radio stations were owned by private businesses. This U.S. model was also the predominant model in Latin America. In Mexico and South America, the radio stations were owned by media barons who handed their broadcasting empires down from one generation to another. The main electronic media company in Mexico was founded by Emilio Azcárraga and was passed on to his son and grandson, also named Emilio. The elder Azcárraga started the company with a chain of radio stations. In the 1930s, he introduced trumpets into Mexican mariachi bands so the music would sound livelier on radio, and today all mariachi bands have horns. In Brazil, the major early radio network was Diários e Emissoras, led by Assis Chateaubriand, but Rádio Globo also held a dominant role. The latter was headed by Roberto Marinho, who would later lead his company to overwhelming dominance in television. Most of these moguls came from the newspaper business, so radio and print journalism were linked. Radio was supported by commercials and featured live news, variety programs, comedies, and **rádionovelas** (serials or soap operas).[5]

Exhibit 14.2
One of the close-ups of a human face from the slaughter on the steps in Eisenstein's *Potemkin*.

(Kobal Collection)

The **public** broadcasting structure was closely aligned to the government and was supported by **license fees** collected from members of the public who owned radios. Britain developed this model, which was then adopted by most of Europe and by British colonies throughout the world. The British Broadcasting Company was formed in 1922 and five years later became the British Broadcasting Corporation (BBC) under a royal charter. This **charter** gave the BBC a monopoly on all radio broadcasting and created a board of governors appointed by the monarch for five-year terms. This board was the policy-making group, while a director-general and staff performed the day-to-day operations of the BBC. The charter stated that there could be no advertisements on the BBC and that it must broadcast daily impartial accounts of the proceedings in Parliament. The license fees were collected by the post office.

Programming on the early BBC was very paternalistic. Designed to upgrade tastes, it was often referred to as programming from "Auntie BBC." The original organization consisted of three national program services, each designed to lead the listener to the next for a higher cultural experience. At the lowest level was the Light Programme, which consisted of quiz shows, light music, children's adventure, and serials. Level two was the Home Service, which included dramas, school broadcasts, and news. The highest level, called the Third Programme, was classical music, literature, talk, and drama. All these services were national, although the BBC later introduced local stations.[6]

Saudi Arabian experiment

private systems

Latin America

public systems

Britain

The **authoritarian** broadcasting structure was closely supervised by the government. It predominated in countries with dictatorships. Russia provides the main model because, from the beginning, broadcasting in Russia was a state monopoly. When the Communists came to power in 1917, one of Vladimir Lenin's first acts was to centralize radio, which was then just developing. In 1924, the Communists founded an organization called Radio for All Society that started radio stations in major cities. Several years later, the Communists formed a state organization that became Gostelradio, the bureaucracy that eventually oversaw all Soviet radio and TV. Radio programming consisted primarily of news and information that the government wanted the people to hear. Educational and cultural programs were evident, but entertainment programs were scarce. The whole system was supported from general tax dollars and did not broadcast commercials.[7]

14.3 The Colonial Era

During the formative days of radio, many European countries had colonies in Africa, the Middle East, and Asia. With Britain's widespread holdings, the saying that "The sun never sets on the British Empire" was true. The British exported their public, noncommercial, charter-oriented model of broadcasting, with its license fees and paternalistic programming, throughout the world. They also established the British Empire Service, a **shortwave** service that sent news and other programming from England to its colonies and dominions in such far-flung places as Australia, Canada, India, and Nigeria.[8]

The British style was to encourage British subjects to use radio to preserve and enhance the local culture. As a result, Britain started the radio organizations with British citizens but focused on training natives to run the stations and networks. After several years, the number of local people working in broadcasting far surpassed the number of British. By the 1940s, some of the colonies even offered services in the vernacular language along with their English services.

This was different from other colonial powers, which exerted stronger governing influence. The French, for example, had a policy known as "assimilation" that used radio broadcasting to tie the colonies firmly to the motherland. Not only were most of the employees French, but many of the programs were produced in France.

When the colonial era ended in the 1950s and 1960s, people in the British colonies continued broadcasting operations because they knew how, while people in the French colonies started over from scratch. The Algerians, for example, forced the French to leave very quickly in 1962, but this left hardly anyone capable of maintaining or operating the broadcasting structure. Almost all the technicians had been French, and when they left, the entire broadcasting system came to a temporary standstill.[9]

14.4 World War II and Its Aftermath

As tensions grew in Europe in the late 1930s, countries used radio for propaganda purposes. In 1937, Benito Mussolini's Fascist Italian army invaded Ethiopia and set up a shortwave station to broadcast anti-British propaganda in Arabic to the Arab world. The British concluded that they could no longer concentrate just on their Empire Service and on building broadcasting systems in their colonies. They needed to counter propaganda transmissions. To do this, they set up an **external service,** the BBC World Service, to broadcast worldwide over shortwave in many different languages. During and after the war, people in many countries listened to this BBC service.[10]

BBC World Service

In 1940, the United States started an external service, which later became known as Voice of America (VOA). Originally, it was aimed at Central and South America because these areas were friendly to Germany, which by then had achieved military victories in Europe. Eventually, VOA broadcast to other areas and became as well known as the BBC. During the cold war period that followed World War II, the United States Information Agency (USIA), which then oversaw VOA, established Radio Free Europe and Radio Liberty, both of which were designed mainly to get America's message into the Soviet Union and its affiliated countries.[11]

VOA

Russia also strengthened its external service, Radio Moscow, during the war. The service was formed in 1925 when a live report of a military parade in Red Square was broadcast to foreign listeners in English, German, and French. During the cold war, Radio Moscow was a widespread Soviet propaganda service that reached many countries.[12]

Radio Moscow

Another American service that developed during the war was the Armed Forces Radio Service (AFRS). Enterprising servicemen in Alaska set up a transmitter and wrote to Hollywood stars asking for radio programs. The stars could not send the programs because of security regulations, so the servicemen contacted the War Department in Washington, which set up AFRS to provide servicemen with American programming they could hear in the field. Some of the broadcasting was accomplished by shortwave, and some was undertaken by troop-operated stations. Several stations actually moved along with the advancing armies, but most were stationary. Studios were set up in Los Angeles so that popular stars could perform on programs, such as *Mail Call,* that were sent to AFRS facilities (see Exhibit 14.3). AFRS was later changed to AFRTS (American Forces Radio and Television Service) and still programs for military personnel and others. Its television programming consists of the best of U.S. network programming, news, sports, and family-oriented material. Radio programming covers sports, news, public radio, and various music formats.[13]

AFRS

Exhibit 14.3

Walter Brennan (*left*) and Gary Cooper perform for AFRS's *Mail Call* about 1943. They read letters to entertain and boost morale of people fighting in World War II.

(Courtesy of True Boardman)

Before the war, Germany had a decentralized radio system. The government wanted to cover all of Germany with just one central radio station, but this plan failed because in the 1920s it was technically impractical to transmit one signal to all the little towns and valleys. As a consequence, Germany

German system was divided into broadcast regions, each operating at least semi-independently. When Adolf Hitler rose to power in the early 1930s, he quickly realized the power of radio. In 1933, the government, which by then was controlled by his Nazi Party, established a Ministry for Public Information and Propaganda and put all radio stations under its power. People who had been top radio administrators were sent to concentration camps, and most employees lost their jobs. All information programming was subject to directives from the Nazis, and entertainment programming was not allowed to be "destructive." This meant that compositions by Jewish composers were forbidden, and jazz disappeared from the airwaves.

After the war, the Allied powers that occupied West Germany reestablished a decentralized form of radio, this time not for technical reasons but for political ones—they did not want radio to aid the resurgence of a strong central government. They set up the ARD, an umbrella organization for stations in nine regions, but it did not broadcast anything nationally. ARD was the only broadcast organization service in West Germany for a number of years. East Germany went its own way with an authoritarian system along the lines of that of its occupying power, the Soviet Union.[14]

Japanese system Japan, before the war, had a government-regulated, nonprofit, private broadcasting system. Private companies (primarily newspapers) owned the stations and collected the license fees, but they were heavily overseen by the government. In 1926, the government formed NHK, a national public-service network that provided much of the programming for the stations. During the war, radio was under state control. After the war, the Allied powers reorganized broadcasting to include both NHK, as a government public-service network similar to the BBC, and

China new private commercial stations that were not allowed to form networks. They were intended to be local stations serving local needs. However, eventually large newspaper companies bought commercial stations in several areas of Japan and played similar programming on all of them, thus creating a semi-network.[15]

As Mao Tse-tung was rising to power in China in the 1940s, he and his revolutionaries set up guerrilla radio stations built with Russian transmitters and scavenged parts. These stations operated out of caves and partially destroyed buildings but were the basis for what eventually became China's centralized radio system, CPBS.[16]

Film, too, was used during World War II for propaganda purposes, especially by Mussolini in Italy. After the war, filmmaking in Europe gradually returned, largely as an artistic experience with the director firmly in charge of style and content. One well-known Italian director was Federico Fellini, whose main star was his wife, Giulietta Masina. With her, he made such films as *La Strada* (1954) and *Night of Cabiria* (1956). Fellini's main theme was sensuality versus spirituality (see Exhibit 14.4).

Exhibit 14.4

Giulietta Masina (*right*) in a scene from *La Strada*.

(©Photofest)

Ingmar Bergman of Sweden was another influential director who probed human emotions and was concerned with spiritual conflict and the fragility of the psyche.[17]

postwar film

Some of the French directors who had made movies before the war picked up again, making the same type of artistic films. A group of film critics, including André Bazin, François Truffaut, and Jean-Luc Goddard, harshly criticized French films, but eventually some of them joined the production process and became directors. England's films of the 1940s and 1950s tended to be somewhat stuffy adaptations of classic literature. The British also shot "quota quickies"—**B pictures** that met the quota of British films that theater owners had to show. These were screened as part of double features to go with American movies that people were more interested in seeing.[18]

14.5 Early Television

Television developed later in most of the world than it did in the United States. One reason was that much of the world was physically recovering from World War II and had other priorities. Also, television was much more expensive than radio, so poorer countries could not easily afford it. TV also was not adopted for social reasons. In South Africa, TV was not introduced until 1976 because the apartheid rulers did not want to disturb the "South African way of life." They feared programming from outside the country because they thought it would create dissatisfaction among the nonwhites who might see how people of their races lived in other countries.[19]

effects of war

South Africa's lateness

In Muslim countries, religion inhibited the introduction of television even more than it had the introduction of radio because the Muslim religion prohibits the creation of graven images—which television was considered to do. In Saudi Arabia, the introduction of TV proved to be fatal to the king. When test TV transmissions started in 1965, Saudi religious conservatives marched on the station. One of the conservatives, Khalid, was killed by a policeman during a struggle with an official of the Ministry of the Interior. Khalid's family appealed to King Faisal to punish the person who shot him, but Faisal said the policeman had acted properly. Ten years later, Khalid's younger brother shot and killed King Faisal and was reported to have said, "Now my brother is avenged."[20]

Saudi Arabia's religion

Television eventually developed almost everywhere, and once again the British model was the most common. The British still had many of their colonies during the early days of TV, but even the countries that were no longer tightly allied with Britain included television in their already established radio structures. Britain had televised the coronation of George VI in 1937—quite a feat for that time. World War II interrupted development, and it was not until 1953, with the televised coronation of Queen Elizabeth II, that the British people became aware of TV and purchased sets.

British influence

The Latin American countries, which like the United States did not experience the ravages of World War II within their borders, started TV relatively early. Part of the reason for this early start was that media companies in the United States, seeking profits, invested in Latin American media. NBC assisted Argentina; ABC was active in Venezuela; CBS invested in Uruguay; and Time-Life

Latin American progress

gave money and advice to Roberto Marinho, who built TV Globo into Brazil's dominant TV network. In general, the same newspaper moguls who dominated radio also dominated television.[21]

14.6 Broadcasting's Development

Between the 1950s and the 1980s, the private, public, and authoritarian systems of broadcasting continued, but they underwent changes and amalgamations. In Britain, the programming's paternalistic nature was criticized. In 1963, **pirate stations** located on ships anchored off the coast of England began broadcasting rock music, which BBC radio disdained. This programming became so popular that when the government planned to suppress it, a huge public outcry prompted the BBC instead to revamp its radio programming from three to four services, one of which played rock and popular music.

pirate stations

ITA and ITVs

A heated 1954 debate in Parliament about the quality and role of television led to the establishment of the Independent Television Authority (ITA), a new television service that was structured very differently from the BBC. The ITA was a regionally based national network that did not produce programs. All programming came from 15 independent companies, called ITVs, selected and franchised by the ITA. Each of these companies served a different area of the country and programmed local material, as well as contributing to the national fare (see Exhibit 14.5). After 1954, a number of new TV services (both BBC- and ITA-operated) were added in Britain, and public broadcasters expanded their offerings in other parts of Europe. Advertising was a controversial subject, with broadcasters trying to maintain services with few or no commercials. But television was too expensive to be supported totally by license fees, and slowly but surely, television commercials were accepted by Europe's public systems.[22]

commercials

Exhibit 14.5

A production scene from *Coronation Street*, one of independent British TV's most popular early programs. Its cast of characters often interacted in the pub, Rovers Return Inn.

(Courtesy of Granada Television)

The public systems adapted in other ways to meet their local needs. Canada provides a good example. Because of its ties to Britain, it has a government-run broadcasting system, the Canadian Broadcasting Corporation (CBC), modeled after the BBC. It also has a private broadcasting system, CTV, similar in structure to that of the United States. Most Canadians live near the U.S. border, and the private stations cannot make money servicing the remote areas, so the CBC's mission includes serving these outlying regions. Canada started cable TV in 1950, two years before it had a broadcast TV network. There were 140,000 TV receivers in Canada before it officially started television broadcasting because cable TV systems were bringing in U.S. stations for subscribers to watch. The presence of American programming in Canada is a thorn in the side of the Canadian government. Politicians are constantly trying to stem the tide of U.S. cultural invasion by setting up rules and **quotas** that are enacted, changed, and opposed, all in an attempt to enable Canada to express its cultural identity.[23]

Exhibit 14.6

In the 1950s, the Russians built this television tower on the outskirts of Moscow next to a monument to conquerors of space.

(AP Photo/Alexander Zemlianichenko)

The authoritarian model of broadcasting continued in full force in the Soviet Union (see Exhibit 14.6) and spread to other Communist countries, most of which were in Eastern Europe. The pattern of central control, however, was not as heavy-handed in these countries as in the USSR. For example, Hungary had local cable TV. Unlike the United States, where cable TV became popular because of the national pay and basic channels, Hungarian cable started because the people saw it as a way of showing locally produced programming that was more liberal in content than the national product.[24]

Canada

Eastern Europe development

The Latin American private systems continued to grow. In the 1960s, they cast off the American companies and assumed all responsibilities, financial and operational. Television, which during its early years was viewed primarily by the economic elite, spread to the masses. The video version of the rádionovela, the **telenovela,** became extremely popular, and Latin American countries (particularly Mexico and Brazil) built worldwide markets for these productions.[25]

Latin America

14.7 The Concerns of Developing Nations

The poorer nations of the world had trouble developing television and filmmaking because of the cost. National pride, however, led them to develop TV systems as best they could. For example, in 1963, Haile Selassie of Ethiopia persuaded the newly formed Organization of African Unity to hold its first meeting in his capital city. A European manufacturer supplied closed-circuit TV for the event. Haile Selassie was so impressed that he ordered a regular television station to be installed in time for the imperial birthday—always a big event. A British firm built the station in six months—just in time for the celebration. Thus it was that Ethiopia, a country with one of the world's lowest per capita incomes, had a national television service in 1964 but hardly any sets on which to view its programs.

Ethiopian TV

Developing countries received outside financial help for their systems because the cold war between the democratic countries and the Communist countries was

cold war help

in full force. Each side was trying to woo as many noncommitted nations as possible; one way was to provide funds or technical expertise for television systems.

Middle East oil

When oil was discovered in the Middle East, this desert area became important to the world economy. In the 1970s, oil prices skyrocketed, and these countries became rich. They used some of their money to build fancy broadcasting systems, but the population of the area was so small that the countries did not have adequate personnel to operate the systems and produce programs.[26]

Indian film

Some of the developing countries that did engage in moviemaking—Brazil, Argentina, Egypt—tended to produce movies of a somewhat pedantic nature. An exception was India, which developed a thriving film production community that produced more movies per year than Hollywood. The films were mainly formula musicals intended to entertain. The people of India loved their homegrown movies; movie stars became celebrities, and the music of the films became very popular. Eventually the term "Bollywood" was coined—short for "Bombay Hollywood"—to refer to the movie industry of India, much of which is in Bombay (now called Mumbai).[27] In 2009, the Mumbai–based film *Slumdog Millionaire* won the Best Picture Oscar.

Whether poor in financial or human resources, these developing countries were happy to accept money and technical expertise from other countries. But they did not want the philosophical and social strings that came with foreign programming. Most of the developing countries had just emerged from colonialism and did not want to be subjected to "electronic colonialism." However, these countries could not afford to produce much of their own TV programming, particularly expensive dramas. They were even less likely to produce movies, which cost even more. As a result, they imported TV programs and movies, particularly the readily available American material. Although the residents of developing nations usually liked the American fare, their governments were uneasy about it. They feared that the lifestyles shown on such programs as *Dallas* would make their people discontent or materialistic. They did not want their youth corrupted by American fashion or mores. Also, the amount of sex and violence in American movies did not sit well with government officials.

electronic colonialism

free flow versus national identity

Some of the major conflict between developed and developing countries manifested itself in the area of news. Most of the democratic developed nations were in favor of the **free flow** of information. They believed that any country or media organization should be able to send information to any other country. The developing nations favored **national identity,** believing that each country should decide what information should be allowed to cross its borders. Radio and television signals do not stop at national borders, so debates raged as to what should and should not be broadcast.

NWIO

The United Nations set up a commission to address the free flow–national identity issue and create a **New World Information Order (NWIO).** The commission tackled the complexities, changing realities, and possible solutions concerning the world's communication structure. Its 1980 report, called the MacBride Report after the commission's chairman, supported the free flow of information but strongly suggested that developing countries be provided with the means to contribute as well as receive. It recommended setting up training programs and giving countries more access to technology. Although the report was hailed when it first came out, eventually it was criticized by both sides, and the issues are still not resolved. Satellite broadcasting further exacerbated the problem because signals from satellites reach much farther than broadcast signals.[28]

14.8 The Coming of Satellites

The first American communication satellite, Telstar I, was launched in 1962, and soon after, the U.S. Congress set up Comsat, an agency to handle American satellite issues. At Comsat's urging, Intelsat was formed in 1964 to handle satellite needs for many countries. At first Comsat managed Intelsat, but in 1973 Intelsat became an independent organization owned by a consortium of countries, each of which paid a yearly fee based on its use of services (see Exhibit 14.7). When a nation wanted to use a satellite for an event, such as telecasting the Olympics or a world soccer championship, it could rent satellite time from Intelsat. In that way, any nation, no matter how small, could have access to satellite technology by paying a modest fee. The organization has changed its structure over the years, but it still handles public-service satellite broadcasting.[29]

Intelsat

One of the first forays into large-scale use of satellites for communication occurred in India. In 1975, the Indian Space Research Organization, in conjunction with the Indian public television organization, Doordarshan, launched the Satellite Instructional Television Experiment (SITE), one of the most ambitious experiments in television history. With the help of the United States, it used a National Aeronautics and Space Administration satellite to beam farm, health, hygiene, and family-planning programs four hours each day to 2,400 villages in rural India. SITE was also used to telecast entertainment programs, such as *I Love Lucy*. Because few people in rural areas had their own TV sets, they watched SITE programs in communal areas (see Exhibit 14.8). SITE's goal of educating people about how they could implement solutions to the country's problems was not met. Tests showed that farmers who watched the SITE programs

SITE

Exhibit 14.7

The lobby of the former Intelsat building in Washington, D.C., displaying some of its early satellites.

were not any more innovative than farmers who didn't. NASA had lent India the satellite for only one year, and when NASA disconnected the satellite in 1976, the SITE experiment ended. It brought the wonders of TV to audiences who otherwise would not have seen them, however, and set the stage for later satellite development. It also resulted in many Indian baby girls being named Lucy, after Lucille Ball.[30]

CNN

Other satellite uses were more successful, including broadcasting directly to homes in Europe and Brazil and the distribution of CNN throughout the world. Although Ted Turner started CNN in 1980 as a cable TV network for American viewers, he had internationalization in mind from the beginning. When the service went worldwide in the mid-1980s, entities in various countries slowly subscribed to (or bootlegged) the satellite feeds, until today most countries air CNN.[31]

STAR

Another satellite success was the Satellite Television Asia Region (STAR) system. Started in Hong Kong in 1991 by property developer Li Ka-shing, it broadcast unscrambled signals of music videos, movies, soap operas, sports, and other fare to countries from Egypt to Japan. It created a huge appetite for its programming. For example, in India, where the only television programming was from the sleepy, paternalistic public system, Doordarshan, enterprising entrepreneurs hooked up rudimentary cable systems to serve large apartment buildings. For a fee, each apartment was wired to a VCR for movies and to a satellite dish for CNN and STAR. The hookup rate was phenomenal. The fate of the characters on *The Bold and the Beautiful* became a main topic of Indian conversation. In 1993, Rupert Murdoch purchased a controlling interest in STAR for $525 million. Li Ka-shing had originally invested $250 million in the service, and after Murdoch's purchase, he still owned one-third of the company.[32]

Satellite programming did not receive instant acclaim from governments that disliked their citizens receiving information and entertainment with views and attitudes different from those espoused by the government. The people, however, liked the diversity, and, in the end, the governments gave in. At one time the Greek government said it would shoot down any satellites that flew over its airspace, but by 1988 the Greek government broadcasting system was airing foreign programs brought in by satellite. China, while trying to position itself as a major manufacturer of satellite dishes, forbid its citizens from owning them. This didn't work, and Chinese now have access to satellite programming.[33]

objections to satellites

14.9 Privatization

Privatization was rampant in the 1980s, particularly in Western Europe where the staid public broadcasting systems seemed to be out of step with the times. Privately owned, advertising-supported stations and networks, organized along the lines of those in the United States and offering similar programming, multiplied during the decade. As might be expected, established government broadcasters fought the introduction of these new services. In many countries, the original private stations were illegal local radio stations that operated without a license. They became so popular that eventually they were legalized. In France, the government suppressed most of them until 1981, when Socialist candidate François Mitterand came out in favor of them. He won the election and passed a law legalizing private radio. In most countries, once private radio was authorized, private TV was not far behind.[34]

illegal radio stations

The advent of satellite services and cable TV aided privatization. By the late 1970s, West Germany had two government TV systems, ARD, the loosely connected state-based system, and ZDF, a more centralized network started in 1963. In 1978, German politicians decided to launch four cable TV pilot projects to improve reception. They also decided to use this pilot to experiment with programming ideas, primarily by allowing private program suppliers to provide material for the cable projects. The cable pilots failed, mainly because of marketing inexperience. For example, one company tried to force all subscribers to sign a 12-year contract. In an attempt to solve the cable problem, the states amended their broadcasting laws so the private programming suppliers could operate in other ways. The most common way was by forming private, advertising-supported TV stations. Eventually they also supplied programming to German satellite services and to revived cable systems.[35]

German cable TV

Privatization affected the public systems, which over the years had become bloated with people and aging equipment. The new entrepreneurs—with slim staffs, new technologies, popular programming, and advertising dollars—stole the audience. Many of the public systems lightened their programming, amid anguished cries about the effects of lowering cultural standards. Others went out of business; the French government in 1987, in a very controversial move, privatized its original government service.[36]

effects on public systems

Even the venerable BBC was attacked for overspending and for forgetting its public-service function. When its charter came up for renewal in the 1990s,

**changes
in Britain**

Parliament scrutinized it carefully; as a result, the BBC streamlined itself and stayed in business, but not as the same old "Auntie." Independent television in Britain underwent its share of turmoil also. All the regional ITV franchises came up for renewal, and some broadcasters who had been in business for years lost their franchises to newcomers.[37]

Privatization changed the nature of broadcasting, not only in Europe but also in Asia, Africa, and other parts of the world that had not previously had a private structure. As a result, many people who 15 years earlier had had one or two program choices now had more than 30.

14.10 The VCR

Adding to the choices were the movies and other programs available on videocassette. **Videocassette recorders (VCRs)** became popular in almost every country. In the early 1980s, much of their use was illegal. Cassettes and VCRs were small enough to be easily smuggled into a country. The programming material on the cassettes was easy to duplicate, so pirated copies of tapes were abundant.

piracy

Many of the countries where early **piracy** abounded were poor countries that did not have **copyright** laws. Even people in the governments did not understand royalties and saw nothing wrong with dubbing copies of programs. Over the years, the United States and other developed countries have persuaded many of them to support the provisions of copyright laws, so the situation is not as bad as it used to be. But the U.S. motion picture industry claims that it still loses about $3 billion a year due to piracy.[38]

censorship

Videocassettes were attractive in some countries because they contained material that was not available otherwise. In Saudi Arabia, for example, companies at first duplicated the tapes legally and brought them into the country despite very strict Muslim censors. The censors did not know what they were dealing with—they couldn't see anything on the physical tapes. When they discovered the forbidden cultural and sexual content of many of the tapes, they outlawed the videos. The result was a thriving underground industry that smuggled, duplicated, and sold tapes any way it could.

popularity

Legal tapes, as well as illegal ones, became very popular. The African country of Cameroon, which had delayed the establishment of television, finally started a television broadcast system in 1986, in part to lure viewers away from videocassettes. In India, video clubs and parlors sprang up, and it became the fashion to ride around in buses that showed videos. Eastern European black markets were a hotbed of pirated tapes during the 1980s. Some people in this part of the world got their first view of Western life from videocassettes. The tapes were instrumental in raising discontent with the Communist lifestyle.[39]

14.11 The Collapse of Communism

When Polish leader Lech Walesa was asked what caused the collapse of Communism, he pointed to a TV set and said, "It all came from there."[40]

**outside
information**

Communist governments tried hard to keep their citizens from obtaining information from outside countries. They jammed signals of external services and even employed wired radio that carried only government stations. But as technology progressed, information on broadcast radio and TV, satellite, and videocassettes did seep in. Most citizens within Communist countries knew they were not getting complete information from their government systems. The desire for more communication with the outside world was part of what led to the uprisings that undermined Communism.

The media also acted as a messenger to show the outside world what was really happening in these countries. Individuals, armed with audiotape recorders and camcorders, taped injustices within their countries and got the material to CNN, VOA, or the BBC. In one 1988 incident, a Czechoslovakian radio hobbyist sent information to Radio Free Europe in Munich describing intervention by Communist troops against people who were praying on Good Friday. Several months later, Czechoslovakian television staffers disobeyed their bosses and aired student protests against Communism. This led to the downfall of Communism in Czechoslovakia.[41]

After Communism fell, Eastern European countries struggled to build a new electronic media system. Political instability racked these countries, and laws could not be drafted fast enough to handle all the societal changes, let alone the changes in broadcasting. But things eventually settled down, and this part of the world also adopted privatization, cable TV, satellite TV, Western-style programming, and advertising.[42]

Even in countries where Communism still survives, such as China, the effects of media can be felt. In 1989, Mikhail Gorbachev, then head of the Soviet Union, visited China, and to demonstrate its new openness, the Chinese government allowed media from throughout the world into the country. The government planned a grand ceremonial event, but China's own citizens had learned about using the media and turned this event into an opportunity to publicly embarrass the leadership by demonstrating in Tiananmen Square (see Exhibit 14.9) for additional

outside communication

new systems

China

Exhibit 14.9

A young man facing down tanks was one of the most memorable shots of the Tiananmen Square incident, which was shown throughout the world.

(AP/Wide World Photos)

freedoms. The media, of course, covered this uprising, and CNN was able to get live pictures out to the world even after the government started restricting media access. Since that time, the Chinese government has periodically become more open. It allowed foreign correspondents to see the devastation of the Sichuan Province 8.3 earthquake in May 2008, and although it was criticized for some of its actions when it hosted the 2008 summer Olympics, it did exhibit more openness than in the past (see "Zoom In" box).[43]

ZOOM IN: China's Olympic Challenges

The 2008 Olympics provided many challenges to China. It wanted to prove to the outside world that it was open to media scrutiny while at the same time controlling potential protests so that they did not dominate the Olympic coverage. It wanted to tell the story of its rapid ascent during the last 20 years and how it had lifted 200 million Chinese people out of poverty, rather than its pollution and bureaucratic red tape. It wanted to showcase the drama of the sports competition and did not want the 20,000 journalists who descended on Beijing to concentrate on political dramas related to Tibet's future, China's relationship to Taiwan, human rights violations, or lack of due process. By spending $40 billion preparing for the games (more than the last three Summer Games hosts combined), it wanted to prove that it was an economic power that had regained its national greatness and rightly deserved the respect of the Western world.

These were daunting challenges for China, but another challenge, rarely mentioned, was the technological challenge of providing the TV equipment and crews to cover the games. As host country, it delivered the main feeds used by other countries. Major international rights holders, such as NBC and BBC, supplemented this footage with some of their own, but China's government-controlled media organization, CCTV, was responsible for the major transmissions. China agreed to make this the first all-HDTV broadcast, even though the country itself had only one HDTV channel and very few viewers with HDTV sets. Of course, it also provided standard-definition feeds for its citizens and those in many other countries. Overseen by Jiang Heping, general manager of the Olympic Channel, also known as CCTV-5, a sports channel

founded in 1995, the Chinese assembled 60 mobile units with multiple cameras and surround-sound capability in order to offer 3,800 hours of Olympics broadcast feed plus 5,000 hours of online material. With more than 50 percent of the earth's population tuning in to the Olympics, CCTV was at the center of the world stage for the 17-day Olympics.

How do you feel about how the Chinese handled the Olympics? Were you impressed? Intimidated? Put off? What did you think of the technical quality of the feeds? Do you think the Chinese government accomplished its social goals mentioned in the first paragraph? Did it clamp down too hard on potential dissidents? Do you think China will continue down a path of media openness?

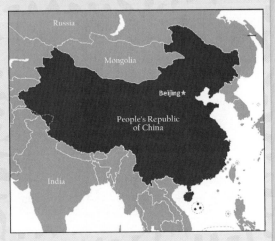

China.

(Courtesy of Brian Gross)

Russian media have taken a slightly different tack. After the fall of Communism, the government-run media services struggled with programming philosophy because their programming had been so highly pedantic. New private companies (primarily joint ventures between Russian companies and Western companies) sprang up—radio stations that played rock music and TV stations that showed American movies and TV series. Eventually a new TV service, NTV, built from scratch by a private Russian company, became very popular. It aired Western material, much of which had not been seen in Russia before, but also began producing slick, gritty, realistic dramas and miniseries as well as some comedies and documentaries. This helped revive the Russian film industry, which had essentially disappeared when American movies were introduced. However, NTV also aired news that dared to criticize the government. When Vladimir Putin became president, he did not take kindly to this approach, and the government started to thwart NTV and other private media operations. In 2001, the government gas monopoly, Gazpron, took over NTV and brought it back to the more staid philosophy of the old authoritarian model. Today both TV and film products are geared mainly to supporting the ideals of the government.[44]

Russia

Another result of the fall of Communism was a change in the Voice of America and similar external services. As previously mentioned, the main mission of these services during the cold war was to send information to the Soviet Union and its satellites. When this was no longer needed, VOA, BBC, and other similar services lost some of their importance. In 1994, the VOA and its sister operations were taken out of the U.S. Information Agency and overseen by an independent government entity, the International Broadcasting Bureau (IBB). After the September 11, 2001, terrorist attacks on the United States, one of the IBB governors, Norm Pattiz, suggested establishing a radio service to target Middle Eastern youths. In 2002, Pattiz oversaw the setting up of Radio Sawa, an Arabic-language music and news station that did, indeed, become popular with the region's young people.[45]

VOA changes

14.12 Indigenous Programming

Most countries have always had the desire to produce radio, television, and film material themselves, but economics and logistics impeded them. Poorer countries found they could purchase imported (mostly American) entertainment much more cheaply than they could produce it themselves. Richer nations, when they first incorporated privatization, had such a great need for programming to fill all the new channels that they, too, relied largely on American fare. American production studios became dependent on exportation. Most TV shows and movies produced in the United States did not make a profit through their national sales; the black ink came only after the products were sold overseas.[46]

American shows

But the media world is changing, and more countries are producing for themselves. Generally, the indigenous radio and TV fare is preferred over imported programming if it is of an entertaining (as opposed to pedantic or propagandistic) nature. Soap operas, game shows, dramas, news (see "Zoom In" box)—all have more appeal if they feature performers speaking the native language without subtitles

local programs

ZOOM IN: Al-Jazeera: A Loud Voice from a Tiny Nation

Al-Jazeera is an indigenous satellite news service established in the Middle East that has taken on an important role in world politics. It was launched in 1996 with $150 million in start-up money from Hamad bin Khalifa al-Thani, the emir of the tiny peninsula country of Qatar (Al-Jazeera means "peninsula" in Arabic). He had overthrown his father in a bloodless coup and decided to make Qatar into a modern democracy, in part by abolishing the country's Information Ministry, which had been responsible for censorship. Although it was pro-Arab, Al-Jazeera was unlike other news programs in the Middle East, which were filled with footage of government officials greeting other government officials; its reporters were encouraged to report on such forbidden subjects as women's rights, polygamy, and the legitimacy of some Arab regimes.

With footage such as that showing the destruction of two ancient Buddha statues in Afghanistan and the burning of an Egyptian flag by Palestinians who felt Egypt was going soft on Israel, it angered neighboring governments that tried to shield their inhabitants from such news. Several Mideast countries briefly recalled their ambassadors from Qatar over stories they didn't like, and many other countries tried, futilely, to keep their citizens from viewing Al-Jazeera.

After the terrorist attack on the World Trade Center and Pentagon on September 11, 2001, Al-Jazeera became well known worldwide. Initially, it was the only network allowed inside the Taliban-controlled areas of Afghanistan, and it is the network Osama bin Laden sends tapes to when he decides to talk about his views of America.

Al-Jazeera has expanded over the years to include more channels, such as ones devoted to sports, documentaries, and children's programming. It also broadcasts in several languages, including English, and can be accessed on the internet and through some American cable TV systems. It has rightly gained an important place in the world media structure.

Qatar and its Middle Eastern neighbors.

(Courtesy of Brian Gross)

or dubbed dialogue and engaging in actions that are familiar to the listeners and viewers. Worldwide, production techniques are flashier and sensationalism is more common than it used to be. In Brazil, for example, the once rock-solid first choice, TV Globo, has been losing audience to upstart services that feature scantily clad women, hostile talk-show hosts, and wrestling.[47]

More countries are also producing movies. New Zealand came onto the film scene in a big way with its involvement in the *Lord of the Rings* trilogy, the last of which won 11 Oscars in 2004 (see Exhibit 14.10). The country has an active national campaign to produce its own quality movies and to encourage production companies from other countries to use New Zealand as a location. China has become a major producer of movies that are well accepted in other countries, bringing profits as well as recognition to China. That country produces more than 200 movies a year, many of which are of high quality with scenic settings and well-choreographed action. In addition, Japan and some Eastern European countries have a history of making quality films, but the filmmaking tradition is crossing new borders. In 2002, more than 50 films were submitted to the Academy of Motion Picture Arts and Sciences for consideration in the foreign-language category. Among the countries submitting films were Albania, Armenia, Croatia, Tanzania, Thailand, and Uruguay.[48]

Exhibit 14.10

New Zealand director Peter Jackson holds the Oscar for best picture for *The Lord of the Rings: The Return of the King.*

(Superstar Images/Globe Photos, Inc.)

Gradually, countries such as Australia, Mexico, Brazil, and Britain, which have always exported some material, are increasing their share of the world market at the expense of U.S. programming. In addition, new countries, such as Poland and the Czech Republic, are entering the TV program distribution field. All these countries find that distributing their own indigenous programming is a harder sell than it used to be because so many countries rely primarily on what they produce internally. As a result, countries that wish to export programming are undertaking new strategies to protect their investments.[49]

movies

distribution

For one, companies are forming alliances and joint ventures in all parts of the world to increase multinational involvement and therefore boost sales because all the countries involved will show the product. The American company MCA and the German company RTL have a deal to co-produce 25 TV series. A French broadcaster, American company, and Korean network have co-productions in the works. Warner Bros. has a joint venture with China Film Group to create films and television programs. Sometimes companies do not simply form temporary

joint ventures

alliances—they buy each other. Sony (Japanese) bought Columbia (American) and renamed it Sony Pictures. A Canadian company owns a New Zealand TV network.[50]

franchise

Some TV production companies **franchise** concepts, and the shows are then produced with local crews and talent. Game shows and soap operas lend themselves to franchising. *Wheel of Fortune* was particularly successful with this strategy. King World, the show's distributor, sent a booklet describing the show and a consultant to every country that wanted to air the program. It required that the format be essentially the same as that shown in the United States, but the "Vanna Whites" and "Pat Sajaks" are local, as are all the contestants. U.S. companies also buy franchises. Both *Who Wants to Be a Millionaire?* and *Survivor* came from concepts developed in Europe. Countries throughout the world have adopted these successful reality formats.[51]

U.S. imports

The United States, which for many years purchased little programming from other countries, now buys more programs. To some degree, this is because minority audiences in the United States are eager to see programming in their native languages. Whole cable channels or independent stations are devoted to programming in Spanish, Vietnamese, Korean, and other languages.

quotas

Despite the recent increase in indigenous programming, American programming still dominates as the most common import (see Exhibit 14.11). It is popular enough that countries institute quotas to protect their local production from this "media imperialism." For example, the European Union requires that 50 percent of programming in its member countries must be from Europe. Some countries

Exhibit 14.11

Not all countries pay the same amount for American products. Generally, poorer countries pay less than richer countries, but because the price is negotiated for each sale, other factors, such as negotiating skills, play a part in the price. This chart shows what different countries might pay for rights to show a typical American made-for-television movie.

What Various Countries Might Pay for a U.S. Made-for-TV Movie

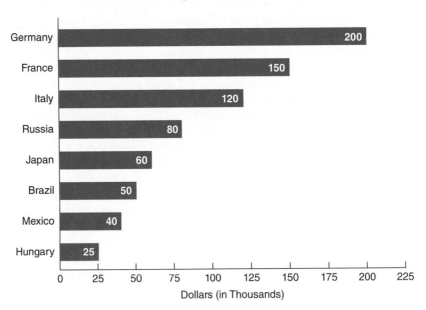

Country	Dollars (in Thousands)
Germany	200
France	150
Italy	120
Russia	80
Japan	60
Brazil	50
Mexico	40
Hungary	25

are even stricter. French law says that 60 percent must be European, and of that amount, 40 percent must be French.[52]

The development of **digital TV** may once again change the programming balance. The many new channels will need additional programming.

14.13 The Digital Age

Digital telecommunication has come to all parts of the globe. The internet has already brought enormous changes to communication. For the first time, theoretically everyone in the entire world has access to the same base of information (or jewelry advertisements or titillating photos). This information is uncensored and free of the **gatekeepers** who decide what people will or will not hear or see with most other media. Perhaps even more important, anyone can be a provider of information or entertainment—again without outside control.

Email between individuals from different countries and **chatrooms** that are not bound by geography greatly enhance interpersonal communications. Most people with access to the internet are fans of it, prompting companies throughout the world to consider it and related services as a big area for potential profit (see "Zoom In" box). Some of the same countries that tried to keep satellite out have also tried to protect their citizens from the internet. China, for example, went to great lengths to police the internet, but hookups in that country are now very common.[53]

Until recently, large areas of the world had antiquated, unreliable rotary-dial telephone systems. They have been able to skip several generations of telephone improvements and go straight to **cellular,** internet, or satellite-based phones. These new phone technologies greatly enhance the ability of people and organizations to communicate internationally.

Most parts of the world are in the process of converting to digital radio and television. Some, such as Canada and Western Europe, are ahead of the United States, particularly in the area of digital radio (see Chapter 5). Others, such as Spain and Greece, are using a go-slow approach and will probably convert later than other countries.[54]

Other aspects of digital technologies are popular throughout the world. **DVDs** and video games (see Exhibit 14.12) have been as much a success in other countries as they have in the United States. European movie producers started shooting movies entirely with digital video well before this practice was common in the United States. Digital technologies are increasing the speed, depth, and breadth of communication.[55]

internet

phones

digital radio and TV

DVDs and games

Exhibit 14.12

One never knows where video game consoles are going to show up—in this case, in a Scottish hospital where Prince Charles was visiting children.

(Courtesy of Globe)

ZOOM IN: **Killer Apps from Estonia**

The internet is truly international, and so are the inventions and applications related to it. For example, the inventors of the internet phone service Skype (see Chapter 2) are Scandinavians who did much of their initial computer engineering work with engineers in Estonia. One of the primary inventors, Niklas Zennstrom, is a native of Sweden, and the other, Janus Friis, is from Denmark. Skype is not their first venture. They developed the music sharing site Kazaa in 2000 and several years later sold it to an Australian company, Sharman Networks.

Then, in 2003, they brought out Skype, which they had been working on simultaneously, and it quickly became quite successful. They sold it to eBay in 2005 for $2.6 billion. They are using some of the money for a new television project called Joost. Their goal is to gather 100-plus channels worth of television material and make it on-demand, so people can access it whenever they wish. And, oh yes, they also paid $100 million to record labels because Kazaa was engaging in illegal music downloading when they developed it. For their television venture, they are making sure they stay strictly legal.

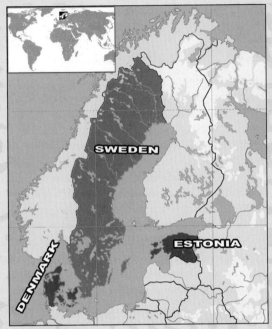

Scandinavian countries of Sweden and Denmark, and nearby Estonia.

(Courtesy of Brian Gross)

global village

rapid communication

14.14 Issues and the Future

In the 1960s, Canadian professor Marshall McLuhan predicted that television would create a "global village" in which all parts of the world could communicate with and perhaps understand each other. As the decades passed, we have come closer to the global village. Satellites and the internet make instant worldwide communication possible, and political changes within countries sometimes make them more willing to communicate.

But rapidity of communication brings with it problems. As recently as the 1960s, President John F. Kennedy had six days to ponder what to do before going public about the Cuban missile crisis. In 1991, during the Persian Gulf conflict, President George H. W. Bush was lucky to have six hours before making a major decision—and then it was with the whole world watching.

With modern communications, uprisings in the most remote areas of the world can be photographed, reported, analyzed, and sometimes blown out of proportion within minutes. Although this ability keeps people informed, it can also aggravate situations that would cool down on their own.

The rapidity with which new forms of communication have swept the world in the past 20 years has also caused problems and societal changes. The Middle East did not have enough people to operate its communication structure. The people of Eastern Europe changed their philosophical and legal mindset to cope with their new media. Public broadcasting systems around the world had to react to the impact of privatization. The effect of rampant advertising is unknown. Will it strengthen the worldwide economy, or will it turn people into avid materialists who are frustrated because they cannot buy all that they see?

problems with rapid change

One major problem is the cost of electronic media. Poor nations and poor people cannot afford every new technological wonder that comes along. The gap between the "haves" and "have-nots" becomes wider. Because much of media consumption relates to obtaining information, the economically poor also become the information poor. Despite the United Nations' attempt to create a New World Information Order, the developed countries still do most of the sending, and developing countries are left to receive information that they do not necessarily want.

developing countries

The barrier to entry caused by the high cost of technology creates a handful of media moguls who extend the reach of their companies into international markets and control a great deal of what is available throughout the world. Can they be trusted to use it in the best interests of the world public?

moguls

Throughout much of the 1900s, the most popular entertainment, whether for a broadcast station, cable system, movie theater, satellite, or the internet, was American. However, many countries are now producing a larger share of their own programming and are successfully exporting it. These programs compete with Hollywood, which by now depends on international sales to make ends meet, so American producers are making adjustments and also looking forward to future exciting international developments.

programs

14.15 Summary

The world market is very important to U.S. media producers in terms of profit, co-productions, and franchises. Those working in global media should know how systems around the world have developed and are structured. One way to examine the basic structures is to look at developments in Brazil, Britain, and Russia.

Brazil's private system of broadcasting is similar to most in Latin America. Early radio was taken over by newspaper moguls, one of whom, Robert Marinho, went on to establish a very successful TV network, TV Globo. Brazil has always had commercials and entertainment-oriented programming. Television started in the early 1950s, with investments from U.S. media companies. In the 1960s, at the same time colonial countries were gaining independence, Brazil and other

Latin American countries cast off the American companies. Brazil, along with Mexico, developed the telenovela into a form that was readily exported. Brazil had one of the early direct broadcast satellite systems. Privatization did not affect Brazil or other South American countries because their media systems were already private, but programming forms have changed as sassier programming services draw audiences away from TV Globo.

Britain, as a public system, led the rest of Europe in media development. BBC radio started in 1922 as a monopoly under a government charter. It was supported by license fees and had no advertisements. Programming was paternalistic. Beginning in the 1930s, Britain spread its public system to its colonies and developed its Empire Service. Unlike France, it believed in local participation in media. During World War II, BBC operated an external service to combat propaganda. After the war, Britain helped the other Allies set up broadcasting systems in Germany and Japan. Television started in the 1930s in England but was interrupted by the war and was not really revived until 1953. In 1963, pirate ships began broadcasting rock music, and the BBC was forced to revamp its staid radio programming to include rock. Independent television started in England in 1954, earlier than in other European countries. Advertisements were allowed on independent TV, much as they were on privatized services throughout Europe. Britain and other European countries received direct broadcast satellite fairly early. As privatization was occurring everywhere, the BBC charter came under review, and independent television was reorganized, with a number of the companies losing their franchises.

Russia is an example of the authoritarian model. Lenin nationalized radio in 1917, and the Communists kept a tight hold on program content through Gostelradio. No advertisements and very little entertainment programming were broadcast. As an ally in World War II, Russia further developed Radio Moscow, which had started in the 1920s. After the war, Russia extended its authoritarian form of broadcasting to East Germany and the rest of Eastern Europe. Illegal videocassettes crept into Eastern Europe and were part of what led to the downfall of Communism, along with access to satellite and broadcast services from the free world. Although other countries in Eastern Europe continue to embrace privatization, Russia has clamped down on and abolished services critical of the government, and most Russian TV and film now reflect the needs of the state.

World communication has developed at a rapid pace, which has created some problems. The tying together of countries through media, however, is bound to continue.

Suggested Websites

http://afrts.dodmedia.osd.mil (site of the American Forces Radio and Television Service)

www.bbc.co.uk (the British Broadcasting Corporation homepage)

www.cbc.ca (the Canadian Broadcasting Corporation homepage)

www.intelsat.com (the site of satellite provider Intelsat)

www.voa.gov (the external service Voice of America homepage)

Notes

1. Gerald Mast, *A Short History of the Movies* (New York: Macmillan, 1986), 207.
2. "A Modern Take on *Dr. Caligari*," *American Cinematographer,* April 2006, 30–33,
3. Jack C. Ellis, *A History of Film* (Englewood Cliffs, NJ: Prentice Hall, 1990), 92–94.
4. Douglas A. Boyd, *Broadcasting in the Arab World* (Ames: Iowa State University Press, 1993), 138–39.
5. "El Tigre," *Los Angeles Times,* November 10, 1991, 26; and "Battle Intensified over Control of Mexican TV," *Los Angeles Times,* January 17, 2005, C-1.
6. "Formation of the BBC," *BBC Fact Sheet 1,* May 1982, 1–2; R. W. Burns, *British Television: The Formative Years* (London: Peter Peregrinos, 1986); and "Britain's BBC-TV," *Broadcasting,* November 3, 1986, 43–53.
7. Bernard Redmont, "Soviet TV: Ballet and Brezhnev, Serials and Symphony," *Television Quarterly,* Spring 1981, 27–35.
8. Louise Bourgault, "Nigeria," in Lynne Schafer Gross, ed., *The International World of Electronic Media* (New York: McGraw-Hill, 1995), 237.
9. Boyd, *Broadcasting in the Arab World,* 205–8.
10. "BBC World Service.com," www.bbc.co.uk/worldservice/index.shtml (accessed August 3, 2008).
11. "Fast Facts about the VOA," www.voa.gov/index.cfm?sectionTitle=Fast%20Facts (accessed August 3, 2008).
12. Sergei V. Erofeev, "Russia," in Gross, *The International World of Electronic Media,* 190.
13. "The GI Joe Network," *Emmy,* August 1990, 70–74.
14. *Radio and Television in the Federal Republic of Germany* (Frankfurt, Germany: Hessicher Rundfunk, 1980), 5–45.
15. Nobuo Otsuka, "Japan," in Gross, *The International World of Electronic Media,* 301–3.
16. Joseph S. Johnson, "China," in Gross, *The International World of Electronic Media,* 278.
17. Ellis, *A History of Film,* 314–17.
18. Mast, *A Short History of the Movies,* 339–49, 380–87.
19. Alan Brender, "After a 25-Year Wait . . ." *TV Guide,* August 14, 1976, 24–26.
20. Boyd, *Broadcasting in the Arab World,* 147–49.
21. Joseph Straubhaar, "Brazil," in Gross, *The International World of Electronic Media,* 63–67.
22. *This Is Independent Broadcasting* (London: Independent Broadcasting Authority, n.d.), 3–14; "U.K.'s Channel 5 Gets Off to a Slow Start," *Electronic Media,* April 7, 1997, 10; and *Advertising on Independent Broadcasting* (London: Independent Broadcasting Authority, n.d.), 3–25.
23. "Canada Wants to Raise CBC's Native Program Levels to 90 Percent," *Broadcasting,* March 9, 1987, 43; "Good Fences," *Hollywood Reporter,* July 7–13, 1998, 12; and "CRTC Mulls Changing the Channel on TV Rules," *National Post,* April 7, 2008, FP-1.
24. "Hungary: A Goulash of Media Activity," *Broadcasting,* July 23, 1990, 81–83.
25. Everett M. Rogers and Livia Antola, "Telenovelas: A Latin American Success Story," *Journal of Communication,* Autumn 1985, 4–12; and "Mexico's Televisa Searches for Global Stardom," *Wall Street Journal,* May 30, 1995, A10.
26. From Drought to Deluge," *Cable Satellite Europe,* December 1993, 20–21.
27. To Make a Big Movie in Less Than a Day, Forget the Script," *Wall Street Journal,* May 6, 1999, 1.
28. Kusum Singh and Bertram Gross, "'MacBride': The Report and the Response," *Journal of Communication,* Autumn 1981, 104–17; and "The Radio Offers Africans Rare Aid in Tune with Needs," *Wall Street Journal,* May 10, 2002, 1.
29. "The Expanding International Horizon of Satellites," *Broadcasting,* July 25, 1988, 36; and "Intelsat Expands with PanAmSat," *Hollywood Reporter,* August 30–September 5, 2005, 12.
30. "From Ahnadabad to Makapura," *TV Guide,* June 19, 1976, 10–12.
31. "CNN: The Channel to the World," *Los Angeles Times,* January 29, 1991, 1.
32. "TV Is Exploding All Over Asia," *Fortune,* January 24, 1994, 98–101; Robbin D. Crabtree and Sheena Malhotra, "A Case Study of Commercial Television in India: Assessing the Organizational Mechanisms of Cultural Imperialism," *Journal of Broadcasting & Electronic Media,* Summer

2000, 364–85; and "Murdoch Star Deal Transforms Asia," *Broadcasting & Cable,* August 2, 1993, 34.

33. "Chinese Wiring the Countryside for Satellite TV," *Los Angeles Times,* September 23, 1999, C-1.

34. "Europe Braces for Free-Market TV," *Fortune,* February 20, 1984, 74–82; and "France: A Revolution in the Making," *Channels,* September/October 1985, 60–61.

35. "Pay TV Comes to Germany," *Broadcasting,* March 25, 1991, 90; and Wolfram Peiser, "The Television Generation's Relation to the Mass Media in Germany: Accounting for the Impact of Private Television," *Journal of Broadcasting & Electronic Media,* Summer 1999, 364–85.

36. " 'Whole Lotta Shakin' Going On in French TV," *Daily Variety,* February 3, 1988, 63.

37. "TV: Revolution Brewing in Thatcher's Britain," *Wall Street Journal,* January 3, 1989, A-9; "Auction Action Shakes Up U.K. TV," *Daily Variety,* October 17, 1991, 1.

38. "DVD Piracy Is Unbelievable Bad, Says Unbelievable Study," http://www.alleyinsider.com/2008/7/dvd-piracy-is-unbelievably-bad-says-unbelievable-study-mvsn (accessed August 3, 2008).

39. Lalit Acharya and Surekha Acharya, "India," in Gross, *The International World of Electronic Media,* 267.

40. "Tuning In the Global Village," *Los Angeles Times,* October 20, 1992, H-1.

41. Ivan Stadtrucker, "The Slovak Republic and the Czech Republic," in Gross, *The International World of Electronic Media,* 162; "US Style Station Is a Hit Among Czechs—And That's a Problem," *Wall Street Journal,* April 30, 1997, 1; and "Cable in Poland," *International Cable,* June 1997, 24–26.

42. Colin Sparks, *Communism, Capitalism, and the Mass Media* (Thousand Oaks, CA: Sage, 1998); and "Maverickski," *Hollywood Reporter,* August 8–14, 2000, 18–19.

43. "Chinese See TV Face-Off as New Page in History," *Los Angeles Times,* June 28, 1998, 1; "Politics at Play, Again," *Denver Post,* August 7, 2008, C-1; "China's Agony of Defeat," *Newsweek,* August 4, 2008, 39–41; and "The Whole World Tuned In," *Hollywood Reporter,* August 11, 2008, 8.

44. Anna Yudin and Michael C. Keith, "Russian Radio and the Age of Glasnost and Perestroika," *Journal of Radio Studies,* December 2003, 246–54; "Russia Alters Direct Hold on Government Channels," *Hollywood Reporter,* December 23–25, 2005, 8; and "Un-Hollywood: In Russia, Films Promote the State," *Wall Street Journal,* January 9, 2008.

45. "Reaching Arabs Via Airwaves," *Los Angeles Times,* August 26, 2002, 1; and "Pentagon Channel: Not Just for the Military," *TV Technology,* May 31, 2006, 12.

46. "Worldwide Opening," *Hollywood Reporter,* July 28, 2005, S1-S7.

47. "Local Heroes," *Hollywood Reporter,* April 2, 2001, S-19–S-43; "U.S. Losing Foreign Airspace," *Hollywood Reporter,* August 6, 1998, 13; and "As 'The Other World' Turns, Brazil Gets Downright Odd," *Wall Street Journal,* September 29, 1999, 1.

48. "The Fate of Middle Earth," *American Cinematographer,* January 2004, 54–61; "How Mr. Kong Helped Turn China into a Film Power," *Wall Street Journal,* September 14, 2005, A1; and "Bumper Crop for Academy," *Hollywood Reporter,* December 3, 2001, 84.

49. "Brits Jack Up Share of Int'l TV," *Hollywood Reporter,* October 7, 1998, 9; and "Eastern Europe Puts Programs into Mix," *Electronic Media,* January 7, 2002, 18.

50. "Titanic Teutonic TV," *Daily Variety,* July 31, 1996, 1; "International TV's Hard Sell," *Electronic Media,* April 12, 1999, 1; and Margie Comrie, "Television News and Broadcast Deregulation in New Zealand," *Journal of Communication,* Spring 1999, 42–54.

51. "Format Fever: The Risks and Rewards," *Broadcasting & Cable,* January 23, 1995, 94; and "How to Use a 'Lifeline,'" *Newsweek,* February 28, 2000, 46–47.

52. "The Going Rate," *Hollywood Reporter,* October 5–11, 2007, S-10; and "The World Turns to U.S. TV," *Broadcasting & Cable,* January 8, 2007, 23.

53. "China Goes One-on-One with the Net," *Los Angeles Times,* January 27, 2001, 1; "Broadband Wonderland," *Fortune,* September 20, 2004, 191–92; and "Trashing the Tube," *Newsweek,* May 14, 2007, 48.

54. "DAB UK," *Diffusion EBU,* 2005/2, 16–19; "Mobile Digital Radio to Target Europe, Asia," *Aviation Week and Space Technology,* September 10, 2001, 28; "Digital Radio: The Way Forward," *Diffusion,* 2002/3, 44–45; and "IBC Looks Beyond HD," *TV Technology,* October 17, 2007, 8.

55. "DVD Becoming a Bigger Player with 247% Rise," *Hollywood Reporter,* January 6, 2000, 51.

GLOSSARY

A

above-the-line Refers to expenses related to administrative and conceptual aspects of a production, mainly those related to producing, directing, writing, and acting. (7.12a)

actor A person portraying someone other than himself or herself. (7.12a)

actual malice In libel suits, taking something that was known to be harmful and saying or printing it anyway, with reckless disregard. (11.7)

adjacencies Commercials that are the first to be aired after programming ends or right before programming starts. (9.9)

advertising agency An organization that decides on and implements an advertising strategy for a customer. (3.6, 5.9, 9.4)

affiliate A station or system that receives programming from a broadcast or cable network. (1.2, 3.2, 8.1b)

aftermarket All the income earned by a movie after its run in American theaters, such as income from DVDs, overseas theaters, and airplane showings. (6.14)

à la carte Offering cable subscribers the opportunity to buy only the channels they want rather than packages of channels the cable system puts together. (4.16)

alternator A device for converting mechanical energy into electrical energy in the form of alternating current (AC). (5.3)

amplitude modulation (AM) Changing the height of a transmitting radio wave according to the sound being broadcast; stations that broadcast in the 535 to 1,705 kHz range. (5.8, 13.5)

analog A device or circuit in which the output varies as a continuous function of the input. (1.2, 4.4, 13.1)

anamorphic A lens that squeezes a wide picture into a frame of film in a camera and then unsqueezes it in the projector. (6.11)

anchor person A person who reads the news on radio or television. (7.12b)

annual flight schedule A plan for running commercials that are aired only near holidays. (9.10)

antenna A wire or set of wires or rods used both to send and to receive radio waves. (13.5)

anthology drama A television play, most commonly associated with the 1950s, that probed character and emphasized life's complexities. (8.6)

ARPANET The original name for what became the internet, developed by the Defense Department's Advanced Research Projects Agency. (2.1)

ascertainment A process stations used to undertake to keep their licenses that involved interviewing community leaders to learn what they believed were the major problems in the community. (11.10)

aspect ratio The relationship of the height of a TV picture to its width. (13.13)

assignment editor A person in a news room who keeps track of the stories that need to be covered and decides which reporters should be sent out to cover each story. (7.12b, 8.3a)

auction To sell something, such as frequency spectrum, to the highest bidder. (4.15, 11.10)

audience flow The ability to hold an audience from one program to the next program shown on the same channel. (8.5b)

audio board A mixer used to combine various sound inputs, adjust their volume, and then send them to other pieces of equipment. (13.2)

audion tube A three-electrode vacuum tube invented by Lee De Forest that was instrumental in amplifying voice so it could be sent over wireless and used for movies. (5.1, 6.7)

audiotape recorder A machine that records and plays sound from a magnetic tape that is usually either 1/4-inch or 1/8-inch wide. (5.11)

auteur The French word for "author," hence the primary creator of a movie; usually used to refer to the director. (6.12)

authoritarian A media system owned and closely overseen by the government and supported by general government funds. (14.2)

Automated Measurement of Lineups (AMOL)   A method that picks up special codes from programs so that Nielsen can determine what programs are on what channels in each market. (10.3)

availabilities (avails)   Times in a station or network schedule when it can air commercials. (9.9)

average quarter-hour (AQH)   A ratings calculation based on the average number of persons listening to a particular station for at least 5 minutes during a 15-minute period. (10.6)

B

backyard satellite   A process through which individual households received satellite signals by putting a receiving dish on their property. (4.3)

bandwidth   The number of frequencies within given limits that are occupied by a particular transmission, such as one radio station; the amount of space (hence, speed) given to particular internet material. (13.5)

banner ad   An advertisement that appears at the top of a webpage. (2.8, 9.7)

barrier   In a communication model, anything that reduces the effectiveness of the communication process. (12.4b)

barter   To receive a program for free because much of the advertising time is sold by the seller. (9.2)

basic cable   Channels, often supported by advertising, for which the cable subscriber does not pay a large extra fee. (4.1)

below-the-line   Refers to expenses for a production that are in the craft

or technical areas, such as salaries for camera operators and editors. (7.12a)

Betamax   The first consumer videotape format. (4.4)

bicycle network   A system in which stations shared program material but did not air it at the same time because the programs were physically transported from one station to another. (3.6, 5.13)

bicycling   A distribution method in which material is mailed, flown, or driven from one place to another. (13.10)

Biltmore Agreement   A settlement between newspapers and radio, worked out at the Biltmore Hotel in the 1930s, that stipulated what radio stations could and could not air in the way of news. (5.10)

blacklisting   A phenomenon of the 1950s when many people in the entertainment business were accused of leaning toward Communism and, as a result, could not find work. (3.5, 6.11, 12.4b)

blackout   Not allowing a particular program, usually sports, to be shown in a particular area because the event is not sold out. (8.12)

blanket licensing   Obtaining the right to use a large catalog of musical selections by paying one set fee. (11.8)

blind bidding   When a theater owner selects a movie to show in theaters without actually seeing it, depending only on publicity and trailers. (9.2)

block booking   A movie distribution technique in which an exhibitor was forced to rent a group of films (some good, some bad) to get any of them. (6.6)

block programming   Airing one type of programming, such as comedies, for an entire evening. (8.5b)

blog   Short for web log, a chronicle placed on the internet with information written by one or just a few individuals, although users may post comments. (2.7, 8.3a)

blogger   A person who places chronicles or other diary-like information on the internet. (7.12b)

Blue Book   The nickname given to a 1946 FCC document that set forth principles regarding license renewal. (11.10)

Bluetooth   A wireless technology that connects portable devices such as cell phones to the internet and to each other by using low-power radio transmission. (2.15)

Blu-ray   Sony's format for high-definition videodiscs that has been accepted as the standard format. (4.11)

boutique agency   An advertising agency that handles only a limited number of tasks for a client. (9.4)

B picture   A movie that was inexpensive to produce and was usually the lesser film of a double feature. (6.8, 14.4)

broadband   High-speed data transmission using relatively wide bandwidths. (1.2, 2.11, 13.9)

broadcasting   The sending of radio and television programs through the airwaves. (1.2, 3.1)

broadcast standards   The department (often called the Department of Standards and Practices) at a network or station that oversees the general standards of acceptability of program and commercial content. (12.1) browser   Computer software that aids in searching the world wide web. (2.6a)

buffer   An area, usually related to a computer, where streamed material stops temporarily so that all of it can be gathered and played as though it were being played in real time. (2.8)

bulletin board   A service that allows people to post electronic messages using the internet. (2.7)

bump system   When a station or network removes an advertiser from a particular time spot because another advertiser has offered to pay more. (9.10)

c

cable modem   An electronic device provided by a cable TV company that provides a high-speed connection to the internet. (13.9)

cable television   A system that delivers TV channels and other services into the home through wires rather than over the air. (1.2, 9.1)

camcorder   A single unit that consists of both a camera and a videocassette recorder. (8.3a, 13.3)

camera   A device that transduces light waves into electrical waveforms. (13.3)

Canon 35   An American Bar Association policy that banned cameras in courtrooms. (11.9)

Canon 3A   An American Bar Association policy that let individual states and judges decide whether or not cameras should be allowed in their courtrooms. (11.9)

capacitance disc   A videodisc system that used a stylus that moved over grooves in a manner similar to a record player. (4.4)

carbon microphone   An early radio microphone that used carbon elements to respond to sound. (5.9)

cardioid   A microphone that picks up sound primarily from one direction, in a heart-shaped pattern. (13.2)

carrier wave   A high-frequency wave that can be sent through the air and is modulated by a lower-frequency wave containing information. (13.5)

cash   Paying a set amount for a program and then being able to sell all the commercial time. (9.2)

cash plus barter   Paying a reduced amount for a program but letting the seller fill some of the advertising time. (9.2)

catharsis theory   A postulate related to violence research that states it is good for people to watch violence because it gives them vicarious excitement that keeps them from inciting real violence. (12.5a)

cathode-ray tube   A special type of vacuum tube consisting of an electron gun that shoots electrons at a phosphor screen in order to create an image. (13.3)

cause and effect   A method of research that tries to determine the relationship between two or more modes of behavior. (12.4b)

C-band   A satellite frequency band between 4 and 6 GHz. (13.7)

cellular phone (cell phone)   A mobile phone that operates by relaying signals from one small area to another through the use of low-powered transmitters. (1.2, 2.15a, 7.12d, 9.1, 13.8, 14.13)

censorship   Not allowing certain information to be disseminated. (11.5)

channel   In a communication model, the element through which messages are transmitted. (12.4b)

charter   A right to operate that is given to a broadcasting corporation (such as the BBC) by the government. (14.2)

chatroom   A computer service that allows many people to write messages that others in the group can see as long as they are online. (2.7, 14.13)

chief executive officer (CEO)   The highest-ranking person within an organization who actively makes decisions for the organization. (7.12f)

chief financial officer (CFO)   A high-ranking person within a company who is in charge of monetary matters. (7.12f)

chip   A tiny piece of silicon with electronic circuits that can store and manipulate data. (13.3)

CinemaScope   A 1950s film experiment that used an anamorphic lens to squeeze a picture so it could be unsqueezed and shown on a large, curved screen. (6.11)

Cinerama   A 1950s film experiment in which three interlocking projectors showed movies on a wraparound screen. (6.11)

clear and present danger   A threatening of the security of the nation if certain information is made public. (11.5)

click   A visit to a website that is counted even if the person or computer has visited the site before. (10.6)

clock   A circle that shows all the segments that appear in an hour's worth of radio station programming. (8.5) closed-circuit TV (CCTV) Television signals that are transmitted via a self-contained wire system, usually within a business or school. (3.17)

clutter   Having a large number of commercials at one time. (9.14)

coaxial cable   A transmission lin in which one conductor surrounds the other, making a cable that is not susceptible to external interference from other sources. (13.9)

commentator   Someone who gives his or her own opinions about news events. (5.10)

commercial broadcasting   A system of disseminating programming that is financially supported primarily by advertising. (1.2)

common carrier   A communication distribution system available for use by those that do not own it. (3.12)

Communications Act of 1934 The congressional law that established the FCC and set the guidelines for the regulation of telecommunications. (4.9, 5.9, 11.1)

community antenna television (CATV)   An early name for cable television. (1.2)

compact disc (CD)   An audio disc on which sound is recorded digitally with optical technology. (13.2)

comparative license renewal   A procedure by which a group could challenge the license of incumbent station owners. (11.10)

competing rights   Different provisions of the Constitution that allow for conflicting actions. (11.9)

composite week   Seven days over a three-year period that were randomly selected by the FCC so that a station's programming record could be judged for license renewal purposes. (11.10)

compression   A process by which redundant information is removed from a digital file, making it smaller so that it can be transmitted more quickly and stored more efficiently. (2.8, 13.1)

compulsory license   A copyright fee that must be paid and that is usually a set fee, such as a percentage of income. (2.11, 11.8)

concept selling   When a station or network uses something unique about itself to sell commercial time to an advertiser, usually because its ratings are so low they are unreliable. (9.3)

content analysis   A research method that involves studying something (such as the number of violent acts) within a group of programs to test a hypothesis. (12.4b)

controller   A person who oversees profit and loss and expenditures for a company or other organization. (7.12f)

control room   An area where the director, technical director, and other crew members select and mix the video and audio coming from a studio. (7.12a)

convergence   The coming together of various media forms and facilities, such as television and computers, so that they merge and share characteristics. (1.7)

copyright   The exclusive right to publication, production, or sale of a literary, dramatic, musical, or artistic work. (2.8, 11.8, 14.10)

corporate video   The use of video by corporations and similar organizations for training, orientation, and similar purposes. (1.2, 3.17, 7.12a, 9.2)

cost per point (CCP)   The cost an advertiser pays per ratings point, calculated by dividing what the advertiser paid by the rating points a commercial accumulates during a week. (10.11)

cost per thousand (CPM)   The price that an advertiser pays for each thousand households or people a commercial reaches (M is from the Latin *mille*, for thousand). (9.9, 10.11)

cover letter   A letter sent with a résumé that highlights aspects of the résumé that are applicable to a particular job. (7.6)

cross-ownership   A situation in which one company owns different media in one market, such as a newspaper owning a TV station. (11.11)

cross-promotion   When one network or station promotes something on another network or station, usually because they have common ownership. (10.1)

cume   The number of households or people that tune in to a particular station at different times. (10.6)

customer service   A department within a company that handles complaints and also sometimes sells services such as cable TV subscriptions. (7.12e)

D

datagram   Encoded delivery information attached to a digital packet when it launches onto the internet, including the senders' and receivers' names and computer addresses. (2.2)

dayparts Segments of the day that reflect the size and composition of the available audience. (8.5, 9.9)

decoder A device that can descramble scrambled television signals. (11.8)

decompression The process of taking file material that has been compressed and getting it back into its original form so that it can be heard and seen through a monitor or TV set. (2.8)

deep focus Shots that have both the foreground and a faraway background in sharp focus. (6.9)

defamation An attack on someone's reputation. (11.7)

demographics Information pertaining to vital statistics of a population, such as age, sex, marital status, and geographic location. (8.2, 10.7)

demo reel A compilation of audio and video material that a person has been involved with that is usually given to a potential employer. (7.6)

deregulation The removal of laws and rules that spell out government policies. (4.6, 5.16)

desensitizing effect A theory that watching violence makes people think that seeing someone commit an act of violence is common and not worth dealing with. (12.5a)

designated market area (DMA) Nielsen's term for the various individual markets it surveys. (10.10)

development The process that a TV program goes through to get from idea to exhibition. (8.2)

diary A booklet used for audience measurement in which people write

down the programs they listen to or watch. (10.5)

digital A device or circuit in which the output varies in discrete on-off steps. (1.2, 13.1)

digital audio radio service (DARS) A form of satellite audio transmission that involves pulse code modulation and CD-quality sound; now often called satellite radio. (1.2, 5.17)

digital cinema A method of projecting movies in a theater that uses digital technology rather than film projectors. (1.2, 13.11)

digital rights management (DRM) Technology that uses encryption for file security. (2.15c, 11.8)

digital subscriber line (DSL) A high-speed telephone-based connection from a computer to the internet that is always on and that can allow for both data and voice communication simultaneously. (13.9)

digital television (DTV) A technical form of TV that allows for high-definition broadcast pictures and/or additional program channels. (4.15, 13.6, 14.12)

digital versatile disc (DVD) A round disc that can be used to store and retrieve audio, video, and data. (4.11, 9.1, 14.13)

digital videodisc The original name for the digital versatile disc, or DVD, because its first use was to store video. (4.11)

digital video recorder (DVR) A hard-disk drive that can record and compress video from off-air, cable TV, or satellite TV and can pause recording and pick up where it left off; often referred to by the name of the popular brand TiVo. (1.2, 4.12, 8.5a, 9.1)

direct broadcast satellite (DBS) A process of transmission and reception in which signals sent to a satellite can be received directly by TV sets in homes that have small satellite receiving dishes mounted on them; now often called satellite TV. (1.2, 4.3)

director The person in charge of the creative aspects of a TV program or movie. (7.12a)

distance learning Instructional programs, usually offered by a college, that are intended to bring education to people who live far from an educational institution or who prefer not to travel to the college. (2.7, 3.17)

distant signal importation The bringing in of stations from other parts of the country by a cable system. (3.3)

distributed network A way of interconnecting computers in a "fishnet" pattern so that data can flow along many paths, rather than to and from a central computer. (2.2)

diversity A set of conscious practices that involve an understanding and appreciation of various cultures and a respect for differing experiences. (7.9)

docudrama A fictionalized production that is based on fact. (8.6)

Dolby A noise reduction technique that raises the volume of the movie sound track elements most likely to be affected by inherent noise during production and then lowers them again during projection so that the noise seems lower in relation to the wanted elements of the sound track. (6.13)

E

earphones Small devices placed over the ears to hear audio signals. (5.9)

E-band The portion of the electromagnetic spectrum from 71 to 76 GHz, 81 to 86 GHz, and 92 to 95 GHz. (13.14)

e-commerce Short for "electronic commerce"; buying and selling goods and services on the internet, often referred to as online shopping. (2.7)

electromagnetic spectrum A continuing series of energies that encompasses frequencies that can carry audio and video signals. (13.4)

electron gun The part of a TV tube that shoots a steady stream of electrons that scan from top to bottom. (13.13)

electronic mail (email) Messages sent over the internet from one person to another person or group, which recipients can read at their convenience. (2.3, 14.13)

electronic scanning A method that analyzes the density of areas to be copied and translates this into a moving arrangement of electrons that can later reproduce the densities in the form of a picture. (3.1)

embedded Mixed in with another group, as reporters were mixed in with troops during the war in Iraq. (8.3a)

encoding Using computer software to translate sound or picture from analog to digital. (2.8)

encryption The process of digitally scrambling a file or document so that only authorized users who have the appropriate software can unscramble the content and use it legally. (5.17)

enhanced underwriting A public broadcasting practice that allows corporate logos and products of companies that contribute to programs to be mentioned on the air. (9.8)

episodic serialized dramas Televised plays that have set characters and problems that are dealt with in each program. (8.6)

equal opportunity Giving the same treatment to political candidates. *See* equal time. (11.12)

equal time A rule stemming from Section 315 of the Communications Act stating that TV and radio stations should give the same treatment and opportunity to all political candidates for a specific office. (11.12)

event video The use of video to record important milestones such as birthdays and weddings. (7.12a)

external service Radio and TV programs that one country produces to influence other countries. (7.12a, 14.4)

F

fairness doctrine A policy that evolved from FCC decisions, court cases, and congressional actions stating that radio and TV stations had to present all sides of controversial issues they discussed. (11.13)

fair trial Making sure someone accused of a crime is treated equitably by the courts. (11.9)

fair use Allowing part of a copyrighted work to be used without copyright clearance or payment. (11.8)

family hour A policy stating that all programs aired between 7:30 P.M. and 9:00 P.M. should be

suitable for children as well as adults. (3.16)

fast national A rating report on national viewing that Nielsen gets to subscribers the morning after the programs air. (10.10)

feedback In a communication model, anything that gives information back to the source. (12.4b)

fiber optic A glass strand through which large amounts of information can be sent. (7.12, 13.9)

field research Studies that are conducted in environments where people naturally are, rather than in special controlled environments. (12.4b)

file sharing An online activity in which computer users engage in digital swapping of content, such as music and video files. (2.9)

film editing A process that involved cutting frames of celluloid film and then positioning and taping them to make an entire movie. (13.3)

filter Software that attempts to identify and block spam. (2.14)

financial interest–domestic syndication (fin-syn) A policy that used to preclude networks from receiving any monetary remuneration from programs aired on the network or any rights to distribute those programs within the country after they had aired on the network. (3.16, 8.1c)

fingerprints Invisible markings on copyrighted goods that travel with them if they are copied illegally in order to try to prevent piracy. (11.8)

fireside chats Radio talks given by President Franklin Roosevelt. (5.9)

firewall   A software barrier designed to keep unwanted information from coming into or going out of a computer network. (2.14)

First Amendment   The part of the U.S. Constitution that guarantees freedom of speech and freedom of the press. (4.16)

first-person shooter   A type of video game in which the screen represents the player's eyes and players see their hands and usually a gun. (2.16c)

first-run syndication   Programs produced for distribution to stations and cable TV rather than for the commercial networks. (8.1d)

fixed buy   When the advertiser states a specific time that a commercial should air and the station or network abides by that time. (9.10)

flash memory   A small form of memory chip that can hold information without having a power source and can be erased and used again. (13.2)

flat-screen TV   A thin television or computer screen that usually incorporates plasma or LCD technology. (13.13)

fleeting expletives   Swear words that someone says in an extemporaneous manner on a live telecast. (11.6)

focus group   An in-depth discussion session to determine what people do and do not like about various aspects of the media. (10.8)

footprint   The section of the earth that a satellite's signal covers. (13.7)

format   The type of programming a radio station selects, usually described in terms of the music it plays, such as contemporary, jazz, rock, and so on. (8.4)

franchise   A special right granted by a government or corporation to operate a facility, such as a cable TV system; the right to produce a TV program that has the format of another TV program. (4.2, 14.12)

free flow   A philosophy that any country or media organization should be able to send information to any other country. (14.7)

freelance   To work on a per-project basis rather than as a full-time employee. (7.7)

freeze   Immobilization or cessation of an activity, such as a stop in the assigning of radio or TV station frequencies. (3.3)

frequencies   The number of recurrences of a periodic phenomenon, such as a carrier wave, during a set time period, such as a second. (11.1, 13.4)

frequency   The average number of times a person is exposed to a particular commercial over a period of time, such as a week. (10.6)

frequency discount   A lessening of cost to an advertiser that airs numerous commercials often on a station or network. (9.9)

frequency modulation (FM)   Placing a sound wave on a carrier wave in such a way that the number of recurrences is varied; stations that broadcast between 88 and 108 MHz. (5.13, 13.5)

frequency of tune (FOT)   An audience measurement that indicates how many times viewers return to a particular network. (10.6)

full-service agency   An advertising agency that handles all the advertising needs for a particular client. (9.4)

G

gatekeeper   A person who makes important decisions regarding what will or will not be communicated through the media. (1.6, 8.11, 12.4b, 14.13)

genre   A categorization of programs, such as drama or documentary. (4.13, 8.6)

Great Debates   The televising of presidential candidates John F. Kennedy and Richard M. Nixon opposing each other face-to-face in 1960. (3.13)

grid   Metal bars that are suspended from a ceiling and are used to hang lights. (13.3)

gridlock   When so many people are trying to access a particular internet site that it cannot handle them all. (13.9)

grid rate card   A list of prices a station or network charges for ads, which vary according to when there is a lot of availability and when the commercial schedule is fairly full. (9.9)

gross average audience (GAA)   The average number of people watching a program over several showings in a market. (10.6)

guild   A unionlike organization for above-the-line people such as actors and directors. (7.10)

H

hammocking   Programming a new or weak program between two successful programs. (8.5b)

hard drive   The part of a computer system that holds most of the information. (13.2)

HD DVD   A high-definition videodisc format that lost out to Blu-ray. (4.11)

HD radio Digital radio transmission that is broadcast in the same frequency band as conventional AM and FM stations; also called in-band on-channel. (1.2, 5.17)

headend The part of a cable TV system where all the network and other programming that is going to be on the system is received. (13.9)

hertz A frequency unit of one cycle per second; abbreviated as Hz. (13.4)

high-definition television (HDTV) Television that scans at approximately 1,000 lines a frame and has a wide aspect ratio. (1.2, 4.15, 7.12, 9.1, 13.6)

high-frequency fluorescent A low-energy, long-lasting light that puts out reds, greens, and blues in a consistent manner. (13.3)

hit Each time a user accesses a website. (10.2)

hoax A deceptive trick, often done in mischief, that, if done on radio or TV, is punishable by fines. (11.14)

Hollywood 10 Ten creative people from the movie industry who refused to answer questions of the House Un-American Activities Subcommittee and were sent to jail. (6.11)

home shopping Programs showing products that viewers can buy instantly by calling a particular phone number. (1.5c, 4.13, 9.6)

host selling A practice that has been outlawed on children's TV of having the host or star of a show also sell products during the show. (9.12)

hotspot The area served by a wi-fi transmitter in which users can receive a wireless signal to connect to the internet. (13.8)

households using TV (HUT) The percentage of homes that have a TV set tuned to any channel. (10.6)

hyperlink A word or image in a document that can be clicked to take the internet user to another document or file automatically. (2.4)

Hypertext Markup Language (HTML) The program language of the world wide web that allows for links and graphics. (2.4)

Hypertext Transfer Protocol (HTTP) A program language of the world wide web that allows documents to be transferred among computers. (2.4)

hypothesis In research, a supposition that is proposed to draw a conclusion or test a point. (12.4b)

I

iconoscope The earliest form of TV camera tube, in which a beam of electrons scanned a photoemissive mosaic screen. (3.1)

identity theft An act by which a hacker steals personal information, such as credit card numbers and passwords, from consumers via computers, telephones, or other telemedia. (2.14)

impressions The average number of people watching a program over several showings in a market. (10.6)

in-band on-channel (IBOC) A radio technology that allows digital signals to be broadcast in the same frequency band as conventional AM and FM stations; often called HD radio. (1.2, 5.17, 13.5)

indecency Language that, in context, depicts or describes sexual or excretory activities or organs in terms patently offensive, as measured by contemporary community standards for the broadcast medium. (5.18, 11.6)

independents TV or radio stations that are not affiliated with one of the major networks; small film and television production companies. (1.2, 3.14, 6.12, 8.1c)

industrial TV The use of video by corporations and similar organizations for training, orientation, and similar purposes. (1.2)

infomercial A program of about 30 minutes that extols the virtues of a particular product. (1.5c, 4.13, 9.6)

information superhighway A metaphor for the internet in which a road system represents the connections among networks and computers, which can be accessed by users through on-ramps. (2.5)

in-house Performing some service within a company rather than hiring outside help. (9.4)

instant messaging (IM) A service that allows users to correspond simultaneously by typing messages into various devices, including computers, PDAs, and cell phones. (2.7)

instructional television fixed service (ITFS) A form of over-the-air broadcasting that operates in the 2,500 MHz range and is used primarily by educational institutions. (3.17)

interactive ad An advertisement that allows people to undertake direct action in order to buy what is advertised. (9.14)

interactive cable Two-way capability that allows interaction between the subscriber and sources provided by the cable TV system. (4.1)

Interface Message Processor (IMP) A small computer that used to be connected to a main computer to translate data from the main computer into a standard protocol for sharing files with other networked computers. (2.1)

interlace scanning A technical method of building a TV picture in which odd lines are laid down and then even lines are laid down in order to build a frame. (13.6)

internet A worldwide computer network that allows people to send and receive email and access a vast amount of information. (1.2, 7.12d)

internet service provider (ISP) A company that facilitates the use of information on the internet by providing easy ways for customers to access the information. (2.6b, 9.1)

internship The chance to work and learn within a company without actually being hired. (7.4)

invasion of privacy Not leaving someone alone who wishes to be left alone, or divulging facts about a person that he or she does not wish divulged. (11.7)

inventory The amount of commercial time that a station or network has to sell. (9.3)

iPod An electronic device developed by Apple to download, store, and play music—and now also video. (2.15c, 9.1)

island A still picture at the end of commercials aimed at children to help them differentiate between the commercial and the program. (9.12)

K

keyword search A type of internet advertising in which links to appropriate sites come up when a person types a request into a search engine. (9.7)

kinescope An early form of poor-quality TV program reproduction that basically involved making a film of what was shown on a TV screen. (3.6)

Kinetoscope An early film viewer that had a peephole for the person to look through in order to see a short movie. (6.1)

Ku-band A satellite frequency band between 11 and 14 GHz. (13.7)

L

laboratory research A form of research conducted in a controlled environment that is not a person's usual environment. (12.4b)

landline A hard-wired telephone line, usually in a home or business. (1.3)

laptop A small-sized computer that includes a keyboard, screen, processing unit, and storage devices all in one unit. (2.15)

laser disc A videodisc system that uses a laser to read the information on the disc. (4.4)

leased access A basic cable channel on which time can be purchased by a business or other organization willing to pay to cablecast a message. (4.1)

length of tune (LOT) An audience measurement that indicates how long viewers watch a particular network. (10.6)

libel To broadcast or print something unfavorable and false about a person. (11.7)

license fee The amount a network pays a producing company or sports franchise to air the program the company produces or the game the sports franchise owns; the amount individuals in some countries pay the government to own a radio or TV set. (14.2)

light-emitting diode (LED) A type of light bulb that consists of tiny electron tubes that put out light. (13.3)

linear editing Editing material from beginning to end by a dubbing method that makes it difficult to make changes once the edits have been completed. (13.3)

line of sight Transmission in which the transmitting agent needs to be lined up with the receiving agent in such a way that there could be visual communication between them. (13.5)

liquid crystal display (LCD) Devices that use chemical elements that show black or a particular color when power is applied. (13.13)

local buying Purchasing advertising time on stations in a limited geographic area. (9.5)

local discount A reduction in cost given to an advertiser who can't profit from the whole coverage area because its place of business is located too far from some of the areas the station reaches. (9.10)

local-into-local The process in which satellite TV transmits local broadcast TV stations into the local areas as part of its service. (4.8)

local origination Programs produced for the local community, particularly as it applies to programming created by cable TV systems. (3.12, 4.1, 8.1a) local peoplemeter (LPM) Nielsen's portable audience measurement device that people carry with them to track their video exposure. (10.5)

log A listing of program and commercial material that is aired on a radio or TV station. (7.12e)

lottery A game or contest that involves chance, prizes, and consideration (money); a random selection method the FCC uses for assigning licenses. (11.10, 11.14)

loudspeaker A device that amplifies sound so that a large group of people can hear it. (5.9)

low angle A camera shot that is taken from below, which makes the person in the shot look powerful and dominant. (6.9)

lowest unit charge The amount that may be charged for a politician's commercial, equal to the lowest rate a regular advertiser would have to pay. (9.13)

low-orbit satellite A satellite that circles the earth at a height of about 485 miles and is not synchronous with the rotation of the earth. (13.7)

low-power TV (LPTV) Television stations that broadcast to a very limited area because they do not transmit with much power. (1.2, 4.5, 11.10)

M

made-for A movie that is specially produced for TV. (3.14)

magazine concept Placing ads of various companies within a program rather than having the entire program sponsored by one company. (3.6)

magazine show A program divided into short segments. (3.16, 4.7)

majors Large film and television production companies. (8.1c)

make-good An electronic media outlet's need to pay back an advertiser some of its money if a promised audience size is not obtained. (9.10)

massively multiplayer online role-playing games (MMORPGs) Video games played on the internet by large numbers of people. (2.16e)

mean world syndrome A theory that watching violence makes people afraid to go out because they think the world is a more dangerous place than it actually is. (12.5a)

mechanical scanning An early form of scanning in which a rotating device, such as a disc, broke up a scene into a rapid succession of narrow lines for conversion into electrical impulses. (3.1)

merchandising Selling the rights to characters or concepts of a TV show or movie that have been purposely included so they can be developed into saleable products, such as toys or clothing. (9.6)

message In a communication model, that which is communicated. (12.4b)

metro survey area (MSA) Arbitron's division for radio ratings of areas that receive radio stations clearly. (10.10)

microphone A device that transduces spoken words or other sounds into electrical waveforms. (13.2)

microwave Radio waves, 1,000 MHz and up, that can travel fairly long distances. (3.18, 13.8)

minidoc A short documentary, often within a newscast. (8.14)

miniseries A dramatic production of several hours shown on TV across a number of different nights. (8.6)

minitheater testing A pretesting procedure that involves bringing participants into a medium-sized room and showing them a movie, program, or commercial in order to get their reactions. (10.8)

modem Short for "modulation-demodulation," a device that allows computer-generated data to be sent over phone lines or cable TV systems. (1.2, 2.8)

modulate To place information from one wave onto another so that the wave that is carrying it represents the signal of the original wave. (13.5)

monaural Sound coming from just one direction. (13.2)

monitor A TV set that does not receive signals sent over the air but can receive signals from a camera or recorder. (13.3)

motion capture Scanning the movements of humans into a computer and then using the scan as a pattern for a computer-generated movie character. (6.15)

MP3 A compression system for audio that is part of the video MPEG compression system and is also used for stand-alone audio applications. (1.2, 2.9)

multiple platforms Related areas that media companies can present to advertisers, such as a TV station and its website, in hopes that the advertiser will buy ads in all the areas. (9.14)

multiple-system operator (MSO) A company that owns and operates several cable systems in different locations. (3.11, 4.2)

multiplex To place more than one service on an allocated band of frequencies. (13.5)

music preference research Determining what music people would be likely to stay tuned to by playing them samples over the phone or in a theater and soliciting their reactions. (10.8)

must-carry A ruling that cable or satellite TV systems must put certain broadcast stations on their channels. (3.11, 4.6)

N

narrowcasting Playing program material that appeals to a small segment of the population, rather than a broad segment. (1.2, 4.1, 8.5a, 12.4b)

national buying Buying advertising time on individual stations throughout the whole country. (9.5)

national identity A philosophy that each country should decide for itself what information should be allowed to enter its borders. (14.7)

National Television Systems Committee (NTSC) The electronic scanning and color system used in the United States and some other parts of the world; the group that set many technical television parameters in the United States. (3.1)

near-video-on-demand (NVOD) A cable TV or satellite service in which the same movie starts on different channels at frequent intervals, such as every 15 minutes, so that subscribers can tune in at a time that is convenient for them. (4.13)

needle drop fee A fee for using music one time that includes as a factor the length of the cut used. (11.8)

network buying Buying ads that are seen everywhere that the network program is seen. (9.5)

network neutrality A proposed legal requirement that all internet sites be treated equally when they are delivered by ISPs; ISPs oppose it because they want to manage traffic on their networks. (2.17)

news agency An organization that provides news to various media entities. (7.12b, 8.3a)

news director The head of a news department. (8.3b)

news producer A person who decides which stories will be presented on a particular newscast. (7.21b, 8.3b)

news release Information put out by a company or other organization that it wishes to have included in the news. (7.12b, 8.3a)

news reporter A person who gathers news and information outside the studio environment. (7.12b)

New World Information Order (NWIO) A United Nations attempt in the 1970s to give developing nations a more prominent role in electronic media. (14.7)

nickelodeon An early movie theater that charged customers a nickel to see a movie. (6.3)

node A computer in a distributed network that relays packets of information from the previous relay point to the next. (2.2)

noise Something physical or psychological that reduces the effectiveness of the communication process. (12.4b)

nonlinear editing A method of editing in which material can be

accessed in any order and easily rearranged. (6.15, 13.3)

O

obscenity Something that depicts sexual acts in an offensive manner, appeals to prurient interests of the average person, and lacks serious artistic, literary, political, or scientific value. (11.6, 12.0)

observational theory In discussions about violence, the idea that people will learn how to be violent by watching it on TV. (12.5a)

off-net Syndicated shows that at one time played on a network. (8.1d)

omnidirectional A microphone that picks up sound from all directions. (13.2)

on-demand Material, usually video or audio, that subscribers can request and store on discs or a computer so they can access it when they want it. (2.8, 4.13)

orbiting Placing ads at a slightly different time each day so that more people have a chance to hear or see them. (9.10)

organizational TV The use of video by corporations, nonprofits, educational institutions, government facilities, hospitals, and the like for training, orientation, and similar purposes. (1.2)

overnight reports Ratings reports that are delivered to customers by noon of the day after the programs air. (10.10)

P

packet switching The method of operation of the internet, in which information is broken down and sent to its destination in small pieces that are then reassembled. (2.2) paradigm shift   A complete change in thinking or a belief system that allows for the creating of something that was previously thought to be impossible. (1.6)

parallel editing   Cutting between two actions or story lines, usually until the two merge. (6.4)

pay cable   A method by which people pay a monthly fee to receive cable programming that is usually free of commercials. (4.1)

payola   The practice of paying a disc jockey under the table so that he or she will air music that might not otherwise be aired. (5.12, 12.0)

pay-per-click (PPC)   A process through which one internet site pays another if someone clicks on a link posted on one site in order to access the other. (9.7)

pay-per-view (PPV)   Charging the customer for a particular program watched. (4.13, 9.1)

pedestal   A mechanism that holds a camera and allows it to be raised or lowered, usually by hydraulic or pneumatic means. (13.3)

PEG   Cable TV channels that are set aside for public, educational, and government access. (4.2)

peoplemeter   A machine used for audience measurement that includes a keypad to be pushed by each person to indicate when he or she is watching TV. (10.5)

people using radio (PUR)   The percentage of people who have a radio tuned to any station. (10.8)

people using television (PUT)   The percentage of people who have a TV set tuned to any channel. (10.6)

performer Q   A measurement that indicates the degree to which

people are aware of and like a particular radio, TV, or movie personality. (10.9)

per-program fee   A right to use music that is obtained by paying a fee for each program that utilizes the music. (11.8)

persistence of vision   A phenomenon in which the eye retains images for a short period of time, enabling fast-moving still pictures to look like constant movement. (6.1)

personal digital assistant (PDA)   A handheld minicomputer that provides convenient functions and can be connected to a computer for full functionality. (1.2, 2.15b)

petition to deny   A process by which citizen groups expressed the desire that a license be taken away from a station. (11.10)

phishing   A kind of spam in which a cybercriminal pretends to be a legitimate company in order to get people to divulge personal information that can be used for identity theft. (2.14)

phosphor screen   A layer of material on the inner face of the TV tube that fluoresces when bombarded by electrons. (13.13)

pilot   A recording of a single program of a proposed series that is produced to obtain acceptance and commercial support. (8.2)

pipes   The physical elements that make up the internet grid, including copper wires, coaxial cables, fiber optics, satellites, and the like. (2.11)

piracy   Obtaining electronic signals or programs by illegal means and gaining financially from them. (6.15, 11.8, 14.10)

pirate stations   Illegal radio stations located on boats off the coast of Great Britain that broadcast rock music during the 1960s. (14.6)

pitch   To try to sell a program idea. (8.2)

pixel   Short for "picture element," one of the small dots that illuminate to make a picture on a television set or computer screen. (13.6)

plain old telephone service (POTS)   Predivestiture voice services provided by phone companies. (4.10)

plasma   A TV set technology that consists of two glass panels with gas between them that is activated by voltage and, in turn, activates colored dots to create a picture. (13.13)

pod   A portable device to which material can be sent through the internet or by other means so that it can be heard or seen using a small screen. (1.2, 2.15c)

podcast   To transmit material so it can be viewed or heard on a portable device such as an iPod. (1.2, 2.15c)

pod deal   A contractual arrangement between a studio or network and a full-fledged production company to have the production company supply a certain amount of programming. (8.1c)

pop-up ad   An ad that shows up on an internet site that can be removed by clicking it off. (9.7)

portable peoplemeter (PPM)   A pager-size audience measurement device that people take with them so that their radio listening is determined for ratings purposes wherever they are. (10.5) portfolio   A compilation of a person's materials, usually of a written nature, that the person puts together in order to show a potential employer his or her skills. (7.6)

postdivestiture   The time after AT&T was broken up in 1980 during which there were many innovations in the phone business. (4.10)

potted palm music   Soft music played on early radio that derived its name because this type of music was usually played at teatime in restaurants by orchestras flanked by potted palm trees. (5.5)

predator   A criminal who abuses the internet to solicit victims. (2.14)

prequel   A movie that is produced after another movie but which tells the story of what happened before the time depicted in the previous movie. (6.13)

press-radio war   A dispute in the 1930s between radio and newspapers over news that could be broadcast on radio. (5.10)

pretesting   Determining ahead of time what programs or commercials people are likely to respond to in a positive manner by having a group of people give written or oral opinions. (10.8)

pretty amazing new service (PANS)   Postdivestiture telephone services, such as call waiting and caller ID. (4.10)

primary channel   The main digital channel of a broadcast station, the one that is high definition. (4.15)

prime-time access rule (PTAR)   A former FCC ruling declaring that stations should program their own material or syndicated material rather than network fare during one hour of prime time. (3.16)

prior restraint   When some entity (such as the government) tries to censor material before it is disseminated. (11.5)

private   A media system in which businesses own the media and finance them primarily by advertising, and the government has only minimal oversight. (14.2)

privatization   A movement of the 1980s in which countries that had public or authoritarian media systems added or changed to privately owned electronic media. (14.9)

producer   The person who is in overall charge of a production, especially the schedule and money. (7.12a)

product integration   Incorporating a particular product within the plot of a TV show in exchange for money from the company that produces the product. (9.6)

product placement   Charging a company to have something that it sells included in a movie, television program, or video game, usually as a prop or set dressing. (1.5c, 9.6)

profanity   Irreverent use of the name of God. (11.6)

program buying   When one company pays the costs of an entire program and has its ads inserted within that program. (9.5)

progressive scanning   A technical method of building a TV picture in which information is laid down one line at a time from top to bottom to create a frame. (13.6)

projection TV   A television screen that is very large, with the signal coming from the rear or front. (13.13)

proliferation   The distribution of media content over an increasing

number of venues and platforms. (1.7)

promise versus performance   A procedure in which broadcasters promised what they would do during a license period and then were judged on the fulfillment of their promises when their license came up for renewal. (11.10)

promotion   Publicizing a program, station, organization, or the like to enhance an image, build goodwill, and ultimately get more customers. (7.12e, 10.1)

promotional spot (promo)   An advertisement for a station's or network's own programs that is shown on its own channel. (10.1)

psychographics   Information pertaining to lifestyle characteristics of a group of people, such as their desire to be involved with new technologies. (10.7)

public   A media system that is closely aligned with the government and is usually supported, at least in part, by license fees. (14.2)

public access   Programming conceived and produced by members of the public for cable TV channels. (3.12, 4.1, 11.6)

public broadcasting   A noncommercial form of broadcasting that gets money from audience members, corporations, and, to some extent, the government. (1.2, 3.15)

public convenience, interest, or necessity   A phrase in the Communications Act of 1934 that the FCC uses as a basis for much of its regulatory power. (11.10)

public domain   Works that are not copyrighted, such as very old material or government publications. (11.8)

public figure   A well-known person who would have to prove actual malice in order to win a libel suit. (11.7)

publicity   Free articles or other forms of enhancement that will garner public attention. (10.1)

public relations   The attempt to build goodwill for a particular organization. (10.1)

public service announcements (PSAs)   Advertisements for nonprofit organizations. (10.1)

pulse code modulation (PCM)   An audio recording and transmitting pattern that records sound digitally. (13.5)

Q

Q score   A measurement method that determines whether people have heard of a particular program or TV talent and whether they like that program or person. (10.9)

qualitative   Research that is not statistical in nature but is based on other forms of analysis, such as anecdotal, historical, or biographical. (10.7, 12.4b)

quantitative   Research that can be analyzed through numerical and statistical methodology. (10.6, 12.4b)

quartz lamp   A lamp with halogen gas and a tungsten filament in a quartz housing. (13.3)

quiz scandals   The discovery during the 1950s that some of the contestants on quiz shows were given the answers to questions ahead of time. (3.10, 8.9)

quota   A limit set on something, such as the amount of foreign programming a country allows on its broadcasting systems. (14.6)

R

Radio Act of 1927   The congressional law that established the Federal Radio Commission. (5.8, 11.1)

rádionovelas   A form of radio soap opera, developed in Latin America, in which stories are serialized for a period of time. (14.2)

radio waves   The electrical impulses of the radio frequency band of the electromagnetic spectrum. (5.1, 13.4)

random sampling   A method of selection whereby each unit has the same chance of being selected as any other unit. (10.4)

rate card   A chartlike listing of what a station, cable system, internet site, or network charges for different types of ads or other services. (9.9)

rating   The percentage of households or people watching or listening to a particular program. (7.12e, 10.6)

rate protection   Guaranteeing a customer a certain advertising fee, even if the rate card increases. (9.10)

raw website return   A type of internet search in which a user enters keywords into a box and the search engine returns a list of websites with those keywords. (2.6c)

reach   The number of different people or households exposed to a particular commercial over a time period, usually a week. (10.6)

reality TV   A genre of programming in which people are actually trying to accomplish real acts, such as getting to a finish line first or winning a singing contest. (4.13)

receiver   In a communication model, the person or group that hears or sees a message. (12.4b)

regional Bell operating company (RBOC)   A local telephone company created as a result of the breakup of AT&T. (4.10)

regional buying   Buying advertising time on stations in a particular part of the country. (9.5)

remote control   A stand-alone device that allows TV set functions, such as channel switching and volume changing, to be achieved from a distance. (1.2, 13.13)

renewal expectancy   An FCC statement that if a station provided favorable service, its license was not likely to be in jeopardy. (11.10)

reregulation   Imposing new rules on some group after rules have been eliminated, usually by the government. (4.6)

resilience   The ability of various forms of electronic media to adapt and continue to exist, even under adverse conditions. (1.7)

résumé   A written document that lists a person's career accomplishments; it is usually used to try to obtain a job. (7.6)

retransmission consent   The ability of TV stations to ask cable systems to pay for the right to carry the TV station. (4.6)

return on investment (ROI)   The value a company or individual gets for the amount of money it spends. (10.7)

ribbon microphone   A microphone that creates representations of sound through the use of a metallic ribbon, a magnet, and a coil. (5.9)

rights fee   Money paid to broadcast something, such as a sports contest. (8.12)

robotic cameras   Studio cameras that do not have individual operators behind them but are run by a person sitting in a remote location. (7.1)

run-of-schedule (ROS)   When a station, rather than the advertiser, decides on the time that a particular commercial will air. (9.10)

S

safe harbor   A period of time when indecent material can be aired because children are not likely to be in the audience. (11.6)

satellite   A human-made object that orbits the earth and can pick up and transmit radio signals. (7.12d, 13.7)

satellite radio   Transmission and reception of audio signals from a satellite that can be received directly by radios designed to receive the signals. (1.2, 5.17, 9.17, 13.7)

satellite TV   A process of transmission and reception in which signals sent to a satellite can be received directly by TV sets in homes that have small satellite receiving dishes mounted on them. (1.2, 4.3, 9.1, 13.7)

scanners   Radio devices that can be used to listen in on police and fire communications. (8.3a)

scanning   The process of producing a video image by illuminating individual pixels one at a time, row by row. (13.6)

scarcity theory   Reasoning that broadcasters should be regulated because there are not enough station frequencies for everyone to have one. (11.3)

scrambling   Changing transmitted electronic signals so they cannot be received properly without some sort of decoding system. (4.8, 11.8)

screener   A tape or DVD sent to Academy members so that they can watch a movie or program in their homes in order to vote for awards. (6.15)

search engine   A type of index to the internet. (2.6c)

secondary channels   Digital signals sent out by broadcast stations that are not their main HDTV channel. (4.15)

Section 315   The portion of the Communications Act that states that political candidates running for the same office must be given equal treatment. (11.12)

segment producer   A person who oversees a particular part of a newscast, such as entertainment news. (7.12b)

self-regulation   Rules that people within the electronic media industry set for themselves. (12.1)

sequel   A second movie that picks up where an earlier, usually successful, movie left off. (6.14, 8.2)

server   A large-capacity computer used to store and place information on the internet, a TV station, or some other entity. (2.5, 13.3)

share   The percentage of households or people watching a particular program in relation to all programs available at that time. (10.6)

shield law   A state law that has the effect of allowing reporters to keep sources of information secret. (11.5)

shortwaves   Radio waves that can travel long distances. (13.4, 14.3)

signal processing   Changing elements of sound, such as loudness or pitch, so that the resulting sound is different from the original sound. (13.2)

slander   To say something false that is harmful to a person's character or reputation. (11.7)

snipe   A pop-up in the lower portion of a TV screen that promotes an upcoming program. (10.1)

social networking   Using internet sites to set up pages where you give information about yourself that others can access and respond to. (1.2, 2.13)

sound bite   A short audio statement about something in the news. (8.3a)

source   In a communication model, the total number of people needed to communicate a message. (12.4b)

spam   Largely unwanted and unsolicited email messages, usually sent to large numbers of people. (2.14)

speaker   A sound box that converts electrical energy back into sound. (13.2)

spin-off   A series that is created out of characters or ideas that have appeared on other series. (3.16, 8.2)

spoofing   A crime in which a hacker gains access to a person's email account and sends messages from that account that the person does not authorize. (2.14)

spot buying   When a company purchases advertising time within or between programs in such a way that the company is not specifically identified with the program. (9.5)

standard definition (SD) Television signals with a 3:4 aspect ratio and 525 lines of resolution. (4.15)

station representative A company that sells advertising time for a number of stations. (9.3)

stereo Sound reproduction using two channels through left and right speakers to give a feeling of reality. (5.13, 13.2)

stereotype A fixed notion about a group of people that is often not true, especially for individuals within the group. (1.5d, 12.5d)

storyboard A chart that contains step-by-step pictorialization of a commercial, TV program, or movie. (9.11, 12.1)

streaming audio Sound that comes over the internet to a computer in real time without being stored on the computer. (2.8)

streaming video Moving pictures that come over the internet to a computer in real time without being stored on the computer. (2.11)

stringer A person who is paid for the stories or footage he or she gathers if that material is used by some news-gathering organization. (5.10, 8.3a)

stripping To air programs, often old network shows, at the same time every day of the week. (8.5b)

studio A space that contains cameras, sets, lights, and microphones where acting and performing takes place. (7.12a)

studio years The period during the 1930s and 1940s when Hollywood movie studios dominated the movie business by having stars under contract and engaging in all phases of production, distribution, and exhibition. (6.8)

stunt broadcasts Radio broadcasts of the 1930s that involved unusual locations, such as gliders and balloons. (5.9)

superstation A broadcast station that is put on satellite and shown by cable systems. (3.18, 4.1)

surround sound Sound that comes from at least six different speakers, giving the effect of sound all around the person who is listening or watching TV or a movie. (4.15)

survey research Study that involves questionnaires that are administered, tabulated, and analyzed. (12.4b)

sweeps The period of time when ratings data are collected for local areas. (10.10)

switcher A piece of equipment used to select the video input that will be taped or shown live. (13.3)

synchronous satellite A satellite whose orbit is synchronized with the earth's rotation, so that it appears to hang motionless in space. (13.7)

syndicated exclusivity A cable TV rule stating that if a local station was carrying a particular program, the cable system could not show that same program on a service imported from a distant location. (3.11)

syndicator A company that produces or acquires programming and sells it to stations, cable networks, and other electronic media. (3.14, 8.1d, 9.2)

T

target To deliver an advertisement only to the people who are most likely to be interested in the product being advertised. (9.14)

Telecommunications Act of 1996 A congressional law that allows many electronic media companies into each other's businesses, deregulates media, and has provisions related to violence and pornography. (4.9, 5.16, 8.16, 11.1)

teleconferencing Transmitting material over satellite from one or various points to other points, usually instead of having a face-to-face meeting or conference. (3.17)

telenovela A form of soap opera, developed in Latin America, in which stories are serialized for several months. (4.7, 8.6, 14.6)

teleprompter A mechanical device that, through the use of mirrors, projects a script in front of a camera lens. (7.12b)

televangelism Using TV to promote religion. (8.15)

tentpoling Programming a successful program between two weak or new programs. (8.5b)

10-watter A public radio station that once broadcast with low power and was used mainly to train students, although now most 10-watters have increased their power to at least 100 watts. (5.15)

testimonials Positive statements about a commercial product usually given by a well-known person. (9.13)

text messaging A system that allows short messages to be sent from one mobile phone to other mobile phone users without necessarily having access to the internet. (1.2, 2.15a)

three-dimensional (3-D) Material that is produced in such a way that it approximates the way the eyes see two images and can then be projected back so that it appears to have multidimensions. (6.11, 13.11)

toll station   A name for the type of radio programming WEAF initiated in 1922 that allowed anyone to broadcast a public message by paying a fee, similar to the way that one pays a toll to communicate a private message by telephone. (5.6)

total survey area (TSA) Arbitron's division for radio ratings of geographically large areas that may not receive all stations equally clearly. (10.10)

toy-based programming Children's programming that was designed around toys as the main characters. (9.12)

trade-out   To give goods or services (such as advertising time) in exchange for other goods or services. (9.10)

trades   Journals and magazines that deal with the industry. (7.3)

traffic   A department that keeps track of programs and commercials that are aired by maintaining a daily log. (7.12e)

trailer   An advertisement for, or preview of, a movie that includes scenes from the movie. (9.2, 10.1)

translator   An antenna tower that boosts a station's signal and sends it to an area that the signal could not reach on its own. (13.8)

Transmission Control Protocol/ Internet Protocol (TCP/IP) The standard computer network language that allows different computer networks to connect to each other. (2.2)

transmitter   A piece of equipment that generates and amplifies a carrier wave and modulates it with

information that can be radiated into space. (13.5)

transponder   The part of a satellite that carries a particular signal. (13.7)

treasurer   A person who handles cash for a company or other organization. (7.12f)

TVQ   A measurement that indicates the degree to which people are aware of and like a particular program. (10.9)

TV receive-only (TVRO)   A satellite dish that can receive but not send signals. (4.3)

twisted-pair   Copper wires used primarily to carry phone conversations. (4.3, 13.9)

U

ultra-high frequency (UHF)   The area in the spectrum between 300 and 3,000 MHz that was used by analog broadcast TV stations above channel 13. (3.3, 13.4)

umbrella deal   An agreement between an independent producer and a studio or network in which the studio provides money, office space, and other amenities to the independent in exchange for the right to distribute its programming. (8.1c)

underwrite   To help pay for public broadcasting in return for a brief mention of the contribution. (9.8)

uniform resource locator (URL) A specific address for a page on the world wide web. (2.4)

union   An organization that establishes pay rates and working conditions for people in a particular part of the industry that production units then need to abide by. (7.10)

unique visitor (UV)   A person who visits a website who has not gone there before. (10.6)

universal resource locator (URL) Same as uniform resource locator. (2.4)

upfront   Ads for programs that are sold before the fall season begins. (9.3)

user-generated content Information or entertainment material put into the media systems by ordinary people who are not usually employed by a media organization. (1.6)

uses and gratifications   A field of research that tries to determine the perceived value people feel by becoming involved with certain actions. (12.4b)

V

vacuum tube   An electron tube evacuated of air to the extent that its electrical characteristics are unaffected by the remaining air. (2.1, 5.1)

vast wasteland   A term coined by FCC chairman Newton Minow in 1961 to refer to the lack of quality TV programming. (3.14)

V-chip   Circuitry in TV sets that allows people to block out programs with violent or sexual content. (4.9, 8.16)

vertical integration   A situation in which one company produces, distributes, and exhibits its products without decision making from any other sources. (6.6)

very high frequency (VHF)   The area in the spectrum between 30 and 300 MHz that was used by analog TV stations that broadcast on channels 2 to 13. (3.1, 13.4)

videocassette recorder (VCR)   A taping and playback machine that uses magnetic videotape enclosed in a container that is automatically threaded when the machine is engaged. (1.2, 4.4, 14.10)

videodisc   A round, flat device that contains video and audio information and can display this information on a TV or computer screen. (4.4)

video game   A form of play that involves interaction between the person or people playing and a TV or computer screen. (1.2, 2.16, 8.9, 9.1)

video home system (VHS)   The consumer-grade half-inch videocassette format developed by Matsushita. (4.4)

video news release (VNR)   A tape given to the media that has information a company, politician, or other person or group wants to have publicized. (3.17, 8.3a)

video-on-demand (VOD)   The ability for subscribers to request a program or other material and then store it so they can access it when they want it. (4.13)

video store   A physical building where movies and other video material can be rented or purchased. (1.2, 4.4)

virtual ad   An advertisement, usually in the form of a billboard, that isn't part of the real scene but is inserted electronically. (9.6)

virus   An intentionally destructive program that can attack computers, erasing files and in other ways damaging them. (2.14)

visual effects   Complicated movie scenes or portions of them that are created in a computer. (13.3)

Voice over Internet Protocol (VoIP)   A telephone technique that uses packet switching to route phone calls via the internet. (2.10)

voice tracking   Making a radio program sound local even though the person doing the talking is in another city. (8.1b)

volume power index (VPI)   A measure that is used when an advertiser pays only for the consumers watching or listening who fit into the advertiser's target audience. (9.10)

W

webcasting   Sending audio or video material over the internet. (1.2, 2.11)

wi-fi   Short for "wireless fidelity," technology that provides broadband connectivity without cables by connecting a low-power radio transmitter to the internet to modulate the data onto radio waves. (2.15, 13.8)

window   The period between the time a movie is shown in a theater and the time it is released for showing on other media, such as pay cable or network TV. (8.1c)

wired   Equipment or distribution methods that use cable, fiber optics, or other wires that hold the signal within them. (13.4)

wireless   Any equipment or distribution method that uses the electromagnetic spectrum. (5.1, 13.4)

wire recording   A form of audio recording that predated tape and used wire that had to be cut and tied in order to edit. (5.11)

wire services   Organizations that supply news to various media. (5.10, 8.3a)

works for hire   Collaborative artistic creations, such as movies or TV programs, for which a company rather than an individual usually owns the copyright. (11.8)

world wide web (www)   Computerized information that is made available from organizations and people to other organizations and people. (2.4)

writer   The person who creates a script for a fictional or news program. (7.12a)

INDEX